Lecture Notes in Networks and Systems

Volume 186

The series "Lecture Notes in Networks and Systems" publishes the latest developments in Networks and Systems—quickly, informally and with high quality. Original research reported in proceedings and post-proceedings represents the core of LNNS.

Volumes published in LNNS embrace all aspects and subfields of, as well as new challenges in, Networks and Systems.

The series contains proceedings and edited volumes in systems and networks, spanning the areas of Cyber-Physical Systems, Autonomous Systems, Sensor Networks, Control Systems, Energy Systems, Automotive Systems, Biological Systems, Vehicular Networking and Connected Vehicles, Aerospace Systems, Automation, Manufacturing, Smart Grids, Nonlinear Systems, Power Systems, Robotics, Social Systems, Economic Systems and other. Of particular value to both the contributors and the readership are the short publication timeframe and the world-wide distribution and exposure which enable both a wide and rapid dissemination of research output.

The series covers the theory, applications, and perspectives on the state of the art and future developments relevant to systems and networks, decision making, control, complex processes and related areas, as embedded in the fields of interdisciplinary and applied sciences, engineering, computer science, physics, economics, social, and life sciences, as well as the paradigms and methodologies behind them.

Indexed by SCOPUS, INSPEC, WTI Frankfurt eG, zbMATH, SCImago.

All books published in the series are submitted for consideration in Web of Science.

More information about this series at http://www.springer.com/series/15179

Tatiana Antipova
Editor

Comprehensible Science

ICCS 2020

Springer

Editor
Tatiana Antipova
Institute of Certified Specialists (ICS)
Perm, Russia

ISSN 2367-3370 ISSN 2367-3389 (electronic)
Lecture Notes in Networks and Systems
ISBN 978-3-030-66092-5 ISBN 978-3-030-66093-2 (eBook)
https://doi.org/10.1007/978-3-030-66093-2

This Springer imprint is published by the registered company Springer Nature Switzerland AG
The registered company address is: Gewerbestrasse 11, 6330 Cham, Switzerland

Preface

This book is based on selected papers accepted for presentation and discussion at the 2020 International Conference on Comprehensible Science (ICCS 2020) that was held on October 30–31, 2020. This conference had the support of the Institute of Certified Specialists (ICS), Russia, and Springer.

The 2020 International Conference on Comprehensible Science (ICCS 2020) is a network of scholars interested in natural and social sciences original research results. The network was established to promote cooperation between scholars of different countries. An important characteristic feature of conference should be the short publication time and worldwide distribution. This conference enables fast dissemination so the conference participants can publish their papers in print and electronic format, which is then made available worldwide and accessible by numerous researchers.

The Scientific Committee of ICCS 2020 was composed of a multidisciplinary group of 110 experts from 38 countries around the world. Thirty-three invited Reviewers, united by science, have had the responsibility for papers' evaluating in a 'double-blind review' process. All the selected papers went through a round of peer review process. The main topics of the included papers are the following: Big Data and Data Mining; Business, Finance and Accounting and Statistics; COVID-19 Impact; Educational Technologies; Innovative Applied Sciences; Innovative Economics; Management Technologies and Systems; Physical and Material Sciences; Medicine, Public Health and Rehabilitation.

The papers accepted for presentation and discussion at the conference are publishing by Springer (this book) and will be submitted for indexing by ISI, SCOPUS, among others. We acknowledge all of those that contributed to the staging of ICCS 2020 (authors, committees, reviewers, organizers, and sponsors). We deeply appreciate their involvement and support that was crucial for the success of ICCS 2020.

October 2020 — Tatiana Antipova

Contents

Big Data and Data Mining

Network Patterns in South African Election Tweets

Aurona Gerber[1,2(✉)] and Stefanie Strachan[1]

[1] Department of Informatics, University of Pretoria, Pretoria, South Africa
aurona.gerber@up.ac.za
[2] Centre for AI Research (CAIR), Pretoria, South Africa

Abstract. Social communities play a significant role in understanding complex societies, from communities formed by support interactions between friends and family to community structures that depict the flow of information, money and power. With the emergence of the internet, the nature of social networks changed because communities could form disassociated from physical location, and social network analysis (SNA) on social media such as Twitter and Facebook emerged as a distinct research field. Studies suggest that Twitter feeds have a significant influence on the views and opinions of society, and subsequently the formation of communities. This paper reports on a study where social network analysis was performed on Twitter feeds in South Africa around the 2019 elections to detect distinct patterns within the overall network. In the datasets that were analysed, a specific network pattern namely Broadcast Networks were observed. A Broadcast Network typically reflects central hubs such as media houses, political parties or influencers whose messages are repeated without interaction or discussion. Our results indicate that there were few discussions and interactions and that messages were broadcasted from central nodes even though the general experience of Twitter users during this time was of intense discussions and differences in opinion.

Keywords: Social network analysis · Graph analysis · Twitter networks · Community clusters · Network visualisation · South african elections

1 Introduction

How communities form plays a significant role in understanding society and society interactions [1, 2]. In the past decade various studies were done on the usage and influence of the internet, technology, and social media in society [3–5]. These studies found that social media became one of the most important means of communication and therefore forms a significant part of what determines the views and opinions of people [6–9]. As a result, social network analysis (SNA) emerged as a research field. SNA analyse network structures in social media networks, and one of these network structures is networked communities and clusters. Understanding the ways in which online communities form and communicate allows us to interpret the flow of information and opinions, power, money and key influencers [10]. This is particularly valuable when analysing political structures within a specific society, an important

T. Antipova (Ed.): ICCS 2020, LNNS 186, pp. 3–13, 2021.
https://doi.org/10.1007/978-3-030-66093-2_1

capability because of the recent developments such as large-scale individualized collective action that can be observed globally and that results in substantial turmoil due to protest action [11–13]. Understanding the formation and dynamics of politics in social networks can provide useful insights on the interactions and changes in political communities, and might assist in the design of interventions [11, 12].

The aim of this study was to detect the social network structures that were formed around specific political events within the South African Twitter community, i.e. whether and how communities or groups were formed and organised within the network, and therefore an analysis of the patterns of information flow within and between clusters. The results of this social network analysis could assist with understanding the South African political landscape.

2 Background and Related Work

This section provides an overview of social networks, social network analysis (SNA), and the application of SNA on social media data.

2.1 Social Networks

A social network is a type of complex network and can be described as a social structure composed of a set of social actors or users, and the inter-relations and social interactions between them. These networks are useful to study the relationships between individuals, groups, social units, or societies [14]. The analysis of social networks is centred around the fundamental theory that the social network is made up of the relations and interaction between users and within units rather than by the properties of the user itself [1, 15]. Network structures are detectable in many different contexts, including biological systems, population and cultural structures, and communication networks. One important aspect of analysing social networks is the detection of clusters or crowds made up of users that are closely interconnected. These structural clusters can be analysed by studying both the links within a cluster as well as the links between different clusters [16]. Depending on the research objectives, there are different general analysis levels of social networks, namely micro-level, meso-level, and macro-level [16]. Micro level analysis generally focuses on an individual or small group and expands to trace the social relationships associated with that user or social unit. Meso-level analysis is used for medium sized groups or to determine how connections are formed between micro and macro level systems. Most human social networks fall within the meso-level category. Macro-level analysis is used when the focus is on the outcomes of interactions and communications of a larger network rather than interpersonal relations within the network. Various network analysis approaches can be followed to study the network structures formed by social entities, as well as the possible patterns found within these structures [8]. Social network analysis is used in the identification of network patterns, the exploration of network dynamics, and the study of information flow within and between, specifically, social media networks [8].

2.2 Social Network Analysis

Social network analysis (SNA) is the general reference to the process of investigating social networks or structures within social media through the use of networks, knowledge graphs, and graph theory. A crucial part of analysing any large dataset is the means to represent the information in a meaningful way – specifically when utilizing the results in decision making processes. Visualisation can prove to be an effective way of conveying results in a manner that is easily understood by users [17]. Some of the social structures that commonly use visualisation through social network analysis include social media networks, information sharing, business networks, general social networks, transmission of disease, and association studies. These types of networks are often visualised by focusing on the social links of persons or users, following the same approach as semantic networks where the significance of an entity is determined by its relation to, and interaction with other entities, rather than its own properties. Entities are represented as nodes or points, and the relationships and interactions between entities are represented by interconnecting lines or edges. Depending on the attribute of interest, the nodes and edges can be partitioned into higher-density subgraphs, resulting in the formation of communities [18].

The Social Media Research Foundation [19] has done various studies on social network analysis and the information about individual behaviour, social relationships, and community value that can be extracted. They subsequently released an open-source tool, NodeXL, that implemented most of the network structure algorithms to support social network and content analysis [20]. In a study on the mapping of Twitter topic networks using NodeXL, Smith et al. identified six distinct archetypical network patterns [21, 22]:

- ***Polarized Crowd*** network pattern usually depict a small number of groups depicting highly divisive and polarized discussions with very little connection between the groups. This is typical of groups that don't argue but ignore each other. The distinct groups therefore rely on different information sources and do not interact [21].
- ***Tight Crowd*** network pattern are characterized by "highly interconnected nodes or people with few isolated participants". Such a pattern is typical of communities formed by conferences, or typically groups associated with professional topics or specific hobbies. Such groups therefore support each other with information flows between members of the group [21].
- ***Brand Clusters*** is a network pattern is highly fragmented and is observable when topics such as well-known products, services or celebrities are discussed in Twitter. The nodes tweet about a topic but not to each other, and the larger the population, the less likely it is that participants are connected to one another. Information flows are this about a topic, and not between members of a group, there is therefore not an exchange of ideas, information is just passed on [21].
- ***Community Clusters*** is a network pattern where multiple smaller groups or hubs are formed each with its own audience, influencers, and sources of information. This pattern therefore reflects multiple centres of activity or community each with its own following and information flows, as well as a fair number of isolates. Such a network would typically represent diverse angles on a subject based on its relevance

to different audiences or it could depict a diversity of opinion and perspective on a topic [21].

- ***A Broadcast Network*** pattern is a distinctive hub-and-spoke that typically depict "Twitter commentary around breaking news stories and the output of well-known media outlets" with many people repeating information such as prominent news and media organizations tweets. Members are often connected only to the hub news source, without connecting to one another. Such a network structure would also depict influencers and agenda setters [21].
- ***A Support Network*** pattern is also a hub-and-spoke pattern that is typically observable where customer services for a major business are handled by Twitter service accounts. In the pattern the "hub account replies to many otherwise disconnected users, creating outward spokes" whereas, in the Broadcast pattern, "the hub gets replied to or retweeted by many disconnected people, creating inward spokes". This is a useful pattern for government, businesses, and groups that provide services and support via social media, because it could serve as a benchmark for evaluating performance [21].

Such social network patterns of Twitter communities can be analysed and visualised to provide insight into society, such as power and political sentiment, as well as information flows about the persons and topics that drive conversations and group behaviour.

The social media data source relevant to this study is Twitter. Twitter, an online news and social networking service created in 2006, is a platform dedicated to personal expression through the minimalist concept of microblogging [23]. Posts are limited to 280 characters, aside from a few language exceptions, and users can follow other users with no mandatory interaction. The simplicity makes it an ideal means to report current events and connect around specific topics, people or interests. The amount and availability of the data makes it possible for researchers to analyse the content based on network maps created by tweets, mentions, retweets, and followers.

3 Approach

The aim of this study was to determine whether any distinct patterns could be found within the social network formed through Twitter conversations on South African political issues, and the data was collected around the 2019 elections. Three specific periods were chosen for the dataset around significant political dates, namely, the last week that political campaigning was allowed (28 April 2019–4 May 2019), election day on 8 May 2019, and inauguration day on 24 May 2019. NodeXL was adopted as the toolset for the network analysis. Data was collected through NodeXL supplemented with data sources directly from the Twitter API. The search terms used for collection of the respective datasets were based on the most active hashtags on Twitter during these periods namely:

- Dataset 1 and 2: SAElections2019, SADecides2019, Xse, XseDay, AfterVotingIExpect, AfterIVoteIExpect.
- Dataset 3: PeoplesInauguration, SAInauguration19

A much larger set of unique Tweets was found for dataset 1 and 2 than for dataset 3. Data wrangling included the removal of duplicates determined by the tweeter's username, the mentioned or replied to username if applicable, and the date and time of the Tweet, or the imported id.

All the datasets were imported into NodeXL, and the unique entries translated into graph vertices, with interactions the edges or links. We adopted the Clauset-Newman-Moore algorithm [24] for the detection of clusters or communities. The network was visualised using the NodeXL graph visualisation features. Because it is Twitter datasets, all the graphs were undirected. The top words and hashtags by frequency of mention were determined for the overall network as well as for each group. The connections within and between the different groups can also be analysed to determine how information flows within the network.

Dataset 1 consisted of 9531 unique entries from 4935 unique entities or users. Dataset 2 consisted of 11 711 unique entries from 8 883 entities or users. Dataset 3 consisted of 3 556 unique entries from 1 938 entities or users. The overall metrics were calculated, and the key properties of the networks are summarized in Table 1.

Table 1. Graph metrics for the datasets

Graph Metric	Dataset 1	Dataset 2	Dataset 3
Vertices	4 935	8 883	1 938
Unique edges	5 753	9 369	2 131
Edges with duplicates	3 778	2 342	1 425
Total edges	9 531	11 711	3 556
Self-loops	2 524	3 032	641
Connected components	457	1 957	151
Single-vertex connected components	333	1 454	111
Graph density	0.000465	0.000200	0.001175

4 Results

Top vertices can be determined by the degree measure, or the betweenness centrality measure. Betweenness centrality would indicate more central, and arguably, the more influential entities, and can assist in determining the key influencers in the network. Few differences were found between the two measures, and the top vertices ranked by betweenness centrality is depicted in Table 2 below.

Table 2. Top vertices for the different datasets ranked by betweenness centrality

Dataset 1	Dataset 2	Dataset 3
IECSouthAfrica	AdvBarryRoux	PresidencyZA
News24	IECSouthAfrica	CyrilRamaphosa
Our_DA	Mandzenga	KhuselaS
SABCNewsOnline	GovernmentZA	GovernmentZA
SimonPGrindrod	tumisole	BiyelaSthe
SizweMpofuWalsh	News24	DIRCO_ZA
aneebh	ThabisoSithole	HloniNyetanyane
ali_naka	MsezaneSifiso	SANDFCorpEvents
AfricaInsights	SABCNewsOnline	TrafficSA

The next aspect we explored is the top hashtags per group, and for dataset 1 the top 5 hashtags per group are listed in Table 3.

Table 3. Top hashtags by frequency of mention for the largest groups in dataset 1

Top hashtags G1	Top hashtags G2	Top hashtags G3	Top hashtags G4
XseDay	SAElections2019	SAElections2019	SAElections2019
SAElections2019	FreedomDay	EFFMayDayRally	Elections2019
8-May	VoteDA	SMWX	ANC
FreedomDay	ANC	TshelaThupa	Elections
SADecides2019	DA	SouthAfrica	ACDPGAUTENG2019

The NodeXL visualization of the network graph for Dataset 1 is depicted in Fig. 1.

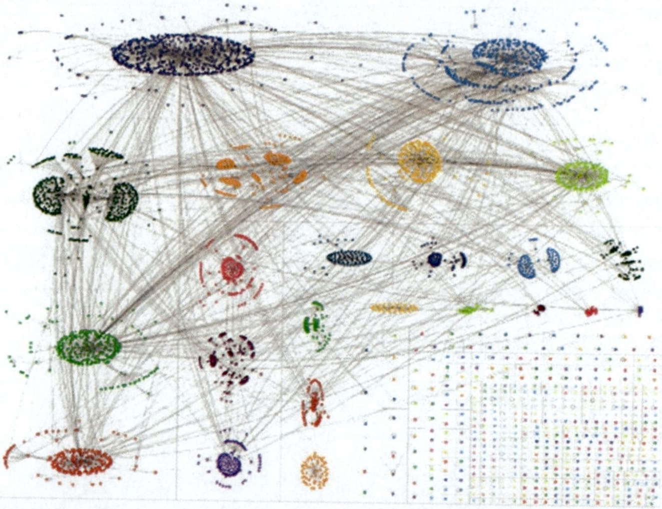

Fig. 1. Network visualisation for dataset 1: last week of campaigning (23–24 April 2019)

The top 5 hashtags per group for dataset 2 are listed in Table 4.

Table 4. Top hashtags by frequency of mention for the largest groups in dataset 2

Top hashtags G1	Top hashtags G2	Top hashtags G3	Top hashtags G4
SAElections2019	SAElections2019	8-May	IAmVotingEFF
XseDay	SADecides2019	XseDay	SAElections2019
SADecides2019	KnowYourDA	SAElections2019	Voting
IAmVotingEFF	OneSAforAll	SADecides2019	SADecides2019
Votingday	sabcnews	Voting	XseDay

The NodeXL visualization of the network graph for Dataset 2 is depicted in Fig. 2 and the network graph for Dataset 3 is depicted in Fig. 3 (Table 5).

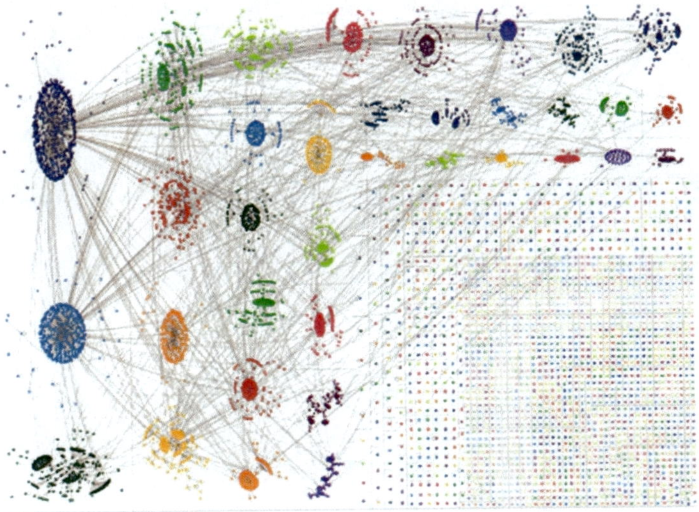

Fig. 2. Network visualisation for dataset 2: election day (7–8 May 2019)

The top hashtags in the five largest groups in all the datasets overlap significantly, and related closely to the top hashtags in the overall networks. It is also possible to determine that some of the smaller groups is centred around a specific political party and therefore referenced that party more prominently. In all the networks the connections within and between groups differs and therefore prominent clusters or groups are distinguishable. A much higher percentage of shared connections are within groups with few connections between groups.

Table 5. Top hashtags by frequency of mention for the largest groups in dataset 3.

Top hashtags G1	Top hashtags G2	Top hashtags G3	Top hashtags G4
PeoplesInauguration	PeoplesInauguration	PeoplesInauguration	SAInauguration19
eSwatini	NewDawn	SAInauguration19	PeoplesInauguration
SAInauguration19	SAInauguration19	NewDawn	NewDawn
WeThePeople	SABCNews	Democracy25	WeThePeople
CyrilRamaphosa	LoftusStadium	WeThePeople	SA

In all the networks, the Broadcast Network pattern is distinguishable, even though there are some similarities to the Community Cluster and Support Network patterns. The distinct hub-and-spoke groups in the graphs, particularly in Fig. 3, differentiates the Broadcast pattern from the Community Cluster, and the fact that we have several such hub-and-spoke groups distinguishes the network from the Support Network. The number of isolates in all the networks also supports a Broadcast Network and we therefore classified the networks therefore as Broadcast Networks. The Broadcast network pattern typically reflects Twitter traffic around breaking news stories or events with many people retweeting posts, or influencers and trend-setters whose tweets are retweeted often [19]. In a Community Cluster there are interactions between members of a group, which does not reflect the hub-and-spoke cores, and the only network that depicts some community clusters are in dataset 2 in Fig. 2 where some of the prominent smaller groups form interactive clusters. Such a cluster would be indicative of discussion between members of a group and not just repeated tweets.

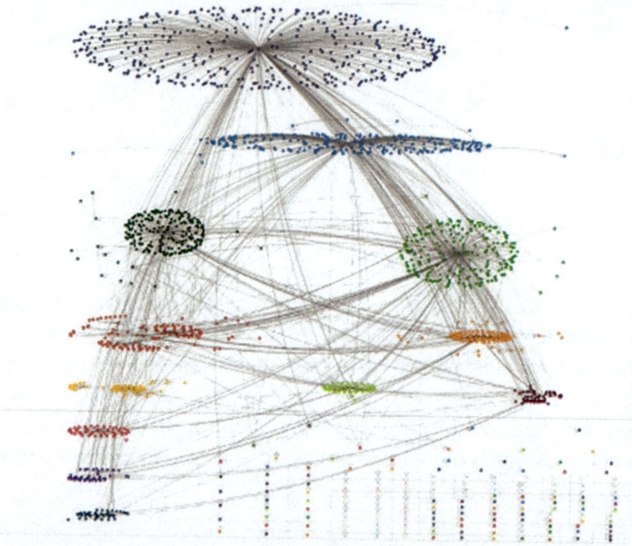

Fig. 3. Network visualisation for dataset 3: inauguration day 23–24 May 2019

5 Findings

All three of the datasets that were analysed can be classified as Broadcast Networks with a number of distinct groups. If we analyse the top hashtags it is clear that these groups were representative of the main political parties as well as the main media accounts that created traffic around breaking news and events. This is very prominent in Fig. 3, inauguration day, where very strong hub-and-spoke groups are discernible.

The social network analysis study done in previous research specifically in the USA found that political communities on Twitter form polarized crowds [11] where, even though both groups are focused on the same topic, there is little conversation between the groups. This study was, however, done in the United States of America where there are only two main political parties feature. Since South Africa has multiple political parties, multiple groups are formed, both on the general subject of the elections, as well as centred around specific political groups or influencers. This is evident in the results of all the datasets where the largest group is centred around the general topic – mostly using hashtags associated with election day, and most of the other groups mention different political parties or specific influential persons. In dataset 3 the groups are not focused on specific political parties but remain segregated. The percentage of connections shared among clusters is also noticeably higher than in the first two datasets. This could be attributed to the fact that the election results were already revealed, and the conversations focused more on the president than on the different political parties that were involved. This is also evident from the hashtags used within the groups – a much larger overlap is seen among groups.

Overall it is surprising that very little discussions are evident from our datasets. The general impression of Twitter users during this time was that there were many discussions and sometimes aggressive differences between individuals, however, when the networks are analysed, it is clear that most interactions were broadcasts from central vertices that represented political role-players.

6 Conclusion

Social media has become the prominent means of communication in recent decades, and conversations on social media create networks with identifiable groups. In Twitter specifically, connections as users reply to and mention one another in conversations form clusters or communities that provide insight. Depending on the subject in question, the groups and conversational structures differ. Mapping social networks can assist in understanding the different ways that individuals form communities and organize online. This information can assist individuals and organisations to make informed decisions on the ways that groups and conversations are formed online. Previous research found that six distinct patterns can be seen in the social structures and conversations of the users, depending on the topic in question [19]. This study investigated the network structures that are formed within the South African Twitter community around the 2019 elections. The NodeXL toolset was used for the social network analysis.

Three datasets were gathered based on significant periods and search criteria. The networks found in all three of the datasets were classified as Broadcast Networks where a large number of disconnected entities contribute to the network by broadcasting key topics and groups do not exhibit strong links to other entities and groups within the network. The network structures are evident when there are central or influential users, typically media houses and political parties. From our datasets very little discussions are evident within groups and most interactions were broadcasts from central vertices that represented political role-players even though the general impression of Twitter users during this time was that there are many more discussions between groups.

References

1. Fu, X., Jar-Der, L., Boos, M.: Social network analysis: interdisciplinary approaches and case studies. CRC Press (2017). https://doi.org/10.1201/9781315369594
2. Scott, J., Carrington, P.J.: The SAGE Handbook of Social Network Analysis. SAGE, London (2011)
3. Hanna, R., Rohm, A., Crittenden, V.L.: We're all connected: the power of the social media ecosystem. Bus. Horiz. **54**, 265–273 (2011). https://doi.org/10.1016/j.bushor.2011.01.007
4. Perrin, A.: Social media usage: 2005–2015. Pew Research Center (2015). (https://www.secretintelligenceservice.org/wp-content/uploads/2016/02/PI_2015-10-08_Social-Networking-Usage-2005-2015_FINAL.pdf)
5. Kaplan, A.M., Haenlein, M.: Users of the world, unite! the challenges and opportunities of social media. Bus. Horiz. **53**, 59–68 (2010). https://doi.org/10.1016/j.bushor.2009.09.003
6. Schivinski, B., Dabrowski, D.: The effect of social media communication on consumer perceptions of brands. J. Mark. Commun. **22**, 189–214 (2016). https://doi.org/10.1080/13527266.2013.871323
7. Wellman, B. (ed.): Networks in the Global Village: Life in Contemporary Communities. Westview Press, Boulder, Colo (1999)
8. Khan, G.F.: Seven Layers of Social Media Analytics: Mining Business Insights from Social Media Text, Actions, Networks, Hyperlinks, Apps, Search Engine, and Location Data. CreateSpace Independent Publishing Platform, Leipzig (2015)
9. Tichy, N.M.: Social network analysis for organizations. Acad. Manage. Rev. **4**, 507–519 (1979)
10. Kietzmann, J.H., Hermkens, K., McCarthy, I.P., Silvestre, B.S.: Social media? get serious! understanding the functional building blocks of social media. Bus. Horiz. **54**, 241–251 (2011). https://doi.org/10.1016/j.bushor.2011.01.005
11. Adamic, L.A., Glance, N.: The political blogosphere and the 2004 U.S. election: divided they blog 8 (2005)
12. Bennett, W.L.: The personalization of politics: political identity, social media, and changing patterns of participation. Ann. Am. Acad. Polit. Soc. Sci. **644**, 20–39 (2012). https://doi.org/10.1177/0002716212451428
13. Trottier, D.: Social Media, Politics and the State: Protests, Revolutions, Riots, Crime and Policing in the Age of Facebook, Twitter and YouTube. Routledge, Abingdon (2014). https://doi.org/10.4324/9781315764832
14. Hansen, et al.: - Do You Know the Way to SNA A Process Model for A.pdf
15. Rainie, H., Wellman, B.: Networked: the New Social Operating System. MIT Press, Cambridge, Mass (2012)

16. Rodrigues, E.M., Milic-Frayling, N., Smith, M., Shneiderman, B., Hansen, D.: Group-in-a-box layout for multi-faceted analysis of communities. In: 2011 IEEE Third Int'l Conference on Privacy, Security, Risk and Trust and 2011 IEEE Third Int'l Conference on Social Computing, pp. 354–361. IEEE, Boston, MA, USA (2011). https://doi.org/10.1109/PASSAT/SocialCom.2011.139
17. Ma, D.: Visualization of Social Media Data: Mapping Changing Social Networks (2012)
18. Hansen, D.L., Rotman, D., Bonsignore, E., Rodrigues, E.M., Smith, M., Shneiderman, B., Capone, T.: Do you know the way to SNA?: a process model for analyzing and visualizing social media data. In: International Conference on Social Informatics, p. 10. IEEE (2012)
19. Social Media Research Foundation. https://www.smrfoundation.org/. Accessed 30 June 2020
20. Fay, D.: NodeXL: Network Overview, Discovery and Exploration in Excel. https://www.microsoft.com/en-us/research/project/nodexl-network-overview-discovery-and-exploration-in-excel/. Accessed 30 June 2020
21. Smith, M.A., Rainie, L., Shneiderman, B., Himelboim, I.: Mapping twitter topic networks from polarized crowds to community clusters. Pew Res. Cent. **20**, 1–56 (2014)
22. Hansen, D.L., Schneiderman, B., Smith, M.A.: Analyzing Social Media Networks with NodeXL: Insights from a Connected World. Morgan Kaufmann, Burlington, MA (2011)
23. Twitter. https://twitter.com/home. Accessed 20 June 2020
24. Clauset, A., Newman, M.E.J., Moore, C.: Finding community structure in very large networks. Phys. Rev. E. **70**, (2004). https://doi.org/10.1103/PhysRevE.70.066111

RusIdiolect: A New Resource for Authorship Studies

Tatiana Litvinova(✉)

RusProfiling Lab, Voronezh State Pedagogical University, Voronezh, Russia
centr_rus_yaz@mail.ru

Abstract. Problems of determining the author of texts (authorship attribution) or her/his characteristics (authorship profiling) using data mining techniques, although often solved disjointly, in fact are related to idiolect identification and should be studied in unified framework with obligatory account for interaction of as much factors of idiolectal variation (both author-based factors – gender, age, cognitive ability, personality traits, etc. and text-based factors – genre, topic, mode, etc.) as possible. Despite an enormous number of papers proposing rigorous methods of authorship analysis and a high social impact of these tasks, practical applicability of the techniques is questioned. This is because of the underestimation of interaction of the above-mentioned factors of idiolectal variation, which may result, for example, in topic, not author identification, in case of lack of topic control in authorship experiments, and other misleading conclusions. A small number of appropriate corpora also hampers progress in idiolect studies. This paper introduces a new freely available resource *RusIdiolect* which allows users to search for factors of idiolectal variation related to both author (person, gender, age, etc.) and text (register, mode, genre) as well as to perform full-text search. Database structure is outlined, as well as its possible applications in idiolect studies. The necessity to further develop corpora supplied with information on factors of idiolectal variation to facilitate this research area is highlighted.

Keywords: Text corpus · Data mining · Idiolect · Factors of idiolectal variation · Text classification

1 Introduction

The problem of assigning a text to the particular author or to class of authors (gender, age groups, etc.) is a task of a great practical value. Revealing information about group identity (gender, psychological profile) of the authors of, for example, online reviews could help companies to develop their marketing strategy more effectively or to aid police in narrowing the potential list of the authors of threatening letters. These tasks, usually referred to as authorship attribution and authorship profiling and solved using data mining techniques, are usually treated separately (see [19] for one the latest reviews). However, they are both related to the analysis of a text as a sample of an idiolect. Idiolect is an individual way of language use which is formed under the influence of a wide range of factors, although by now there is no agreement on the list,

T. Antipova (Ed.): ICCS 2020, LNNS 186, pp. 14–23, 2021.
https://doi.org/10.1007/978-3-030-66093-2_2

typology and magnitude of effect of these factors (see [1] for discussion). Some factors (i.e., sources of idiolectal variability) such as individual sociolinguistics history (languages a person speaks, linguistic communities he/she lives in, etc.), education, gender, age, personality traits, cognitive ability and so forth can be proposed as author-based stable factors of interindividual idiolectal variation. This does not mean that these factors are constantly stable, though – of course, they can change with time, but rather slowly, unlike people's moods, feeling, states and other author-based dynamic factors which could be accounted for by both intraindividual (if texts by the same person produced in different states are compared) or interindividual idiolectal variation (if texts written by persons in different states, for example, persons with and without depression are compared). Not only author characteristics, however, are reflected in idiolects, but also the context of a particular text production which define mode, register, topic, genre of the text. These are texts-based factors of idiolectal variation. Another important factor of this type is a factor of an audience. A type of device used for text production (cellphone, notebook, etc.) and its characteristics (for example, keyboard type) as well as a type of messenger or another program used for communication are other important factors whose effect is yet to be assessed. All the above-mentioned text-based factors could be also considered both as factors of intraindividual or interindividual idiolectal variation depending on the task at hand.

All the factors of idiolectal variation are related to each other, and lack of account of their interaction could result in misleading conclusions. For example, success of authorship attribution predictive models could be in fact that of topical classification if all authors in the dataset writes about different topic (which is often the case when, for example, only genre is controlled for) bearing in mind that state-of-the-art approaches to text classification imply the same features set for both tasks [12]. Another example is a well-known distinction in usage of certain function words which is often attributed as male/female writing styles distinction which was later shown to be genre-dependent (however, only in fine-grained scale, i.e. after manual dividing texts labelled as blogs into subgenres – diaries and filter, with diaries containing more "female" features irrespective of the real gender of the author) [3], so that it is obviously necessarily to take into account a wide range of factors and their possible interactions to draw really sound conclusions about authorship of the questioned text.

State-of-the-art in authorship analysis is characterized by the prevalence of engineering approach since nowadays computer scientists are mostly engaged in solution of these problems. This approach is aimed at gaining the highest possible accuracy of predictive models not in the close examination of datasets, which often requires manual work, for example, for fine-grained genre and topic labelling, notwithstanding achievements in automated genre and topic classification. Research is needed to deepen our understanding of the complex picture of interaction of as many factors of idiolectal variation as possible and to develop methods for authorship analysis in accordance with practical needs. To this end, it is crucial to have datasets which contain metadata in the form of different factors of idiolectal variation. Development of such resources is a time- and labor-consuming task, so they are rather scarce. A new resource *RusIdiolect* is introduced which allows users to search for different factors of idiolectal variation related to both author (person, gender, age, etc.) and text (register, mode, genre) simultaneously as well as to perform a full-text search. Its structure is outline and the metadata it contains are described in detail. I also highlight opportunities it provides to study the interaction of factors. To the best of my knowledge, this is the first resource specially designed to study interactions of idiolectal variation factors.

2 Corpora for Automated Idiolect Identification

2.1 Corpora of "Experimental" Texts

To control as many different factors of idiolectal variation as possible (by ensuring sociodemographic homogeneity of authors and designing writing tasks to control for genre or/and topic), as well as to obtain information about authors (e.g., to assess their personality profile) and to be sure that texts are produced by authors themselves (i.e., not copy-pasted), researchers (especially at the early stage of automated idiolect identification research) have made a great amount of effort to carefully construct corpora of "experimental" texts, i.e. texts produced by respondents (usually university students) by the researcher's request. One of the examples of such corpora used for individual idiolect identification are Dutch Authorship Benchmark Corpus (ABC-NL), which consists of 72 Dutch texts by 8 authors [2], or "essays" dataset [14] of stream-of-consciousness texts (about 2400, one for each author/user), labelled with personality classes, considered as the gold-standard dataset of this type and widely used for personality prediction.

This type of corpora is very useful for idiolect identification research; however, a compilation of such datasets is related to high time and labor costs. Besides, such corpora are usually small-sized and restricted (typically) by the texts of one social group (university students), which does not allow one to generalize conclusions made on their basis.

2.2 Corpora of "Real-World" Texts

Corpora of "real-world" texts, i.e. texts produced in different situations of real life (in contrast to experimental conditions), are widely used in idiolect identification studies. Two sources of such corpora can be allocated. The first source is "professional" texts, i.e. texts produced by the advanced writers as a product of their professional activity (fictions, scientific papers, journalist articles, etc.). Such corpora are easy to obtain, texts are typically lengthy, and a lot of texts per author can be gathered. The example of such corpora is Reuters Corpus of over 800,000 manually categorized for topic newswire stories made available by Reuters, Ltd. Many of such news have also author label which made this corpus widely used for authorship identification since work by [18]. Another example is corpora of fiction texts derived from Gutenberg project [16]. Such corpora are suitable both for individual and group (mostly gender) idiolect identification, although they are not free from drawbacks, which needs to be taken into consideration when reporting the results of authorship experiments. First, methods developed with a lot of textual data from the author could not be applicable to forensic cases typically involving small amount of short texts. Second, advanced writers could have more stable and discriminative idiolect markers than less experienced ones.

He/she can have a list of favorite topics, plots, etc., which makes them easily recognized than naïve writers. Third, "professional" texts could be exposure to different types of editing, which is a sort of noise in authorship studies.

The second source of "real-world" texts is social media texts (blogs, tweets, comments) frequently used in recent studies for both individual and group (mostly gender) idiolect identification [17]. Such corpora are easy to collect automatically. A typical example of the process of collection of a social-media text corpus (namely, comments left on Reddit, social news platform) is given in [11]. The authors state that such a corpus is an example of a multi-genre, multi-topic and multi-language database for idiolect studies. However, topic labelling is taken from platform structure which does not guarantee high-quality (users can discuss things far from news that started an initial discussion). The possibility of multiple accounts of the same person, trolls, unmarked citations and other "noises" should also be kept in mind when working with corpora of this type.

Overall, as I briefly outlined, each an existing type of corpora has both advantages and drawbacks which must be taken into account in idiolect studies. The biggest drawback is the lack of labelling (with at least a partial manual check) for the factors of idiolectal variation. I argue that when one works with idiolect identification, the quality not just quantity of data matters.

3 RusIdiolect: A New Resource for Idiolect Research

3.1 General Overview of the Resource

RusIdiolect is a database containing both texts in Russian (all authors are native Russian speakers) and rich metadata related to the information about authors as well as about texts. The database contains several corpora that have been collected by the author of this paper since 2012 up to present. Several corpora included in *RusIdiolect* have previously been described in [5] and used in a series of research in authorship profiling (gender [6] and age detection [7], personality prediction [9]) and deception detection [4]. Besides, *RusIdiolect* contains texts from RusIdioStyle corpus designed for research in stability of idiolectal markers over texts by the same authors. It was used for experiments in authorship attribution in [13].

The database is freely accessible via webpage https://rusidiolect.rusprofilinglab.ru and allows users to search for a wide range of attributes (i.e. factors of idiolectal variation) related to both author and text (see below for details). The results are saved in Excel format (texts+metadata; current limit for saving is 1050 texts). Full-text search is also available (for word forms, lemmas and part-of-speech tags). User can also choose a corpus to search in.

All the texts were either taken from open sources (i.e., blogs) or were produced in the course of experiments conducted by research team of Corpus Idiolectology Lab (https://rusprofilinglab.ru). RusIdioStyle also contains both "experimental" and "real-world" (provided by the authors themselves) texts by the same respondents. Informed consent was obtained from all authors whose texts and metadata were received during the experiments.

All personal information (except for taken from open sources) was removed from texts, so that identification of the author identity is impossible.

The metadata were obtained in the course of experiments (respondents filled questionnaires; psychological tests were interpreted by the professional psychologist). The metadata for authors of "real-world" texts were taken from their profiles and checked manually. In case of inconsistences (for example, author reported male in his blog profile, but writes as if he was a woman, which is indicated by gender endings typical for Russian) author and his/her texts were removed from database. Blogs were plagiarism-checked, and texts consisted mostly of plagiarism fragments were removed.

3.2 Metadata

3.2.1 Factors Related to Author

Currently, the database contains labels for the following factors of idiolectal variation.

ID (Author Identifier). Each text in the results of search over the database is accompanied by the unique identifier of its author coded in filename after # (see Fig. 1). When viewing a text, a user has an option to view all the texts by the same author (button "All texts by this author").

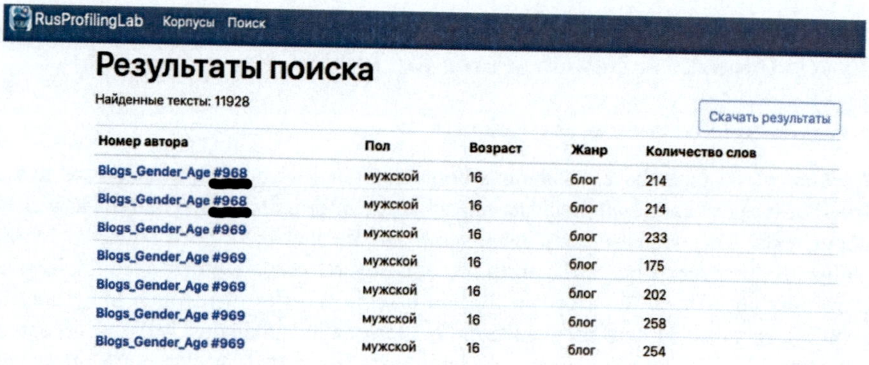

Fig. 1. Example of search results in RusIdiolect (author ID is highlighted as an example)

To date, overall, database contains *22675 texts by 5024 authors.* The mean number of texts per author varies significantly across corpora: from 1 in the **Corpus for controlled authorship attribution** to 9 in **RusIdioStyle** (Fig. 2).

Gender. This attribute, like ID, is provided for all the authors whose texts are included in the database (except for 3 authors who did not report their gender). Gender is coded as a binary variable. Currently, database contains 10741 texts by female authors (i.e., 47.36% of total number of texts) and 11928 texts by male authors.

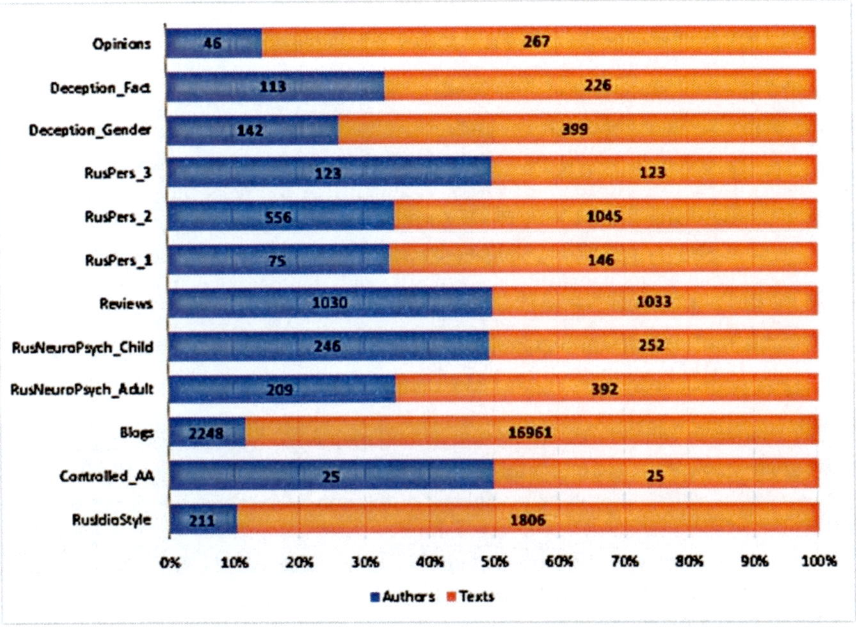

Fig. 2. Author and texts distribution over corpora included in RusIdiolect

Age. Age is provided as a continuous variable (user can set age value in the form 'from… to…') for 20092 texts (or 88.6% of the total number of texts). Age distribution of the authors is presented in Fig. 3.

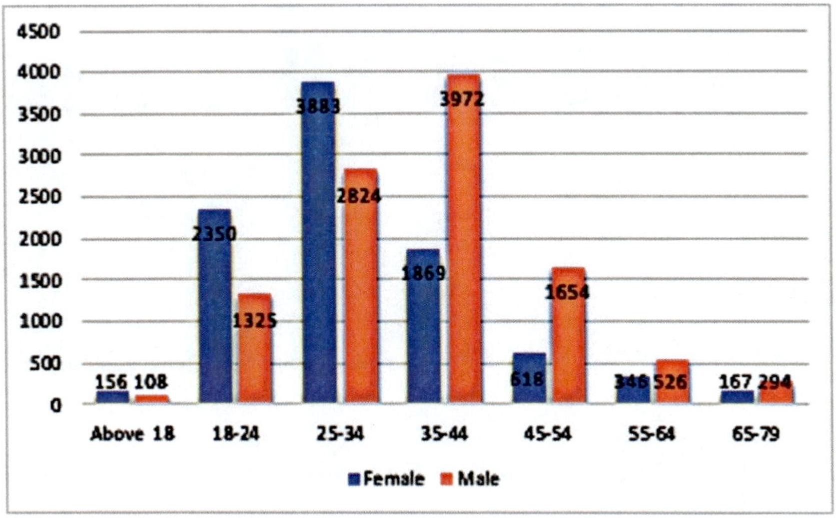

Fig. 3. Distribution of texts over age and gender groups of their authors

Personality Traits. A considerable part of author-based metadata in *RusIdiolect* belongs to personality profile. Different personality tests were proposed to respondents. Since tests took different time, some respondents chose to take only one, while others several. The selection of a particular test was motivated by the necessity to compare the results with the ones obtained by researchers for English texts (Big Five Personality Test - “Big-5”) or their ability to reveal potentially dangerous characteristics (Freiburg Personality Inventory, form B - FPI-B). A full list of the tests the respondents took include, among Big-5 and FPI-B, Bem Sex–Role Inventory, IQ test “Domino”, questionnaire “Styles of self-regulation” by V.I. Morosanova [10], Leonhard’s test for revealing attenuated personality traits [15], Belov’s temperament test. All these tests measure traits, i.e. some stable personal characteristics. Another line of tests the respondents took is related to states. The examples are Hospital Anxiety and Depression Scale (HADS) and method for the self-assessment of the emotional states by Hans Eysenck.

The database allows one to search simultaneously over different scales of the above-mentioned tests, for examples, predefine values for Extraversion (from… to …) in Big-5 test or select all non-zero values (i.e. to choose all texts with Big-5 scores for authors).

Along with personality trait tests, some authors were proposed to take neuropsychological assessment (see [8] for details).

The distribution of the authors with respect to tests they took is shown in Fig. 4.

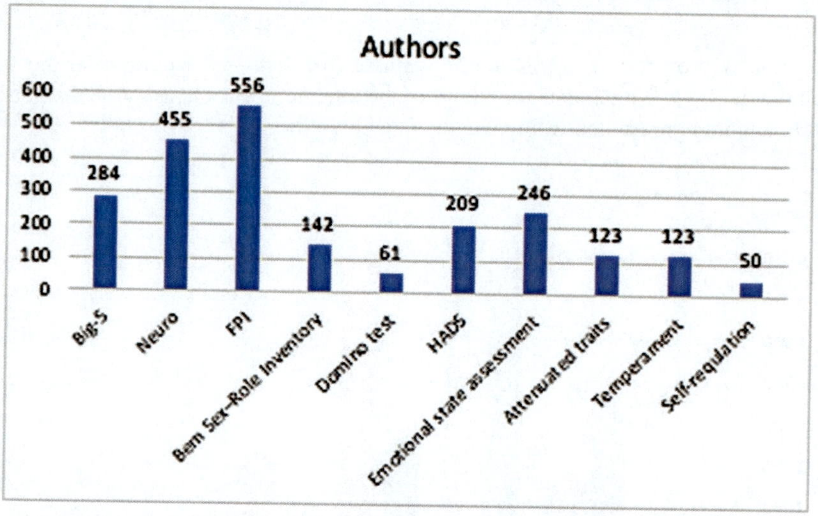

Fig. 4. Number of authors with different types of personality profile metadata

3.2.2 Factors Related to the Context of Text Production

Another zone of factors of idiolectal variation supplied in *RusIdiolect* is the one related to context of text production (named as **Text**). The process of manual labelling of this group of factors is ongoing, and the structure of this zone is subject to change.

In the database, user has an option to search for the following factors related to the context.

Mode. This attribute can have 1 of 3 values: oral (539 texts), written (4076) and electronic (18060). Our collection contains oral texts in the form of their orthographical transcriptions with segmentation on elementary discourse units (EDUs). First, speech-to-text software was used, then manual examination and correction with a follow-up segmentation have been performed by the members of Corpus Idiolectology Lab. By introducing oral texts to idiolect studies, I would like to facilitate research in cross-modal of authorship attribution which is now an understudied problem. Label “written” was assigned to texts which originally were written by hand and afterwards converted into electronic form (this work was also done by Corpus Idiolectology Lab team members) as well as to the texts typed on computer but not in the situation of electronic communication (writing assignments, fragments of dissertation, etc.). ‘Electronic” label was assigned to the texts originally produced in the course of computer-mediated communication (blogs, vlogs, etc.).

Situation. Texts produced by the researcher request are marked as “experimental” (4067), while “real-world” texts are labelled as “natural” (either produced in internet-communication (18343) or in live dialogues (265)).

Intent. Since *RusIdiolect* contains two corpora for deception detection studies, it provides information for presence/absence of an intent to deceive a text recipient for some part of the collection (I recommend first to choose search in corpus “Intention_Lie_Author” of “Intention_Lie_Fact”, since for other texts this information is unavailable). Two types of deceptive information are present in the texts: either intent to deceive the reader regarding author identity (“stylistic deception”) or regarding information on the situation described in texts (“factual deception”).

Genre. As it have been already mentioned, genre detection is a complicated problem requiring manual labeling. The situation becomes even more complicated if the absence of a generally excepted genre typology is taken into account. Manual labeling has already been performed for RusIdioStyle corpus. The process of labeling the other corpora is ongoing. Currently, fine-grained genre labelling has been performed (for example, currently there are texts labelled as “Day”, “Day in reverse order”, “Day written later”), but the plans include combining the some existing genre labels as well as to label texts in accordance with the register. Users will be allowed to search for mode (written/oral), form (monologue or dialogue), register and genre of the texts.

Users can also search inside a particular corpus, search for word forms, lemmas and part-of-speech-tags throughout the database.

After finishing the works on text labelling, the users will be provided with a manual on the database usage and a detailed description of texts falling into this or that register of genre label.

3.2.3 Factors Related to the Process of Text Production

Some texts (*Opinion corpus* and some texts from *RusIdioStyle*) are accompanied by the video of their production made by the authors with screen recording software on different types of devices. Videos are available by request.

Notwithstanding the *RusIdiolect* database is not fully functional yet (works on manual labelling for genre is still in progress), a plethora of existing metadata allows one to conduct multiple research in idiolect identification. For example, research in gender identification under genre+topic control, interaction of gender and personality traits, authorship identification with account for personality traits with genre control, cross-modal authorship identification, and so forth.

4 Conclusions

Research in idiolect identification (both on individual and on group level) using stylometric approach is of a high theoretical and social impact. However, in order to be applied in real-world cases, methods of idiolect identification should be theoretically grounded, the results interpretable, which is currently not the case due to the trend on purely computational (engineering approach) to this problem associated with the ultimate goal of obtaining the highest possible accuracy of predictive models. *RusIdiolect* database, whose structure and content were briefly outlined in this paper, will hopefully facilitate research in this area since it allows us to examine the effect of several factors of idiolectal variation on stylometric markers simultaneously, thus providing a framework for unified research in authorship attribution and profiling. In the course of database collection, the list of the types of texts representing idiolect will be expanded by introducing oral texts (both monologue and dialogue) and dialogue texts (both written and oral) in addition to monologue written texts – a typical object of authorship research. The share of such texts in the database is, however, small, since this work is time-consuming and/or related to privacy issues, but at least the existing data provide researchers with opportunities to conduct initial work in this direction.

Further manual labelling for texts genre and register as well as expansion of the database with a special attention to the fixation of the process of text production by the same individuals with different technical means of communication (notebooks, mobile phones, tablets, etc.) will be performed since this type of factors is believed to affect idiolectal variation significantly, but this influence is yet to be studied.

Acknowledgement. The research has been performed in Voronezh State Pedagogical University under the support of Russian Science Foundation, grant number 18-78-10081, which is gratefully acknowledged.

The author expresses her gratitude to Bulat Yaminov and Ildar Yaminov for their invaluable contribution to database design and construction. The author also thanks two anonymous reviewers for their insightful comments.

References

1. Grant, T., MacLeod, N.: Resources and constraints in linguistic identity performance–a theory of authorship. Lang. Law/Linguagem e Direito **5**(1), 80–96 (2018)
2. Van Halteren, H., Baayen, H., Tweedie, F., Haverkort, M., Neijt, A.: New machine learning methods demonstrate the existence of a human stylome. J. Quant. Linguist. **12**(1), 65–77 (2005)

3. Herring, S.C., Paolillo, J.C.: Gender and genre variation in weblogs. J. Sociolinguist. **10**(4), 439–459 (2006)
4. Litvinova, O., Seredin, P., Litvinova, T., Lyell, J.: Deception detection in Russian texts. In: Proceedings of the Student Research Workshop at the 15th Conference of the European Chapter of the Association for Computational Linguistics, pp. 43–52 (2017)
5. Litvinova, T., Litvinova, O., Zagorovskaya, O., Seredin, P., Sboev, A., Romanchenko, O.: "RusPersonality": a Russian corpus for authorship profiling and deception detection. In: Proceedings of the International FRUCT Conference on Intelligence, Social Media and Web (FRUCT 2016), pp. 1–7 (2016)
6. Litvinova, T., Pardo, F.M.R., Rosso, P., Seredin, P., Litvinova, O.: Overview of the RUSProfiling PAN at FIRE track on cross-genre gender identification in Russian. In: FIRE (Working Notes), pp. 1–7 (2017)
7. Litvinova, T., Sboev, A., Panicheva, P.: Profiling the age of Russian bloggers. In: Conference on Artificial Intelligence and Natural Language, pp. 167–177. Springer, Cham (2018)
8. Litvinova, T., Seredin, P., Litvinova, O., Ryzhkova, E.: Estimating the similarities between texts of right-handed and left-handed males and females. In: Jones, G., et al. (eds.) Experimental IR Meets Multilinguality, Multimodality, and Interaction. CLEF 2017. Lecture Notes in Computer Science, vol. 10456, pp. 119–124. Springer, Cham (2017)
9. Litvinova, T.A., Seredin, P., Litvinova, O., Zagorovskaya, O.: Profiling a set of personality traits of text author: what our words reveal about us. Res. Lang. **14**, 409–422 (2016)
10. Morosanova, V.I.: Self-regulation and personality. Procedia-Soc. Behav. Sci. **86**, 452–457 (2013)
11. Murauer, B., Specht, G.: Generating cross-domain text classification corpora from social media comments. In: Crestani, F., et al. (eds.) Experimental IR Meets Multilinguality, Multimodality, and Interaction. CLEF 2019. Lecture Notes in Computer Science, vol. 11696. Springer, Cham (2019)
12. Panicheva, P., Litvinova, O., Litvinova, T.: Author clustering with and without topical features. In: Salah, A., Karpov, A., Potapova, R. (eds.) Speech and Computer. SPECOM 2019. Lecture Notes in Computer Science, vol. 11658, pp. 348–358. Springer, Cham (2019)
13. Panicheva, P., Litvinova, T.: Authorship attribution in Russian in real-world forensics scenario. In: Martín-Vide, C., Purver, M., Pollak, S. (eds.) Statistical Language and Speech Processing. SLSP 2019. Lecture Notes in Computer Science, vol. 11816, pp. 299–310. Springer, Cham (2019)
14. Pennebaker, J., King, L.A.: Linguistic styles: language use as an individual difference. J. Personal. Soc. Psychol. **77**(6), 1296–1312 (1999)
15. Pukrop, R., Steinmeyer, E.M., Woschnik, M., Czernik, A., Matthies, H., Sass, H., Klosterkötter, J.: Personality, accentuated traits and personality disorders. A contribution to dimensional diagnosis of personality disorders. Der Nervenarzt **73**(3), 247–254 (2002)
16. Qian, C., He, T., Zhang, R.: Deep learning based authorship identification. Department of Electrical Engineering, Stanford, CA (2017)
17. Rocha, A., Scheirer, W., Forstall, C., Cavalcante, T., Theophilo, A., Shen, B., Carvalho, A., Stamatatos, E.: Authorship attribution for social media forensics. IEEE Trans. Inf. Forensics Secur. **12**(5), 5–33 (2016)
18. Stamatatos, E.: Author identification using imbalanced and limited training texts. In: Proceedings of the 4th International Workshop on Text-Based Information Retrieval, September 3–7, Regensburg, Germany (2007)
19. Tai, K.Y., Dhaliwal, J., Shariff, S.M.: Online social networks and writing styles – a review of the multidisciplinary literature. IEEE Access **8**, 67024–67046 (2020)

Some Peculiarities of Big Data Legal Regulation in the Digital Age

A. A. Andreasyan[1], Artem Balyakin[2], Marina Nurbina[2](✉), and S. B. Taranenko[2]

[1] Otkrytoe Pravo Ltd., 103, 6, 4th Dobryninsky, Moscow 119049, Russia
[2] NRC Kurchatov Institute, 1, ac. Kurchatov sq., Moscow 123182, Russia
Nurbina_MV@nrcki.ru

Abstract. The paper discusses the legal features of digital economy formation. It is shown that the digitalization should be treated as a complex phenomenon, which includes both the formation of knowledge economy and the modernization of existing institutions, including the institution of law. A number of socio-economic phenomena associated with digitalization are described. The idea of eventual degradation of a number of institutions is formulated. The possible transformation of law towards its fragmentation is discussed. The risk of loss of state control over the field of law due to the high degree of commercialization of digital technologies is formulated. It is proposed to use the practice of legal regulation of megascience facilities to form a stable regulatory system taking into account the relationship between man, science, the state, society and nature. Examples of soft law and hoftlaw in the field of legal regulation of scientific activity are presented. The idea of the need for an integrated approach to the development of digital technologies is formulated. These efforts should include the development of existing institutions, taking into account the emerging risks and challenges.

Keywords: Digital technologies · Legal regulation · Institute of law · Law of science · Smart content · Big data · Digital future

1 Introduction

The pace to digital future is inevitable all over the World. For many countries the introduction of digital technologies is currently considered as one of the tools to ensure technological independence and successful modernization development, supporting the transition to the next technological mode. The policy of introducing new technologies is also combined with existing peculiar national-wide problems that are not objectively dependent on the development of digital technologies: disengagement of territories and complex logistics; uneven economic development of regions, socio-political challenges, etc.

The introduction of digital technology is reflected in such a social institution as law. Its transformation (both forced and conducted by the state), has a serious impact on the possibility of sustainable economic growth, and produces a number of new risks and challenges that are institutional in nature. Moreover, many risks and challenges are not

T. Antipova (Ed.): ICCS 2020, LNNS 186, pp. 24–32, 2021.
https://doi.org/10.1007/978-3-030-66093-2_3

new and often repeat those already existing in human history [1]: some negative aspects of the development of digital technologies make transparent what has taken place and developed independently previously.

2 Main Objectives

This paper contains the results of the scientific research carried out within the framework of the RFBR project No. 18-29-16130. This work is devoted to the study of the peculiarities of legal regulation associated with the introduction of digital technologies (including big data, artificial intelligence technologies, etc.).

3 Socio-economic Aspects of Digitalization

It is too early to state which institutions precisely will be transformed most in the near future under the influence of innovative technologies. The analysis enables one to identify a number of main trends. However, it is impossible to conclude which course will be preferred (chosen) by the society. Obvious manifestation of this fact is the unformed glossary of digital technology and the "vulgarization" of many terms (e.g., one to be mentioned – "artificial intelligence").

Within the framework of our research, we have formulated the following series of trends that accompany the digitalization processes that are important from the point of view of this study.

Firstly, various methods and approaches used to gather, process and store data have a significant impact on the formation of the digital ecosystem. As an example, it is worth to note the big data technology in relation to which a single conceptual and regulatory apparatus has not been formulated yet: it is technically customary to characterize "big data" by a set of properties [2]. From a business point of view, big data technology is a process that provides insights into decision making [3]. In jurisprudence, disputes continue [4] whether to understand "big data" as a process or an object (for example, one considers the opportunity to treat "big data" as an anomalously large database with specified properties).

Secondly, digital technologies are often understood as ready-made solutions (in extreme cases, as identified correlations according to criteria a priori). The "knowledge environment" is replaced by the "environment of ready-made solutions": the proposed solutions are notorious for their apparent universality of answers. In fact, we are talking about hidden subjectivity [5] and the degradation of scientific thinking skills [1] in favor of correlation approach.

Thirdly, two opposing trends have been outlined: on the one hand, data depersonalizing (i.e., reduction of data knowledge to methods for ranking and sorting data based on self-learning algorithms). This phenomenon is called "Smart Content", which is characterized by fewer than the current number of variables, along with the ability to convolution and data recovery. On the other hand, there is now an active request for the creation of the reverse algorithms (data de-anonymization ones). This process is initiated both by private companies in order to maximize profits, and by state institutions in order to gain and persist control. The latter trend is most remarkable in Chinese social rating policy.

Fourth, in the social sphere, attempts to combine universal digitalization and human rights led to the "open data" concept. This approach (traced from CERN experiments) suggests that all scientific knowledge and the achieved results of scientific activity, financed from public funds and/or from the state, should be in the public domain [6]. In relation to personal data (not involving personal identification), it is proposed to use the concept of "no one's data" ("common data") [7]. The idea of maximum openness of the society and the economy (open data) is currently being seriously discussed as an alternative to the concept of maximum protection of intellectual rights and the right to protect personal information [8].

Fifthly, the policy of localizing data centers is politically biased. The authorities strive to control physical entities based on the positive experience of "property law". This leads to the need to develop the appropriate infrastructure of the digital economy for example [9–11], which includes both physical data centers and numerical and analytical methods and approaches [12], including big data, artificial intelligence technologies, etc. Thus, in Russia the President's decree on the development of high technologies in Russia was approved, which provides for the construction and reconstruction of the five largest facilities of the mega-science class [13]. A similar policy is carried out in other countries, too [5]. Also, China's success in the field of big data regulation, thanks to the centralized collection of data on a large number of objects, has led to a close examination of the experience of the Chinese model in other countries. Especially this practice was in dire demand in experiments on managing the situation during the Covid-19 pandemic.

Sixth, socially at the international level, digitalization is seen as another manifestation of globalization, since there is no division among countries, there is a high interconnection of technologies and approaches.

Seventh, digitalization is increasingly understood as a process identical with commercial success and cost reduction. As a result, digital giants, repeatedly increase their capitalization and influence, become monopolists in providing access to information, distorting the functioning of the market [14].

4 Digitalization and the Institution of Law

Authors claim that the change in the institution of law in the course of digitalization is characterized by the following aspects.

Firstly, there is a decrease in the role of the state as a lawmaker and an increase in the importance of NGO institutions in the formation of regulatory approaches. Accelerated digitalization is largely explained by the commercial success of digital technologies, which leads to the fact that the regulation of digitalization issues is transferred to the relevant processes actors [10]. They, for their part, establish rules on the fact (based on market interest); the rules they introduce become mandatory not only for business, including competitors, but also for the average citizen ("consumer" in modern terminology). The introduced changes have a fundamental impact both on the structure of the economy and on the institutional structure of society. The visible effect of such processes can be called "digital expropriation" [15].

The seemingly universality of the proposed solutions, their apparent isolation from subjective preferences leads to the expropriation of human rights as an institution. Relying on a "computer solution" as an explanatory factor in case of a denial of law is a common feature of our time. The computerized decision-making process becomes a component of the 21st century state economy [16].

Secondly, there is a deformation of law due to changes in the nature of production processes in the digital economy [17]. A new reality is accompanied by pronounced processes of degradation: the development of digital technologies displaces high-tech (and, therefore, costly) production processes in favor of low-tech (cheaper) ones. At the same time, new players in the market are trying to create their own platforms - the "blue oceans" in the terminology of the Club of Rome, using digital technology to circumvent current legislation, working conditions and fiscal systems. Accordingly, the development of digital business without proper consideration of public interest is a source of bias in the entire structure of the economy in favor of easily profitable low-tech processes (including the widespread adoption of platform solutions), and with it the structure of the economy and society as a whole.

It is believed that companies such as Uber or food delivery services distort the existing structure of economic entities and, in their current form, do not meet sustainability criteria (i.e. their users do not share the general costs, e.g., for the urban infrastructure used by drivers, etc.) These business projects may contain a real danger of uncontrolled development, an increased risk of monopolization and unethical use of technology, and it is not yet clear how to avoid these threats [18].

The presence of "old" norms and rules comes into conflict with the development of digital businesses: digitalization supporters are trying to remove established institutional barriers. Among these barriers there are the basic institutions of society, such as law. Thus, the law acquires not only a previously unused source, but it should also be expected that the institution of legalization of "corporate" law will be developed, having in mind not a legal term, but a law dictated by corporate interests and aimed at all those who have relations with the corporation: including not only consumers, but also other interested parties. This process forms new externalities or the so-called "External effects of the corporation" [19].

With that said, thirdly, we should expect the effect of fragmentation of law. This suggests (due to the multiplicity of sources) that the law will be different in various parts: it will be distinguished both by geographical criterion, by technological or sectorial characteristics, etc. It is noteworthy that this is not only about the formation of new branches of law, but about the formation of various zones of law enforcement and jurisdiction.

Note that over the past few years there has been a discussion about the need to legalize a new set of human rights - digital rights. The idea of expanding the concepts of an individual and a legal entity is discussed: for example, the idea of a digital profile is proposed as a set of information about citizens and legal entities contained in the information systems of state bodies and organizations that exercise separate public powers in accordance with federal laws, as well as in a unified identification system and authentication. In this case, the difference between digital identity and real is allowed [20]. And –

fundamentally – this "right" ceases to be a right subordinate to public control, including in the form of state institutions.

In our opinion, there is a challenge to the loss of state control, its transfer to other agents (both authorized by the state and those who received these opportunities in an explicit manner). Overcoming the difficulties is possible by creating regulatory institutions that provide control over the development of digital business, the digital economy as a whole and in its separate parts, including the creation of comfortable legal conditions for the introduction of new technologies in the economy. In practice, we are talking about the transition from law in the form of requirements to law in the form of management.

In the context of the digital economy, one should expect acceleration of the process of shifting the law from the model of "rules" to the model of "operational management" (arising seignorate). It is this model that removes the obstacles to the regulation and operational tuning of digital processes being built. Furthermore, digital technology is an effective tool for such a senior. And in this "state" version, there is a pronounced risk of fragmentation of the right to certain legal seignorates, equivalent to the reincarnation of territorial feudalism in the form of guild feudalism. There is a perceptible risk of a return to previous generations of law, a rejection of a group of human rights and freedoms (new in content and in the form of realization that arose in the second half of the 20th century – Fig. 1). The flip side of the coin threatens to be the regulation of all legal processes to the complete impossibility of implementing effective business management of private companies (such as tightening regulatory measures under the GDPR [21]).

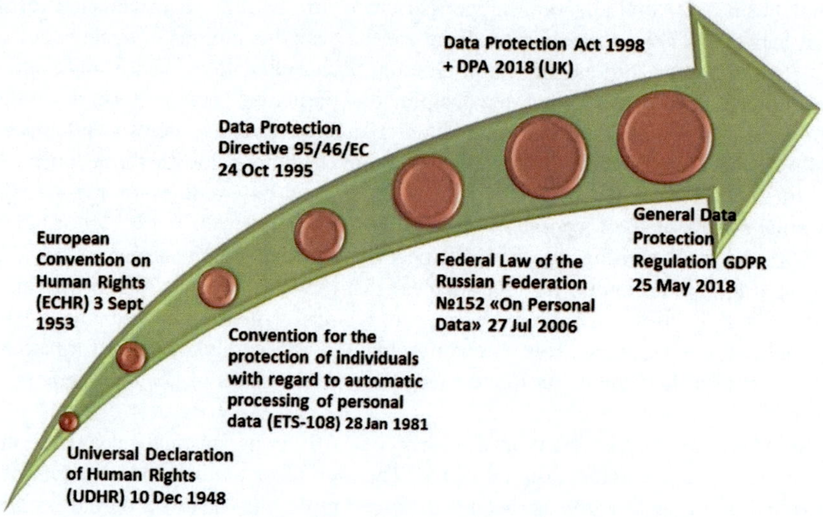

Fig. 1. Legal chronicles of big data

5 The Role of Science in Future Basics of Law Formation

The expected development of digital technologies (artificial intelligence, big data, etc.) no longer refers to purely scientific issues, but represents an important component of both business processes and government management. The main problem of new technologies is an attempt to "embed" them into the existing structure of society without taking into account possible societal consequences. This approach is explained by the commercial success of digital technologies, but expectations to transfer the experience of using big data technology from retail, social networks, financial sector to the life of society as a whole seem to be overestimated and unrealizable in full in the near future.

Awareness of current trends, risks and threats associated with digitalization should be accompanied with the understanding that society (represented by the state) sets the rules and regulations for business, and not vice versa. That is why it becomes socially necessary to formulate a thoughtful and effective regulatory system concerning the relationship between man, science, the state, society and nature. Although the prohibition and restriction policy seems to be the simplest solution to emerging problems, this approach is not fruitful. In the applied aspect, it is possible that part of such a system of future regulation could belong to the newly formed integrated branch of law – research law (the law of science) [22].

Currently, most countries understand the importance of high-tech for the development of society. Mega-science projects based on international cooperation are actively being created in the world for basic research in various fields. Main traits of such projects together with peculiarities of international scientific cooperation shape a number of features in the regulation area (such as risk sharing, financing, responsibility, etc.).

In the scientific sphere, soft law mechanisms currently dominate. Examples of soft law implementation in science are CERN (Conseil Européen pour la Recherche Nucléaire, European Organization for Nuclear Research) and NICA (Nuclotron-based Ion Collider Facility) projects. For instance, there are 18 participants in the mega-science project NICA [23], and questions about the financing of the project and the allocation of time are defined in internal regulatory documents. The rights to the result are determined by the participants in a separate memorandum of understanding, which avoids the so-called "over-regulation".

For Large Hadron Collider (LHC) to operate, the CERN bodies, in the framework of their organizational and administrative powers, issue binding normative acts that determine the general rules for the creation and functioning of international scientific collaborations [24]. LHC collaborations have their constituent documents under the general name "Memorandum of Understanding", which is a soft law mechanism that allows members of the collaboration to agree on their positions.

A similar approach was adapted for the International Experimental Thermonuclear Reactor—ITER [25]. The basic documents establishing the legal status of the ITER mega-science facility are the Agreement on the Establishment of the ITER for the joint implementation of the ITER Project and the Agreement on the Privileges and Immunities of the ITER for the joint implementation of the ITER Project. They formulated key conditions for the creation and operation of the project, determined the basic principles for financing the ITER mega-science facility and the distribution of costs between project partners.

We note that all these examples prove scientific groups to work comfortably and interact profitably in the absence of strictly legal regulation. Main aim of the collaboration (to achieve some scientific goal) force participants not to violate their obligations. This (disinterest in violation of obligations), however, does not completely exclude a possible dispute between partners. At the moment, there is already a practice of satisfying the claim of violation of the moral obligation of the party [26], which gives grounds to assert soft law to turn binding, and actually we face new form of regulations – so called "hoftlaw".

6 Conclusions

The technologies considered in this paper are not socially neutral – they, at the same time, as they offer society new opportunities and solutions, are a source of serious social risks and challenges. The search for ways to accelerate the development of innovative digital technologies as much as possible should make them routine, a lag in this matter can seriously undermine the country's security. There is a high probability that in the field of the prominent digital technologies (including artificial intelligence) we will soon see a real digital race – similar to nuclear, space and supercomputer ones. But, in general, the continued total implementation of digital technologies carries a number of risks, seriously changing existing social practices. There is a high probability of deep deformation (up to a complete shift) of the basic institutions of the society in favor of institutions that provide more effective and uncontested control of business over society and its institutions.

It is necessary to seek other methods and approaches to regulation, including a clear ethical framework for interaction with regard to digital technologies. In general, deep socio-theoretical models of the basic institutions of society in the era of digital transformation are needed. One of the possible theoretical and economic tools seems to be the idea of an institutional trap as a factor determining the direction and limitations of the society development [27].

Another requirement (in the moral and ethical area) is the anthropocentrism principle. This approach is already being implemented in the European Union and, taking into account the situation with the coronavirus pandemic (the importance of human life to be strictly articulated), one should expect the spread of this practice in territories with similar values [28].

In the field of law, it seems possible to develop adequate mechanisms of legal regulation in the digitalization era (taking into account the high role of technology in society), drawing on the experience of international scientific groups working with mega-science projects based on soft law and hoftlaw. The authorities will constitute and thrust their purposes, and the role of the governmental institutions (or society's structures associated with the state interests) will doubtlessly increase.

Ongoing international cooperation (and, in particular, joint financing, distribution of risks and rights to the results of work) require great flexibility in regulatory matters.

The high degree of interdependence of participants, the real globalization of all aspects of life, leads to the fact that the use of soft law and semi-rigid law (whose main strength lies in the high degree of persuasiveness and effectiveness of their inherent scientific objectivity of logic) seems more relevant in modern conditions than the use of binding law [22].

Totally, in order to overcome the emerging risks and challenges in connection with the digitalization processes, it is necessary to form a comprehensive attitude to technological development. It requires not only formal rules and regulatory mechanisms, but also the development of public understanding of the processes of introducing innovative technologies.

Acknowledgments. This work was supported by the RFBR grant No. 18–29-16130.

References

1. Balyakin, A.A., Malyshev, A.S., Nurbina, M.V., Titov, M.A.: Big data: Nil Novo Sub Luna. In: Antipova T. (eds.) Integrated Science in Digital Age. ICIS 2019. Lecture Notes in Networks and Systems, vol. 78, pp. 364–373, Springer, Cham (2020)
2. Lynch, C.: How do your data grow? Nature **455**, 28–29 (2008)
3. Kalyvas, J.R., Overly, M.R.: Big Data A Business and Legal Guide, p. 232. CRC Press Taylor & Francis Group, (2015)
4. Perspectives in Law, Business and Innovation. Springer Nature, p. 341. Singapore Pte Ltd. (2017)
5. Mayer-Schönberger, V., Cukier, K.: Big Data: A Revolution That Will Transform How We Live, Work, and Think. Mann, Ivanov, Ferber, Moscow (2014)
6. Balyakin, A.A., Mun, D.V.: Formation of an open science system in the European Union information and innovations. In: Proceedings of the Conference “Sciencemetry and bibliometry”, pp. 33–37 (2017)
7. Manuka, J., Chui, M., Brown, B., Bughin, J., Hobbs, R., Roxburgh, C., Byers, A.: Big Data: The Next Frontier for Innovation, Competition, and Productivity. McKinsey Global Institute (2011)
8. Florio, M., Sirtori, E.: Social benefits and costs of large scale research infrastructures. Technol. Forecast. Soc. Change **112**, 65–78 (2016)
9. Decree of the President of the Russian Federation dated December 01, 2016 No. 642 “On the Strategy for Scientific and Technological Development of the Russian Federation”
10. Order of the Government of the Russian Federation of July 28, 2017 No. 1632-r Program “Digital Economy of the Russian Federation”. https://static.government.ru/media/files/9gFM4FHj4PsB79I5v7yLVuPgu4bvR7M0.pdf. Accessed 02 Oct 2019
11. Forecast of the long-term socio-economic development of the Russian Federation for the period until 2030 (developed by the Ministry of Economic Development of Russia). https://www.consultant.ru/document/cons_doc_LAW_144190/. Accessed 07 Feb 2017
12. Ivanov, K.V., Balyakin, A.A., Malyshev, A.S.: Big data technologies as a tool for ensuring national security. St. Petersburg State Polytech. Univ. J. Sci. Tech. Sheets Econ. Sci. **13**(1), 7–19 (2020)
13. Decree of the President of the Russian Federation of July 25, 2019 No. 356 “On measures to develop synchrotron and neutron research and research infrastructure in the Russian Federation” (2019)

14. Mau, V.: Uberization of everything. RSMD (2020). https://russiancouncil.ru/analytics-and-comments/comments/uberizatsiya-vsego/?sphrase_id=40996163#detail. Accessed 02 June 2020
15. Kagarlitsky, B.Y.: Marxism: not recommended for education. In: Petrova, B. (eds.) Publishing house Algorithm. Series: Philosophical Bestseller, Moscow (2006)
16. Sadowski, J.: When data is capital: datafication, accumulation, and extraction. Big Data Soc. **6**(1), 1–12 (2019)
17. Zhulego, V.G., Balyakin, A.A., Nurbina, M.V., Taranenko, S.B.: Digitalization of society: New challenges in the social sphere. Bull. Altai Acad. Econ. Law **9–2**, 36–43 (2019)
18. Von Weizsaecker, E., Wijkman, A.: Come On! Capitalism, Short-termism, Population and the Destruction of the Planet, p. 46. Springer (2018)
19. Pigou, A.C.: The Economics of Welfare. Macmillan, London (1920)
20. Kornev, A.V.: Digital technologies, social processes, modernization of law and the possible future of Russian legal education. Lex Russica **4**(149), 23–30 (2019)
21. The EU General Data Protection Regulation (GDPR). https://gdpr-info.eu. Accessed 18 Nov 2019, Accessed 17 Mar 2020
22. Kashkin, S.Y.: The formation of the law of science as a new integrated branch of law. Bull. Univ. Named After O.E. Kutafina, Moscow (5), 16–27 (2018)
23. Nuclotron-based Ion Collider Facility. https://nica.jinr.ru/ru/megaproject.php. Accessed 3 oct 2019
24. Chetverikov, A.O.: The Large Hadron Collider as a Legal Phenomenon (Part 1). Lex Russica **2019**(4), 151–169 (2019). (In Russian)
25. THE ITER ORGANIZATION. https://www.iter.org/org. Accessed 17 Apr 2020
26. Balyakin, A.A., Nurbina, M.V., Taranenko, S.B.: Big data technologies as a factor in the integration of the EAEU. Mod. Eurasian Stud. **1**, 6–17 (2019)
27. Polterovich, V.M.: Institutional traps and economic reforms. Econ. Math. Methods **35**(2), 3–20 (1999)
28. Balyakin, A.A., Taranenko, S.B., Nurbina, M.V., Titov, M.A.: Social aspects of big data technology implementation. ICS. J. Digit. Sci. **1**(1), 15–24 (2019)

Citation Recommendation with Random Walking

Quang-Vinh Dang(✉)

Industrial University of Ho Chi Minh City, Ho Chi Minh City, Vietnam
dangquangvinh@iuh.edu.vn

Abstract. Citation recommendation is a research problem that has a lot of potential impact in both academia and practice. In this paper, we present a method of citation recommendation based on the content of documents. Our method does not require the metadata of the documents. We evaluate our method against some real-world datasets. The experimental results claim the advantages of our proposed approach compared to the other techniques.

Keywords: Citation recommendation · Natural language processing · Privacy preserving machine learning

1 Introduction

In recent years, we observe several major improvements in digital library technology [1]. However, due to the rapid development of research studies all over the world, it is difficult for a typical researcher to keep up with the state-of-th-eart research works [42]. The researchers are usually faced with the problem of finding the relevant information, or particularly, for their citations [20]. The citation recommendation task is defined as suggesting a small number of relevant of citations for the current research study of the researchers [27].

The citation recommendation task can be defined as follows. Given a query document d_q and a large number of candidate documents C. The task is to rank the documents in C in the order that the highest ranked documents should be cited in the d_q [5, 18].

The citation recommendation is an important task for researchers. Many solutions have been proposed over years. For instance, several studies suggest using the metadata such as author names or venue for the recommendation [45, 54]. Other studies suggested using a list of *seed* citations provided by the authors [28, 34, 39]. The authors of [5] argued that this information might not be available during some periods of the research process, such as in the early stage of the project.

It is worth to notice that the citation recommendation is different from the paper recommendation [4]. The paper recommendation aims to recommend users relevant papers to read [3] but not necessarily to cite for a study [33, 47].

In this paper, we present our method of relying on the *content* of the research study to find the relevant citation, like the work of [5, 27, 55]. However, we differ from these

T. Antipova (Ed.): ICCS 2020, LNNS 186, pp. 33–39, 2021.
https://doi.org/10.1007/978-3-030-66093-2_4

methods by using the random walk with restarts to measure the similarity of documents. We evaluate our method using PubMed and DBLP datasets (Fig. 1).

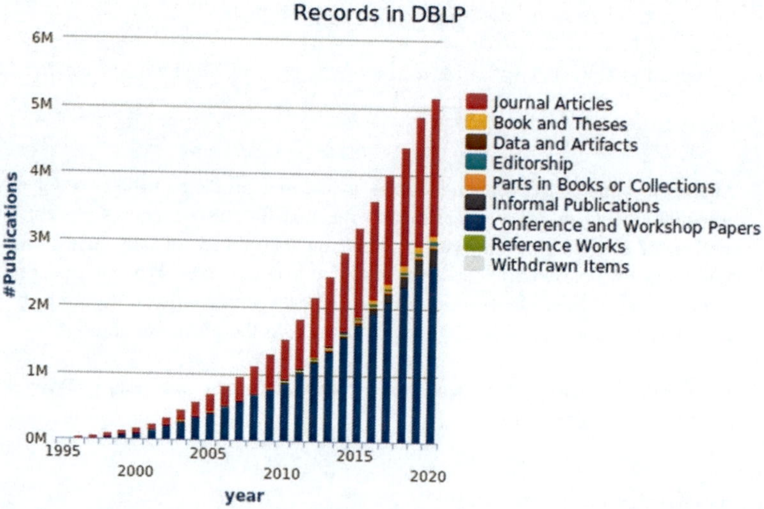

Fig. 1. Number of papers indexed by DBLP over year. Data source: https://dblp.org/statistics/recordsindblp

2 Related Works

The authors of [24] classified the citation recommendation task into the *local* and *global* recommendation. A local recommender looks into a specific context, such as a paragraph of a paper [22, 25] or even an abstract [29] while a global recommender take into the consideration the entire content, including the metadata of the paper [30, 48]. In this paper we follow the global recommendation approach [2]. According to the recent survey [18], the global recommendation is the most common approach in citation recommendation.

Several early research works have focused on hand-crafted feature engineering combining with an off-the-shelf classification algorithm for the citation recommendation, such as SVM [36] or the cosine similarity of TF-IDF of the documents [16]. Some authors used the more complicated machine learning algorithms, mostly ensemble methods based on decision trees [23, 37]. However, as the authors of [13] argued, hand-crafted feature engineering is a cumbersome task and might not be easy to transfer cross-domain.

Several authors use the topic modelling techniques [24, 30] based on the idea of LDA [6, 7]. The authors of [24] proposed a probabilistic model using Gleason's theorem [19] while the authors of [30] reused the LDA-variations Link-LDA and Link-

PLSA-LDA [41]. However, the major disadvantage of the approach is that the computationally expense is quite high [40, 49].

An interesting approach in citation recommendation is using machine translation for the task [22, 25]. The core idea of the approach is to *translate* the context of the query document into the reference, or the citation. The machine translation approach is suitable for both global [38] and local recommendation [22, 25]. The authors of [38] translated the entire original document into the cited documents, while the authors of [22] translated each part of the documents. The authors of [25] improved the approach by only translating to the identifier of the citation, saving a lot of the computation cost.

Due to the rapid development of deep learning [12, 32], it is not surprised that many recent research works focus on neural network models [51, 53]. The citation recommendation can be designed specifically or can be put under the umbrella of hyper-link prediction [11, 14, 21]. The authors of [17] designed a convolutional neural network with attention mechanism [52]. Negative sampling strategy is incorporated as well [31].

We would like to note a few approaches that exploits the metadata of the documents, such as the user's information. Particularly, the work of [5, 35] also allows the users to leak a self-defined proportion of personal information.

3 Our Approach

In this paper, we utilize the random walk [44] to measure the similarity between documents. Random walk has been applied successfully in many learning tasks, such as learning the similarity between graph nodes [8, 11, 14].

Assuming that we are currently at the *doc1* and we are moving to another node, either *word1* or *word2*. At the moment we do not have any preference so the probability of moving to any node of these two nodes is 0.5. By extending the process following the idea of PageRank [26, 43], we can generalize the transition matrix between nodes. The transition matrix states the probability of moving from one node to the other and can be interpreted as the similarity between documents controlling the words. The graph is modified regarding the citation information we gathered from the dataset (Fig. 2).

Overall, our idea is inspired by the work of [5]. The process can be divided into three phases:

- Candidate generation.
  - In this phase, we use pretrained BERT model [15] to embed the document context into the vector space. However, we differ from the work of [27] as we use solely the content of the document, while the authors of [27] also utilized the metadata of the text.
  - We measure the similarity of documents using the random walk [44]. At this step, we differ from the works of [5] and [27].

- The candidate list is established by using the nearest neighbors. The list can be extended by the rule as *citation in citation* under some constraints that are pre-defined. We used the metric of cosine similarity.

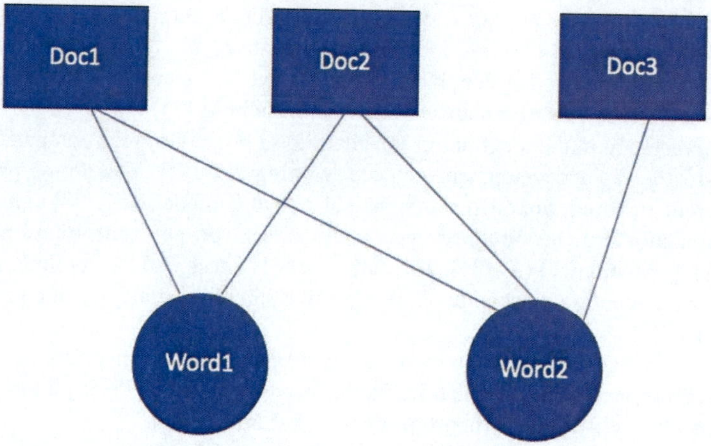

Fig.2. Documents with sharing words

– Ranking.
  - In this phase, each candidate is measured against the query document. – Recommendation.
  - In this phase, the final recommendation list is made.

4 Experimental Results

We used the DBLP [50] and PubMed dataset [45] for the evaluation. The latest DBLP dataset version as the time of writing is the version 12, that contains 4,894,081 articles with 45,564,149 citations[1]. The PubMed dataset contains more than 29 million citations.

We used the Mean Reciprocal Rank (MRR) and F1@20, consistency to the work of [5] for evaluation. We evaluated against two baseline models of BM25 [46] and ClusCite [45].

Table 1. F1@20 score

	DBLP dataset PubMed dataset	
BM25	0.121	0.213
ClusCite	0.242	0.268
RW	0.267	0.305

[1] https://www.aminer.org/citation.

Table 2. MRR score

	DBLP datasetPubMed dataset	
BM25	0.0.419	0.587
ClusCite	0.552	0.586
RW	0.589	0.652

We presented our model in Table 1 and Table 2. We observed our model achieved higher scores in both performance metrics, but there is still room for improvement.

5 Conclusions

In this paper, we present our approach on content-based citation recommendation. We improve the previous research works by introducing the random walk. In this study, we assume all honest citations between papers. In the future research, we would like to incorporate malicious citations into the consideration [9, 10].

References

1. Alzahrani, A.I., Mahmud, I., Ramayah, T., Alfarraj, O., Alalwan, N.: Modellingdigital library success using the delone and mclean information system success model. J. Librarianship Inf. Sci. **51**(2), 291–306 (2019)
2. Ayala-Gómez, F., Daróczy, B., Benczúr, A., Mathioudakis, M., Gionis, A.: Globalcitation recommendation using knowledge graphs. J. Intell. Fuzzy Syst. **34**(5), 3089–3100 (2018)
3. Bai, X., Wang, M., Lee, I., Yang, Z., Kong, X., Xia, F.: Scientific paper recommendation: a survey. IEEE Access **7**, 9324–9339 (2019)
4. Beel, J., Gipp, B., Langer, S., Breitinger, C.: Research-paper recommender systems: a literature survey. Int. J. Digital Libr. **17**(4), 305–338 (2016)
5. Bhagavatula, C., Feldman, S., Power, R., Ammar, W.: Content-based citation recommendation. In: NAACL-HLT, pp. 238–251. Association for Computational Linguistics (2018)
6. Blei, D.M., Ng, A.Y., Jordan, M.I.: Latent Dirichlet allocation. In: NIPS, pp. 601–608. MIT Press (2001)
7. Blei, D.M., Ng, A.Y., Jordan, M.I.: Latent Dirichlet allocation. J. Mach. Learn. Res. **3**, 993–1022 (2003)
8. Dang, Q.: Trust assessment in large-scale collaborative systems. (Evaluation de la confiance dans la collaboration à large échelle). Ph.D. thesis, University of Lorraine, Nancy, France (2018)
9. Dang, Q.: Studying machine learning techniques for intrusion detection systems. In: FDSE. Lecture Notes in Computer Science, vol. 11814, pp. 411–426. Springer (2019)
10. Dang, Q.V.: Active learning for intrusion detection systems. In: IEEE Research, Innovation and Vision for the Future (2020)
11. Dang, Q.V.: Link-sign prediction in signed directed networks from no link perspective. In: Lecture Notes in Networks and Systems. pp. 291–300. Springer (2020)
12. Dang, Q.V., Ignat, C.: Quality assessment of wikipedia articles: a deep learning approach. SIGWEB Newslett. **2016**(Autumn), 5:1–5:6 (2016)

13. Dang, Q.V., Ignat, C.: Quality assessment of wikipedia articles without feature engineering. In: JCDL, pp. 27–30. ACM (2016)
14. Dang, Q., Ignat, C.: Link-sign prediction in dynamic signed directed networks. In: CIC, pp. 36–45. IEEE Computer Society (2018)
15. Devlin, J., Chang, M., Lee, K., Toutanova, K.: BERT: pre-training of deep bidirectional transformers for language understanding. In: NAACL-HLT, no. 1, pp. 4171– 4186. Association for Computational Linguistics (2019)
16. Duma, D., Klein, E.: Citation resolution: a method for evaluating context-based citation recommendation systems. In: Proceedings of the 52nd Annual Meeting of the Association for Computational Linguistics (Volume 2: Short Papers), pp. 358–363 (2014)
17. Ebesu, T., Fang, Y.: Neural citation network for context-aware citation recommendation. In: SIGIR, pp. 1093–1096. ACM (2017)
18. Färber, M., Jatowt, A.: Citation recommendation: approaches and datasets. CoRRabs/ 2002.06961 (2020)
19. Gleason, A.M.: Measures on the closed subspaces of a hilbert space. J. Math. Mech. **6**, 885–893 (1957)
20. Guo, L., Cai, X., Qin, H., Guo, Y., Li, F., Tian, G.: Citation recommendation with a content-sensitive deepwalk based approach. In: ICDM Workshops, pp. 538–543. IEEE (2019)
21. Han, J., Song, Y., Zhao, W.X., Shi, S., Zhang, H.: hyperdoc2vec: distributed representations of hypertext documents. In: ACL, no. 1, pp. 2384–2394. Association for Computational Linguistics (2018)
22. He, J., Nie, J., Lu, Y., Zhao, W.X.: Position-aligned translation model for citation recommendation. In: SPIRE. Lecture Notes in Computer Science, vol. 7608, pp. 251–263. Springer (2012)
23. He, Q., Kifer, D., Pei, J., Mitra, P., Giles, C.L.: Citation recommendation without author supervision. In: WSDM, pp. 755–764. ACM (2011)
24. He, Q., Pei, J., Kifer, D., Mitra, P., Giles, C.L.: Context-aware citation recommendation. In: WWW, pp. 421–430. ACM (2010)
25. Huang, W., Kataria, S., Caragea, C., Mitra, P., Giles, C.L., Rokach, L.: Recommending citations: translating papers into references. In: CIKM, pp. 1910–1914. ACM (2012)
26. Jardine, J., Teufel, S.: Topical page rank: a model of scientific expertise for bibliographic search. In: Proceedings of the 14th Conference of the European Chapter of the Association for Computational Linguistics, pp. 501–510 (2014)
27. Jeong, C., Jang, S., Shin, H., Park, E.L., Choi, S.: A context-aware citation recommendation model with BERT and graph convolutional networks. CoRR abs/1903.06464 (2019)
28. Jia, H., Saule, E.: An analysis of citation recommender systems: beyond the obvious. In: ASONAM, pp. 216–223. ACM (2017)
29. Jiang, Z., Lu, Y., Liu, X.: Cross-language citation recommendation via publicationcontent and citation representation fusion. In: Proceedings of the 18th ACM/IEEE on Joint Conference on Digital Libraries, pp. 347–348 (2018)
30. Kataria, S., Mitra, P., Bhatia, S.: Utilizing context in generative bayesian models for linked corpus. In: AAAI. AAAI Press (2010)
31. Kobayashi, Y., Shimbo, M., Matsumoto, Y.: Citation recommendation using distributed representation of discourse facets in scientific articles. In: JCDL, pp. 243– 251. ACM (2018)
32. LeCun, Y., Bengio, Y., Hinton, G.: Deep learning. Nature **521**(7553), 436–444 (2015)
33. Li, S., Brusilovsky, P., Su, S., Cheng, X.: Conference paper recommendation for academic conferences. IEEE Access **6**, 17153–17164 (2018)
34. Liu, H., Kong, X., Bai, X., Wang, W., Bekele, T.M., Xia, F.: Context-based collaborative filtering for citation recommendation. IEEE Access **3**, 1695–1703 (2015)

35. Liu, X., Yu, Y., Guo, C., Sun, Y., Gao, L.: Full-text based context-rich heterogeneous network mining approach for citation recommendation. In: JCDL,. pp. 361–370. IEEE Computer Society (2014)
36. Liu, Y., Yan, R., Yan, H.: Guess what you will cite: personalized citation recommendation based on users' preference. In: AIRS. Lecture Notes in Computer Science, vol. 8281, pp. 428–439. Springer (2013)
37. Livne, A., Gokuladas, V., Teevan, J., Dumais, S.T., Adar, E.: Citesight: supporting contextual citation recommendation using differential search. In: SIGIR, pp. 807– 816. ACM (2014)
38. Lu, Y., He, J., Shan, D., Yan, H.: Recommending citations with translation model. In: CIKM, pp. 2017–2020. ACM (2011)
39. McNee, S.M., Albert, I., Cosley, D., Gopalkrishnan, P., Lam, S.K., Rashid, A.M., Konstan, J.A., Riedl, J.: On the recommending of citations for research papers. In: CSCW, pp. 116–125. ACM (2002)
40. Nallapati, R., Ahmed, A., Xing, E.P., Cohen, W.W.: Joint latent topic models for text and citations. In: KDD, pp. 542–550. ACM (2008)
41. Nallapati, R., Cohen, W.W.: Link-PLSA-LDA: a new unsupervised model for topics and influence of blogs. In: Icwsm, pp. 84–92 (2008)
42. Orduña-Malea, E., Ayllon, J.M., Martín-Martín, A., López-Cózar, E.D.: About thesize of google scholar: playing the numbers. CoRR abs/1407.6239 (2014)
43. Page, L., Brin, S., Motwani, R., Winograd, T.: The pagerank citation ranking: bringing order to the web. Technical report, Stanford InfoLab (1999)
44. Perozzi, B., Al-Rfou, R., Skiena, S.: Deepwalk: online learning of social representations. In: KDD, pp. 701–710. ACM (2014)
45. Ren, X., Liu, J., Yu, X., Khandelwal, U., Gu, Q., Wang, L., Han, J.: Cluscite: effective citation recommendation by information network-based clustering. In: KDD, pp. 821–830. ACM (2014)
46. Robertson, S.E., Walker, S.: Some simple effective approximations to the 2-poissonmodel for probabilistic weighted retrieval. In: SIGIR, pp. 232–241. ACM/Springer (1994)
47. Sharma, R., Gopalani, D., Meena, Y.: Concept-based approach for research paper recommendation. In: PReMI. Lecture Notes in Computer Science, vol. 10597, pp. 687–692. Springer (2017)
48. Strohman, T., Croft, W.B., Jensen, D.D.: Recommending citations for academic papers. In: SIGIR, pp. 705–706. ACM (2007)
49. Tang, J., Zhang, J.: A discriminative approach to topic-based citation recommendation. In: PAKDD. Lecture Notes in Computer Science, vol. 5476, pp. 572–579. Springer (2009)
50. Tang, J., Zhang, J., Yao, L., Li, J., Zhang, L., Su, Z.: Arnetminer: extraction and mining of academic social networks. In: KDD, pp. 990–998. ACM (2008)
51. Tang, X., Wan, X., Zhang, X.: Cross-language context-aware citation recommendation in scientific articles. In: SIGIR, pp. 817–826. ACM (2014)
52. Vaswani, A., Shazeer, N., Parmar, N., Uszkoreit, J., Jones, L., Gomez, A.N., Kaiser, L., Polosukhin, I.: Attention is all you need. In: NIPS, pp. 5998–6008 (2017)
53. Yin, J., Li, X.: Personalized citation recommendation via convolutional neural networks. In: APWeb/WAIM, no. 2. Lecture Notes in Computer Science, vol. 10367, pp. 285–293. Springer (2017)
54. Yu, X., Gu, Q., Zhou, M., Han, J.: Citation prediction in heterogeneous bibliographic networks. In: SDM, pp. 1119–1130. SIAM/Omnipress (2012)
55. Zhang, Y., Ma, Q.: Doccit2vec: Citation recommendation via embedding of content and structural contexts. IEEE Access **8**, 115865–115875 (2020)

Business, Finance & Accounting and Statistics

Digitization of Investment-Innovation Development of Ukrainian Economy: Empirical Analysis

Svitlana Tkalenko[1], Natalya Sukurova[2(✉)], Liudmyla Kudyrko[3], and Natalia Litvin[4]

[1] Department of European Economics and Business, Kyiv National Economic University named after Vadym Hetman, pr. Peremogy 54/1, Kiev 03057, Ukraine
sv.tkalenko@gmail.com

[2] Department of International Management, Kyiv National Trade and Economic University, Kioto Street 19, Kiev 02156, Ukraine
natalill@ukr.net

[3] Department of World Economy, Kyiv National Trade and Economic University, Kioto Street 19, Kiev 02156, Ukraine
l.kudyrko@knute.edu.ua

[4] «KROK» University, Tabirna Street, 30-32, Kiev 03113, Ukraine
nlitvin@gmail.com

Abstract. This article reveals the principles of research of investment-innovation development of the country in terms of digitization of the economy, which will lead to economic growth and increase the country's competitiveness on the world market. The article reveals the global trend of digitization, analyzes approaches to digitization, identifies the place of information and communication and digital technologies at the current stage of economic development, and focuses on the need of digital transformation of Ukraine. The authors focus on the multifaceted processes of digitization and digital transformation of the world's economies. Current trends of digital transformation of the world's economies and various processes have necessitated this study. In particular, the authors pay more attention to the digitization of investment-innovation development of Ukrainian economy on the basis of empirical analysis. The role and importance of foreign direct investments and the need of financing innovations in the digital transformation of Ukraine's economy to accelerate economic growth and increase global competitiveness are proved. Based on the analysis of works of domestic and foreign scientists, author's research and generalization of world economic development trends and Ukraine's for the last twenty years, the leading place of foreign direct investments and innovations in transformation processes that promote rapid economic growth have been proved. The econometric model constructed by the authors showed the relationship between macroeconomic factors (GDP, foreign trade, inflation and employment) and their influence on investment-innovation development in terms of digitization of economy.

Keywords: Digitization · Globalization · Information economy · Information and digital technology · Foreign direct investments · Innovations · Economic growth

T. Antipova (Ed.): ICCS 2020, LNNS 186, pp. 43–60, 2021.
https://doi.org/10.1007/978-3-030-66093-2_5

1 Introduction

The process of integration of Ukraine to the world community involves the development of investment-innovation activity of society, which radically changes economic, technological, social, organizational and other terms of this process and determines the competitiveness and stability of the country taking into account current global trends. Processes of globalization, scientific-technical and technological revolution, the dynamic development of information and communication technologies determine the importance of the need of digital transformation of Ukraine's economy, including investment-innovation development.

The purpose of our research is to identify changes of determinants of investment-innovation development of the country in terms of digital transformation as a general trend of further development of world economy. Based on this goal, the main tasks are to research the basic principles of digitization of the country and to identify its influence on the process of attracting foreign direct investments and the costs on innovations in the economy on the example of Ukraine.

The novelty of research in the presented article is the substantiation and econometric testing of functional relationships between the attraction of foreign direct investments and realization of costs on innovations and important macroeconomic indicators of the national economy. This approach will provide a more objective understanding of interdependencies between external and internal factors of development of Ukraine's economy and will form methodological bases for expert assessment in substantiating the priorities of digital and investment-innovation development in terms of global challenges of the XXI century.

Theoretical and applied aspects of digital transformation and development have been revealed in the works of famous Ukrainian scientists, such as Bilyk O [1], Dulska I. [2], Kraus N. [3], Rudenko M. [4], Fedulova L. [5] and others. The analysis of scientific works prepared by these authors showed that they paid attention to the digital transformation of the world's economies; to the processes of production and provision of services, which are one of the leading trends; to the issues of identifying the main drivers of digitization of the economy and their importance for social development; to the peculiarities of high-tech development in terms of globalization, the need for the introduction of digital technologies, which will contribute to the realization of the synergetic effect and economic growth.

Important achievements in the research of digitization of economy have made foreign scientists, the last of which are Stjepić A.-M. and et al [6], Isaacson W. [7], Grimes A. [8], Grabis J. [9], Niebel T. [10], Chinese researchers such as Young-Chan Kim and Pi-Chi Chen [11], etc.

Thus, digitization is of great importance, which is a tool for economic growth, increasing the competitiveness of investment-innovation sphere, creating new jobs and improving the lives of the population. That is why the authors prove the relevance of the digital transformation of Ukraine's economy and its influence on investment-innovation development. The role of foreign direct investments (FDI) in the digital economy, the determinants of attracting investments in the national economy and factors influencing on the level of costs on innovations are identified based on econometric modeling, using E-Views program.

2 Materials and Methods

An active phase of world development today is the global trend of digitization. Developed countries are successfully digitizing their economies, giving preference to innovative technologies, including digital platforms. The XXI century, which is characterized by rapid changes, has actualized the field of digitization, which is caused by the rapid development and spread of information-communication and digital technologies. Versatility and debatability of some issues related with digitization and digital transformation lead to further research of the impact of information-communication and digital technologies on economic development. Theories of information-communication and digital technologies have been studied by D. Bell [12], Castells M. and Himanen P. [13], A. Toffler [14], Wallerstein I [15] and others. In their researches, they revealed the principles of information economy formation, determined the role and influence of information-communication and digital technologies on economic development.

Consulting company (USA) Gartner, which specializes on information technology markets, defines digitization as “the use of digital technologies to change a business model and provide new revenue and value-producing opportunities” [16]. This definition focuses on changing business models and does not cover all areas of public life and economic development of the country. American scientists J. Scott Brennen and Daniel Kreiss consider digitization as “the way in which many domains of social life are restructured around digital communication and media infrastructures” [17]. Digital transformation is a broader term than digitization. This term includes strategic transformation and introduction of information-communication and digital technologies [18]. According to the authors’ views, digitization of the economy involves the creation of information and digital platforms by disseminating information-communication and digital technologies, which will solve more quickly and effectively strategic tasks.

In Ukraine, the need to form a digital economy and society is recognized at the state level, and digital technologies are considered as one of the key drivers of sustainable development [19]. The rapid and profound consequences of the transition to the digital economy will be possible only when digital transformation becomes the basis of life of Ukrainian society and economy, becomes a commonplace and everyday phenomenon, will be “our key agenda on the path to prosperity and the basis of well-being of Ukraine” [20]. The development of the national economy, modernization of existing technologies, digitization necessitate structural reforms.

Innovative technologies require large volumes of financing and investments. For Ukraine, whose economy has lack of national investments the question of attracting FDI is acute [21].

To achieve the setted goal it is necessary to research the current stage of digital transformation and current trends of national economy development; identify factors influencing the digitization of investment-innovation development; prove the importance of FDI and innovations for economic growth; using the econometric model to show the influence of the most significant factors on the volume of FDI attractions and investing in innovations and to prove their influence on growth in terms of digital transformation of the economy.

Due to the digital transformation, it becomes possible to develop innovations more rapidly, support startups, teach everyone the basics of programming and implement digital technologies in the field of economics. However, according to experts of The Boston Consulting Group by 2020, about 25% of the world economy will move to the introduction of digital technologies that allow government, business and society to function more efficiently and 16% of job losses [22].

Our hypothesis is that in terms of digital transformation of the Ukrainian economy the role and necessity of attraction of FDI and realization of bigger volumes of innovations grows; so, with the help of econometric modeling we will check the influence of exogenous and endogenous macroeconomic factors on investment-innovation development.

This study is based on authors' empirical analysis for the period 1999-2018 of economic development of Ukraine. This approach made it possible to identify those factors that have the greatest influence on investment-innovation development of the country in the era of digital globalization. Factors that affect the quantitative value of investment-innovation development (the volume of attracted FDI and volume of costs on innovations) are: GROWTH, FT, I, LM. The general view of such model illustrates the relationship of selected factors and is described by the following equation:

$$\mathrm{FII}(\mathrm{FDI} + \mathrm{SI}) = \mathrm{F}(\mathrm{GROWTH},\ \mathrm{FT}, \mathrm{I}, \mathrm{LM})$$

Where,

FII = the sum of attracted foreign direct investments in the economy of Ukraine and the costs made on innovations, mln. USD;

GROWTH = Growth represents National economic growth measured by the growth rate of real Gross Domestic Output (Ukraine's GDP), mln. USD;

FT = foreign trade of goods and services of Ukraine, USD mln. (export, import);

I = inflation index for the year;

LM = employed population, million people.

The authors built a multifactor regression model based on statistical data for the last two decades. The hypothesis concerning the influence of selected factors on the investment-innovation development of the country in the modern terms of economic digitization is tested with the use of software product E-Views. The absence of autocorrelation and heteroskedasticity was checked using various tests, the high quality of this model was proved, and the causal relationships between the selected factors were investigated using the Granger test.

However, it should be noted that authors used data before the current situation related to Covid-19, which may make adjustments to the economic development of the country. Such a "Black Swan" will make significant adjustments to the modern economic development of the world. We consider a faster transition to online technologies in most spheres of society as a positive consequence, in particular, revenues will increase in e-Commerce, pharmaceuticals, delivery services, etc; the negative consequences are projected in the near future to fall world GDP (in particular, according to JP Morgan - the loss will be $ 5.5 trillion by the end of 2021 [23]), loss of production, declining consumer demand, expected decline in revenues.

3 Results

Digitization of investment-innovation development involves the use of modern technologies to increase the volume of FDI and investments in innovations [24]. To build the model, we analyzed statistics for the last twenty years, such as the volume of foreign investment in Ukraine, the costs made on innovation (excluding foreign investors), GDP, exports and imports of goods and services, the rate of inflation and the rate of employed population [25]. Let's check how the factors we choose influence on the amount of FDI and investments in innovations. The authors constructed a correlation matrix (Table 1).

Table 1. Correlation matrix of selected variables

	FII	GROWTH	FT	I	LM
FII	1	0.1653	0.8397	0.03352	-0.6635
GROWTH	0.1653	1	0.2766	-0.1981	-0.0774
FT	0.8397	0.2766	1	-0.1301	-0.3443
I	0.0335	-0.1981	-0.1301	1	0.0019
LM	-0.6635	-0.0774	-0.3443	0.0019	1

Source: authors' development

Table 1 presents a correlation matrix that explains the relationship between selected variables and shows their influence on foreign direct investments and incurred costs for innovations. Matrix constructed by us confirms the success of the model.

The correlation matrix of variables shows a strong correlation between attracted foreign direct investments and the costs made on innovations and foreign trade - 84%; negative correlation (66%) between FII and the employed population (66%) and the presence in the model of a negative correlation between other variables. In general, an acceptable result of the correlation between variables actually confirms that our model is successful. The results of multifactor regression are shown in Table 2.

Table 2. The results of multi-factor regression

Variable	Coefficient	Std. Error	t-Statistic	Prob.
GROWTH	-0.007656	0.016551	-0.462595	0.6503
FT	0.280631	0.038253	7.336194	0.0000
I	21131.13	16048.06	1.316740	0.2077
LM	-3584.295	812.2249	-4.412934	0.0005
C	43745.64	25921.55	1.687617	0.1122
R-squared	0.881596	Mean dependent var		26690.03
Adjusted R-squared	0.850021	S.D. dependent var		18405.75
S.E. of regression	7128.008	Akaike info criterion		20.79377
Sum squared resid	7.62E+08	Schwarz criterion		21.04270
Log likelihood	-202.9377	Hannan-Quinn criter.		20.84236
F-statistic	27.92120	Durbin-Watson stat		2.258566
Prob(F-statistic)	0.000001			

Source: authors' development

Table 2 shows the results of multi-factor regression using the least squares method and different coefficients, which as a whole prove the significance of the equation.

Regarding the quality requirements of variables, we will apply the accepted average of 5-10% of the significance level. According to Fisher's F-statistics, all the coefficients of the regression equation simultaneously are not equal to 0. In our equation, foreign trade and the employed population are less than 5%; therefore, these factors are the most significant. As a rule, there are no strict requirements for the constant, in our case it is also not significant (more than 5-10%). The variables GROWTH (GDP) and I are insignificant because their probability is more than 10%.

R2 shows the extent to which the selected variables and their quantity explain investment-innovation development, which the authors statistically examine through the amount of FDI attracted to Ukraine and investments made in innovations. In the author's model, the researched variables explain the investment-innovation development of Ukraine by 88.2%; adjusted R-squared is 85%.

The presence of a sufficiently strong correlation indicates a correlation coefficient, which is 0.88; F-statistic = 0.000001, the probability of accepting the null hypothesis, confirms the need to take an alternative hypothesis, which certifies the significance of the equation as a whole.

We check the higher-order autocorrelation using the Breusch-Godfrey test. The idea of this test is as follows: if there is a correlation between adjacent observations, then we should expect that in the equation:

$$e_t = \rho \cdot e_{t-1} + v_t, \; t = 1, \ldots\ldots n,$$

where e_t – the regression residue obtained by the usual least squares method; coefficient - will be significantly different from zero.

The results of the test are shown in Table 3. We check with the help of diagnostics: at lag 2, the probability of accepting the null hypothesis is much greater than 0.05, which means that this lag is not significant. At lag 1, at 5% significance level, the probability of accepting the null hypothesis RESID (−1) = 0.6393 and RESID (2) = 0.8919, which is more than 5%, and means that we can accept the null hypothesis of no autocorrelation, i.e. no autocorrelation. The probability of accepting the null hypothesis is 86%, i.e. we claim that in our model there is no autocorrelation of random deviations.

Table 3. Breusch-Godfrey serial correlation LM test:

F-statistic	0.151902	Prob. F(2,13)	0.8606
Obs*R-squared	0.456717	Prob. Chi-Square(2)	0.7958

Test Equation: Dependent Variable: RESID; Method: Least Squares; Sample: 1999 2018 Included observations: 20;
Presample missing value lagged residuals set to zero.

Variable	Coefficient	Std. Error	t-Statistic	Prob.
GROWTH	0.000620	0.017620	0.035210	0.9724
FT	-0.000680	0.040803	-0.016667	0.9870
I	1747.407	18036.44	0.096882	0.9243
LM	-150.3835	988.6197	-0.152115	0.8814
C	871.3210	30440.79	0.028623	0.9776
RESID(-1)	-0.150210	0.313075	-0.479790	0.6393
RESID(-2)	0.043524	0.313926	0.138645	0.8919

R-squared	0.022836	Mean dependent var	-6.34E-12
Adjusted R-squared	-0.428163	S.D. dependent var	6333.402
S.E. of regression	7568.781	Akaike info criterion	20.97067
Sum squared resid	7.45E+08	Schwarz criterion	21.31917
Log likelihood	-202.7067	Hannan-Quinn criter.	21.03870
F-statistic	0.050634	Durbin-Watson stat	2.029886
Prob(F-statistic)	0.999252		

Source: authors' development
Table 3 shows the presence or absence of autocorrelation.

Let's analyze the information criteria Akaike, Schwarz, Hannan-Quinn, Durbin-Watson. Criteria Akaike (20.79), Schwarz (21.04), Hannan-Quinn (20.84) are not significant, which confirms the success of the model. We check the equation for the presence of autocorrelation using the Darbin-Watson test (d = 2.25). From the Darbin-Watson statistics table, we determine the significant points d_L and d_U. For the number of observations of 20 and 4 variables at a significance level of $\alpha = 5\%$ $d_L = 0.9$ and d_U

= 1.83; at 1% significance level $d_L = 0.68$ and $d_U = 1.57$. In our case, DW is greater than d_U, which means no autocorrelation.

We check the model for heteroskedasticity, using the following tests: Breusch-Pagan-Godfrey, Harvey, Glejser, ARCH, White and check the heteroskedasticity (non-constant variance) of random errors of the linear regression model. The test results are shown in Table 4, 5, 6, 7, and 8.

The probability of accepting the null hypothesis by the Breusch-Pagan-Godfrey test (Table 4) is 0.3926, which is more than 5% and confirms the absence of heteroskedasticity.

Table 4. Heteroskedasticity test: Breusch-Pagan-Godfrey

F-statistic	1.099571	Prob. F(4,15)	0.3926
Obs*R-squared	4.534714	Prob. Chi-Square(4)	0.3385
Scaled explained SS	2.674106	Prob. Chi-Square(4)	0.6138

Test Equation:	Dependent Variable: RESID^2	Method: Least Squares
Sample: 1999 2018	Included observations: 20	

Variable	Coefficient	Std. Error	t-Statistic	Prob.
C	-2.43E+08	2.04E+08	-1.191828	0.2518
GROWTH	-21.97924	130.0934	-0.168950	0.8681
FT	587.8557	300.6744	1.955124	0.0695
I	1.13E+08	1.26E+08	0.898619	0.3831
LM	5124186.	6384210.	0.802634	0.4347

R-squared	0.226736	Mean dependent var	38106378
Adjusted R-squared	0.020532	S.D. dependent var	56611406
S.E. of regression	56027223	Akaike info criterion	38.73289
Sum squared resid	4.71E+16	Schwarz criterion	38.98182
Log likelihood	-382.3289	Hannan-Quinn criter.	38.78149
F-statistic	1.099571	Durbin-Watson stat	1.758367
Prob(F-statistic)	0.392561		

Source: authors' development
Table 4 shows the presence or absence of heteroskedasticity

The Harvey test in Table 5 also shows the absence of heteroskedasticity, the probability of accepting the null hypothesis is 0.1539.

Table 5. Heteroskedasticity test: Harvey

F-statistic	1.952104	Prob. F(4,15)	0.1539
Obs*R-squared	6.846961	Prob. Chi-Square(4)	0.1442
Scaled explained SS	13.63282	Prob. Chi-Square(4)	0.0086

Test Equation:	Dependent Variable: LRESID2	Method: Least Squares
Sample: 1999 2018	Included observations: 20	

Variable	Coefficient	Std. Error	t-Statistic	Prob.
C	3.172557	10.67429	0.297215	0.7704
GROWTH	-5.85E-06	6.82E-06	-0.857891	0.4045
FT	3.51E-05	1.58E-05	2.225578	0.0418
I	-3.658406	6.608464	-0.553594	0.5880
LM	0.715873	0.334468	2.140334	0.0492

R-squared	0.342348	Mean dependent var	15.36130
Adjusted R-squared	0.166974	S.D. dependent var	3.216007
S.E. of regression	2.935257	Akaike info criterion	5.203785
Sum squared resid	129.2360	Schwarz criterion	5.452718
Log likelihood	-47.03785	Hannan-Quinn criter.	5.252379
F-statistic	1.952104	Durbin-Watson stat	2.044076
Prob(F-statistic)	0.153939		

Source: authors' development
Table 5 shows the presence or absence of heteroskedasticity

The following check for heteroskedasticity is by using the Glejser test for our variables (Table 6). All variable models were also tested separately by the test. All variables are not statistically significant, as the probability of accepting the null hypothesis is more than 5%, ie it can be accepted. The presence of regression in general is also high, ie in our model there is no heteroskedasticity, ie the model residues do not depend on the selected variables.

Table 6. Heteroskedasticity test: Glejser

F-statistic	1.610237	Prob. F(4,15)	0.2232
Obs*R-squared	6.008082	Prob. Chi-Square(4)	0.1985
Scaled explained SS	5.671248	Prob. Chi-Square(4)	0.2251

Test Equation:	Dependent Variable: ARESID	Method: Least Squares
Sample: 1999 2018	Included observations: 20	

Variable	Coefficient	Std. Error	t-Statistic	Prob.
C	-16942.15	14662.43	-1.155480	0.2660
GROWTH	-0.004867	0.009362	-0.519876	0.6107
FT	0.053611	0.021638	2.477676	0.0256
I	5439.481	9077.529	0.599225	0.5580
LM	560.7568	459.4321	1.220543	0.2411

R-squared	0.300404	Mean dependent var	4547.380
Adjusted R-squared	0.113845	S.D. dependent var	4283.102
S.E. of regression	4031.932	Akaike info criterion	19.65420
Sum squared resid	2.44E+08	Schwarz criterion	19.90313
Log likelihood	-191.5420	Hannan-Quinn criter.	19.70279
F-statistic	1.610237	Durbin-Watson stat	1.681205
Prob(F-statistic)	0.223173		

Source: authors' development
Table 6 shows the presence or absence of heteroskedasticity

The ARCH test (Table 7) also shows the absence of heteroskedasticity and the probability of accepting the null hypothesis.

Table 7. Heteroskedasticity TEST: ARCH

F-statistic	2.404302	Prob. F(1,17)	0.1394
Obs*R-squared	2.354206	Prob. Chi-Square(1)	0.1249

Test Equation: Dependent Variable: RESID^2 Method: Least Squares
Sample (adjusted): 2000 2018 Included observations: 19 after adjustments

Variable	Coefficient	Std. Error	t-Statistic	Prob.
C	25706243	15662977	1.641211	0.1191
RESID^2(-1)	0.353007	0.227661	1.550581	0.1394

R-squared	0.123906	Mean dependent var	39861904
Adjusted R-squared	0.072371	S.D. dependent var	57600668
S.E. of regression	55477229	Akaike info criterion	38.60014
Sum squared resid	5.23E+16	Schwarz criterion	38.69956
Log likelihood	-364.7014	Hannan-Quinn criter.	38.61697
F-statistic	2.404302	Durbin-Watson stat	1.743029
Prob(F-statistic)	0.139418		

Source: authors' development
Table 7 shows the presence or absence of heteroskedasticity

Another test using the White test, which is universal. By value of accepting the null hypothesis, all variables are not statistically significant; in addition, prob. F = 68%, which indicates the absence of heteroskedasticity (Table 8).

Table 8. Heteroskedasticity test: White

F-statistic	0.766654	Prob. F(14,5)	0.6827
Obs*R-squared	13.64400	Prob. Chi-Square(14)	0.4766
Scaled explained SS	8.045819	Prob. Chi-Square(14)	0.8869

Test Equation:	Dependent Variable: RESID^2	Method: Least Squares
Sample: 1999 2018	Included observations: 20	

Variable	Coefficient	Std. Error	t-Statistic	Prob.
C	1.92E+09	1.09E+10	0.176764	0.8666
GROWTH	80322.68	115039.3	0.698219	0.5162
GROWTH^2	-0.016848	0.019646	-0.857574	0.4303
GROWTH*FT	-0.000751	0.102203	-0.007351	0.9944
GROWTH*I	-84594.06	110139.1	-0.768066	0.4771
GROWTH*LM	1023.592	1400.684	0.730780	0.4977
FT	-87501.24	132810.4	-0.658843	0.5391
FT^2	0.020337	0.115164	0.176592	0.8668
FT*I	96144.39	115707.1	0.830929	0.4439
FT*LM	-1300.350	1697.071	-0.766232	0.4781
I	-4.62E+09	1.53E+10	-0.301549	0.7751
I^2	-5.10E+08	4.54E+09	-0.112256	0.9150
I*LM	2.80E+08	3.06E+08	0.915547	0.4019
LM	77673021	3.80E+08	0.204560	0.8460
LM^2	-9837937.	8373992.	-1.174820	0.2930

R-squared	0.682200	Mean dependent var	38106378
Adjusted R-squared	-0.207641	S.D. dependent var	56611406
S.E. of regression	62211810	Akaike info criterion	38.84369
Sum squared resid	1.94E+16	Schwarz criterion	39.59049
Log likelihood	-373.4369	Hannan-Quinn criter.	38.98948
F-statistic	0.766654	Durbin-Watson stat	2.357996
Prob(F-statistic)	0.682714		

Source: authors' development
Table 8 shows the presence or absence of heteroskedasticity

The model is verified on explanatory capability, since it should accurately reflect FDI inflows using available independent variables (*see Fig. 1*). As can be seen from the graph, the fitted values (Fitted) have the same trends with actual values (Actual), so the model is fully acceptable under this criterion.

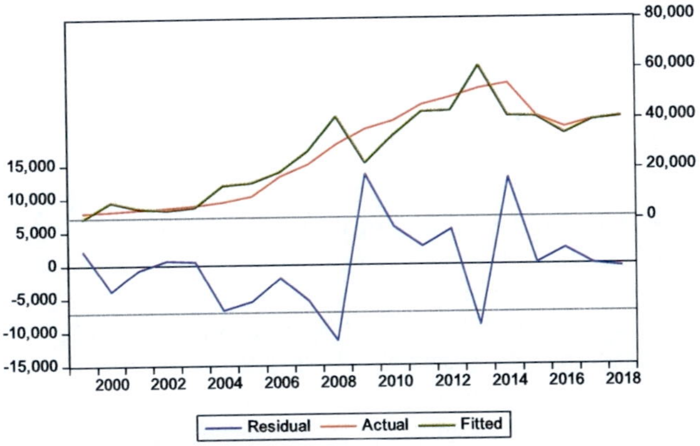

Fig. 1. Explanatory capability of the model *Source: authors' development*

We check the model for predictive quality: we pay attention to the MAPE criterion. For each integrated risk indicator, an approach is defined that improves the minimum forecast error. In our case, the criterion MAPE = 29.66, which means acceptable (but not high) accuracy of the forecast (see Fig. 2).

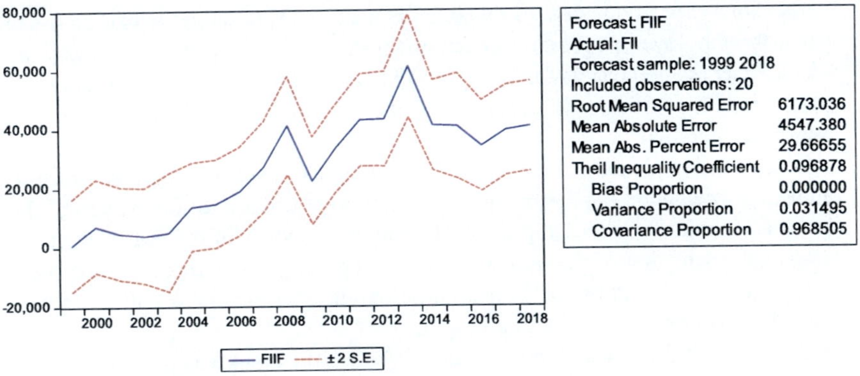

Fig. 2. Forecast of models *Source: authors' development*

The study of causal relationship between the selected variables and the inflow of FDI and the realization of financing for innovations was implemented using the Granger test. The idea of such test is as follows: if a change in variable A is the cause of change B, then change A precedes change B; that is, that the primary A or B. The regression equation of the two hypotheses of the directions of causality is as follows:

$$A_i = \alpha_0 + \sum_{i=1}^{n} \alpha_1 B_{t-i} + \sum_{j=1}^{n} \beta_j A_{t-j} + \varepsilon_t,$$

where,

A i *B* – dependent and independent variables, respectively; α_0, α_i, β_j – coefficients calculated as a result of regression; ε_t ε_t – the final member of the regression; *ii*, *j* = 1, 2, …, *n* – the time delay.

As a result of the Granger test, the null hypothesis A is tested and is not the cause of changes B. The criterion for accepting the hypothesis is Prob. (the value of the probability of accepting the hypothesis). At the same time, the opposite causal relationship is also checked. If two coefficients are statistically significant at the same time, the dependence is two-way, or feedback-dependent. The existence of a two-way causal relationship may mean the existence of a third variable, which is the real cause of the changes in the two variables considered in the equation. If two coefficients are statistically zero at the same time, there is no cause-and-effect relationship (this is possible even if there is a statistically significant correlation). The tests were performed for lags 2–6. The results of the test are shown in Table 9.

According to the results of the Granger test (Table 10) we make a number of generalized conclusions:

- during the first year, the employed population and the volumes of attracted FDI and innovations expenditures are in dynamic interaction. This suggests that, on the one hand, the employed population affects investments and innovations development, on the other hand, the situation with the volume of attracted FDI and investments in innovations determines the dynamics of the employed population (including the return of migrants);
- change of the employed population at a certain time (+5 years) affects the process of attracting foreign direct investments and increasing the costs for innovations, which in turn will contribute to the growth of national economy in the near future;
- unilateral orientation of the influence of the employed population from the second period (+5 years) also has an impact on inflation. Such influence will promote the growth of volumes of production output;
- with a delay in the future, foreign trade will affect the employed population, which indicates a growing trend in services sphere.

The general form of the multi-regression model of dependence of investment-innovation development of Ukraine from independent variables can be described by the following equation:

Substituted Coefficients:

FII = −0.00765641794241 * GROWTH + 0.280631482649 * FT + 21131. 1275458 *I − 3584.29483973 * LM + 43745.6381445

Table 9. The results of Pairwise Granger Causality Test on cause-effect dependence for the period 1999-2018

	Null hypothesis	lag 2 (Obs 18)			lag 3 (Obs 17)		lag 4 (Obs 16)		lag 5 (Obs 15)		lag 6 (Obs 14)	
		F-Stat.	Prob.	Conclusion	F-Stat.	Prob.	F-Stat.	Prob.	F-Stat.	Prob.	F-Stat.	Prob.
1	GROWTH does not Granger Cause FII	0.042	0.958	accept	0.315	0.814	0.193	0.934	0.518	0.756	7.726	0.268
	FII does not Granger Cause GROWTH	0.253	0.780	accept	0.183	0.904	0.676	0.629	1.039	0.499	6.577	0.289
2	FT does not Granger Cause FII	2.617	0.110	accept	1.978	0.181	0.823	0.549	0.592	0.712	2.064	0.487
	FII does not Granger Cause FT	0.328	0.725	accept	0.451	0.721	0.442	0.775	0.524	0.752	0.680	0.729
3	I does not Granger Cause FII	2.225	0.147	accept	1.517	0.269	1.664	0.260	7.258	0.038	1.673	0.531
	FII does not Granger Cause I	0.609	0.558	accept	1.304	0.326	0.646	0.647	0.312	0.882	1.946	0.499
4	LM does not Granger Cause FII	5.559	0.018	reject	9.109	0.003	7.946	0.009	9.284	0.025	24.065	0.154
	FII does not Granger Cause LM	4.005	0.044	reject	2.092	0.164	2.417	0.145	3.342	0.132	1.385	0.571
5	FT does not Granger Cause GROWTH	0.184	0.833	accept	0.142	0.932	0.637	0.652	0.753	0.625	2.342	0.462
	GROWTH does not Granger Cause F	0.012	0.987	accept	0.020	0.995	0.112	0.974	0.445	0.800	11.679	0.220
6	I does not Granger Cause GROWTH	2.960	0.087	accept	2.372	0.131	10.365	0.004	22.011	0.005	1.821	0.513
	GROWTH does not Granger Cause I	0.032	0.968	accept	0.072	0.973	0.505	0.734	1.031	0.502	0.615	0.750
7	LM does not Granger Cause GROWTH	0.022	0.977	accept	0.084	0.966	2.065	0.189	0.921	0.546	38.092	0.123
	GROWTH does not Granger Cause LM	0.002	0.997	accept	0.012	0.997	0.005	0.999	0.077	0.992	0.502	0.791
8	I does not Granger Cause FT	0.689	0.519	accept	0.383	0.767	0.568	0.694	0.399	0.829	0.940	0.657
	FT does not Granger Cause I	1.052	0.377	accept	0.554	0.656	0.193	0.934	0.160	0.964	8.497	0.256
9	LM does not Granger Cause FT	0.305	0.741	accept	0.548	0.660	0.716	0.606	0.908	0.552	1.780	0.518
	FT does not Granger Cause LM	1.607	0.237	accept	1.386	0.303	1.437	0.316	652.656	7.E-06	854.48	0.026
10	LM does not Granger Cause I	2.575	0.114	accept	4.616	0.028	4.989	0.032	5.673	0.058	6011.6	0.009
	I does not Granger Cause LM	0.147	0.864	accept	0.870	0.488	0.791	0.566	1.560	0.343	0.962	0.652

Source: authors' development

Table 9 shows the results of Pairwise Granger Causality Test on cause-effect dependence with 2, 3, 4, 5 and 6 lags, which show the effect of one factor on another in our model.

Table 10. Interpretation of Granger test results

lag 2	lag 3	lag 4	lag 5	lag 6
GROWTH there is no connection FII	GROWTH there is no connection FII	GROWTH there is no connection FII	GROWTH there is no connection FII	GROWTH there is no connection FII
FT there is no connection FII	FT there is no connection FII	FT there is no connection FII	FT there is no connection FII	FT there is no connection FII
I there is no connection FI	I there is no connection FI	I there is no connection FI	I ⟶ FI	I there is no connection FI
LM ⟷ FII	LM ⟶ FII	LM ⟶ FII	LM ⟶ FII	LM there is no connection FII
FT there is no connection GROWTH	FT there is no connection GROWTH	FT there is no connection GROWTH	FT there is no connection GROWTH	FT there is no connection GROWTH
I there is no connection GROWTH	I there is no connection GROWTH	I ⟶ GROWTH	I ⟶ GROWTH	I there is no connection GROWTH
LM there is no connection GROWTH	LM there is no connection GROWTH	LM there is no connection GROWTH	LM there is no connection GROWTH	LM there is no connection GROWTH
I there is no connection FT	I there is no connection FT	I there is no connection FT	I there is no connection FT	I there is no connection FT
LM there is no connection FT	LM there is no connection FT	LM there is no connection FT	LM ⟵ FT	LM ⟵ FT
LM there is no connection I	LM ⟶ I	LM ⟶ I	LM ⟶ I	LM ⟶ I

Source: authors' development

Table 10 shows the interpretation of the Granger test from the point of view of the direction of causal relationships between selected variables studied in the model.

4 Conclusion

Researched problem proves the significant role of the era of digital globalization, which is characterized by objectivity and irreversibility of the process of formation of information economies, continuous and rapid data flows that contain knowledge and innovations, ideas and information. Positive effects of digitization will soon be manifested through increased efficiency, faster receipt of necessary information and the establishment of mutually beneficial relations with other entities, optimization and scaling of operations, and so on. For Ukraine, this will be manifested through the growth of the activity of TNCs, which will lead to the growth of FDI attraction and to increase the competitiveness of the national economy on the world market to increase investments in innovations.

In this article, the authors analyze the determinants of investment-innovation development of the country's economy, which are necessary for digital transformation. The constructed model showed a high level of sensitivity of FDI and costs made on innovations from the rate of employment and inflation. Predictions are confirmed that digitization will help to release jobs, increase unemployment and rethink the role of the employee in the society and in the company. The growth of the level of foreign trade by 1% will contribute to the growth of the sum of FDI and investments in innovations, which is a positive trend for information transformation of Ukraine's economy. GDP in our model does not play a significant role however, we consider that increasing the volume of innovative competitive products will have a positive impact on foreign investors and innovators.

Prospects for the study should focus on the creation and promotion of digital platforms, on the support of innovative companies, on creating an attractive environment for investment and innovation development of the country. At the same time, further steps are studies that will reveal the following important questions: How will Covid-19 affect the economic development of the world and determine the trends of information transformation? What incentives should be given to companies to support them, which invest in innovative, high-tech products? What steps should the country take to accelerate digital transformation reforms?

References

1. Bilyk, O.: The impact of the digital economy on reducing the negative effects of social risk. Probl. Econ. Manage. Issue **4**, 8–16 (2019). (in Ukrainian)
2. Dulska, I.: Digital technologies as a catalyst for economic growth. Econ. Forecast. **2**, 119–133 (2015). (in Ukrainian)
3. Kraus, N.: Digital economy: trends and prospects of avant-garde development. Efficient economy. http://www.economy.nayka.com.ua/pdf/1_-2018/8.pdf №. 1. (2018). (in Ukrainian)
4. Rudenko, M.: Digitization of economy: new opportunities and perspectives. Ekonomika ta derzhava **11**, 61–65 (2018). https://doi.org/10.32702/2306-6806.2018.11.61. (in Ukrainian)
5. Fedulova, L.: Digital transformation (digitization) of the regions of Ukraine. Analytical note (2018). http://academy.gov.ua/pages/dop/198/files/4ba4c1b4-cefe-4f27-b58b-3aee7c8cf152.pdf. (in Ukrainian)
6. Stjepić, A.M., Ivančić, L., Vugec, D.S.: Mastering digital transformation through business process management: Investigating alignments, goals, orchestration, and roles. J. Entrep. Manage. Innov. (JEMI) **16**(1) (2020)
7. Walter, I.: The Innovators: How a Group of Hackers, Geniuses and Geeks Created the Digital Revolution, 561 p. (2014)
8. Grimes, A., Ren, C., Stevens, P.: The need for speed: impacts of internet connectivity on firm productivity. J. Prod. Anal. **37**(2), 187–201 (2012)
9. Grabis, J., Kampars, J.: Application of micro services for digital transformation of data-intensive business processes. In: ICEIS, no. 2, pp. 736–742 (2018). https://www.scitepress.org/papers/2018/68052/68052.pdf
10. Niebel, T.: ICT and Economic Growth: Comparing Developing, Emerging and Developed Countries. ZEW Discussion Paper 14–117, Germany (2014)
11. Young-Chan, K., Pi-Chi, C.: The Digitization of Business in Chins, London, UK, 197 p. (2018)
12. Bell, D., Inozemtsev V.L.: The Age of Disunity: Reflections on the World of the 21st Century, Moscow (2007). (in Russian)

13. Castells, M., Himanen, P.: The Information Society and the Welfare State: The Finnish Model. Oxford University Press, Oxford (2004)
14. Alvin, T., Heidi, T.: Revolutionary Wealth, Moscow (2008). (in Russian)
15. Wallerstein, I.: The end of the familiar world. Sociology of the XXI century, Moscow (2003). [in Russian]
16. Gartner Glossary. https://www.gartner.com/en/glossary
17. Scott Brennen, J., Daniel, K.: Digitization (2016). https://onlinelibrary.wiley.com/doi/epdf/10.1002/9781118766804.wbiect111
18. Bloomberg, J. Digitization, Digitization, And Digital Transformation: Confuse Them at Your Peril. https://www.forbes.com/sites/jasonbloomberg/2018/04/29/digitization-digitization-and-digitaltransformation-confuse-them-at-your-peril/#1e1fe-3d02f2c. Accessed 05 Nov 2018
19. The concept of development of the digital economy and society of Ukraine for 2018–2020. http://zakon3.rada.gov.ua/laws/show/67-2018-%D1%80 [in Ukrainian]
20. Digital Agenda of Ukraine – 2020. Conceptual principles (version 1.0). Priority areas, initiatives, projects of «digitization» of Ukraine until 2020 (2016). https://ucci.org.ua/uploads/files/58e78ee3c3922.pdf. (in Ukrainian)
21. Svitlana, T., Natalya, S.: The influence of FDI on sustainable economic development of Ukraine in terms of global digitization Journal of Digital Science, vol. 2, no. 1, pp. 40–54 (2020). https://ics.events/journal-of-digital-science/
22. Website The Boston Consulting Group. https://www.bcg.com/d/news/8june15-projected-twenty-five-percent-increase-in-use-of-robotics-15337
23. Website JP Morgan. https://www.jpmorgan.com/global/insights/coronavirus-update
24. Tkalenko, S., Sukurova, N., Honcharova, A.: Determinants of the foreign direct investments in terms of digital transformation of the Ukrainian economy. In: Antipova, T., Rocha, Á. (eds.) DSIC 2019. AISC, vol. 1114, pp. 148–164. Springer, Cham (2020)
25. Website State Statistics Committee of Ukraine. www.ukrstat.gov.ua/

Russian Accounting Practice in the 17th Century and Its Transformation by Peter the Great

Svetlana Karelskaia(✉) and Ekaterina Zuga

St. Petersburg University, Saint Petersburg, Russia
s.karelskaya@spbu.ru

Abstract. The paper is devoted to the development of accounting in Russia at the turn of the 17th–18th century. Authors prove the necessity of consideration for particular types of documents of business practice to regulate accounting rules, - acting letters, memories and decrees. Their study allows to explain the similarity of accounting systems of various economic entities, including state institutions, state-owned enterprises and monasteries, which help to clarify the content of transformations of Russian accounting provided by Peter the Great. They are systematized and classified as innovative for Russia or practice-based. The research is based on the analysis of national and local regulations published in the Complete collection of laws of the Russian Empire, as well as documents of accounting practice described in the works of researchers of the history of state records management and accounting.

Keywords: Accounting · Reform · Peter the great · State-owned enterprise · Accounting books

1 Introduction

In Russia the reign period of Peter the Great (years of life: 1672–1725, years of rule: 1682–1725, including the reign without co-rulers: 1696–1725) accounting historians characterize as the beginning of government regulation of accounting rules, the introduction of special accounting terminology into practice, the appearance of the position of accountant in state bodies and enterprises. In our opinion, these changes should be considered by dividing it into two groups: completely innovative for Russia and caused by the economic practice of the previous century. Peter I and his reforms always attract the attention of many researchers. They study transformations in Russian accounting, but there are regular discoveries, which confirm the relevance of this research area.

The methodological basis of this paper is works on the accounting history of Sokolov [1], as well as history of records management - Pronina [2], Sirenov [3], etc., and regulation of accounting rules and accounting practices in state-owned enterprises of the 17th century – Lvova [4], Motyka [5], Platonova [6], Sidorova and Nazarov [7], Sokolov et al. [8], Shirokiy [9], Mazdorov [10], etc. The approaches used to study the

T. Antipova (Ed.): ICCS 2020, LNNS 186, pp. 61–69, 2021.
https://doi.org/10.1007/978-3-030-66093-2_6

accounting practice by Russian scientists E. Sievers and A. Galagan were described by Kuter et al. [11].

The purpose of this paper is to analyze Peter the Great innovations in accounting regulation and assess their impact on accounting practice. The key task is to study the use of a special type of documents of economic practice - orders. The subject of the research is the evolution of accounting regulations in Russia during 17th and 18th centuries.

2 Methodological Framework

The paper is aimed at the analysis of changes in the accounting regulation at the turn of the 17th–18th century, implemented during Peter the Great reforms of the state structure, as well as the development of public sector accounting, the charge and discharge accounting [12–14].

Context studies based on different theories (Foucault, Habermas, Bourdieu, etc.) are widespread among recent papers on the accounting history. But this study is more closely related to the traditional technical research according to the classification of Broadbent and Guthrie [15]. It is devoted to the technical accounting, which according to the classification of historical works by Jacobs [16], belongs to the group that does not contain a theory. Almost half of the works on the history of public sector accounting are written in such way.

Various historical sources are used to study Russian accounting practices at the turn of the 17th and 18th centuries. Special accounting terms appeared in the Russian language, the structure of accounting registers was completely transformed, simplifications and uniformity were introduced into office work. This made possible the perception of double entry accounting, which was applied in Europe. Three decades after Peter the Great reforms it was introduced in Russia at the legislative level.

3 History and Structure of the Administrative and Financial Systems of Moscow Monarchy

In the 17th century Russia[1] was a centralized Moscow state that began to form after the liberation from the Tatar-Mongol yoke (1240s–1480). Centralization was carried out by uniting the scattered Russian principalities and free cities into a single state around Moscow. The Central government relied on military force instead of commercial and business elite, as it was in Western Europe [17].

Russian industry was poorly developed. By the end of the 17th century there were no more than 30 manufacturies [19]. Almost all of them were state-owned and covered a narrow range of industries, their development was determined by the state needs and interests of the Royal court (arming troops, coining money, etc.). One of the first such

[1] In this paper Russia is understood similarly to Motyka [5] as the territories that were originally the territories of the state of Veliky Novgorod and the Moscow state before the 18th century. Since the 18th century they belong to the Russian Empire.

enterprises was the foundry Gun yard built in 1479 in Moscow. There were also a Printing House, a Khamovny[2] House, a Mint, an Armory, a Silver chamber, brick factories. Similar state-owned enterprises were also common in Western Europe (manufactories of Henry of Bourbon and Colbert in France, Elizabethan manufactories in England) [20]. The needs of the Royal court were served by the Palace manufactories produced luxury goods: velvet, fine linen, fine leather, morocco, glass, and etc. [17]. All the products of the manufactories went to the state Treasury and controlled by state authorities. Supplies to the Treasury were carried out mainly at cost and only its surplus came to the market for sale [18]. But state participation extended not only to unprofitable enterprises. Also, regalia - state monopolies on trade in certain types of goods - were established for the most profitable types of trade: vodka within the country; hemp, bread, resin, etc. – out of the country.

The centralization of the state, which began under the Grand Duke of Moscow Ivan III (years of life: 1440–1505, years of rule: 1462–1505), was accompanied by the formation of the state administration apparatus. Since the end of the 15th century the state was governed through a system of Orders (*prikazi*) - permanent state institutions with functional or territorial jurisdiction. In the 17th century there were almost 40 of them. The Orders (*prikazi*) had its own Treasury and the right to collect various taxes and bear expenses, each within its own jurisdiction. There were budgets for each province and Chancellery. Each Chancellery usually exercised financial control over its subordinate services. Thus, the financial system of the state was decentralized [8]. Accounting Order (*prikaz*) was mentioned in documents from 1656 to 1701. Its main task was the implementation of state external financial and property control [21].

Unified Russian state legislation began with the adoption the Law Code (*sudebnik*) of Ivan III[3] in 1497. It was a set of laws of the Russian state, which established rules of legal proceedings uniform for the entire territory of the centralized Moscow state, as well as uniform legal provisions on offenses and measures of legal responsibility [23]. Later it was replaced by the Law Code (*sudebnik*) dated 1550 and 1589. In addition the legislation of the 17th centuries included other legal acts issued by the czar, state and local authorities in the form of letters, sentences, and resolutions on a wide range of issues. The statutory and specified book of Orders (*prikazi*) (Assault, Local, Rural, etc.) were compiled. They were significant in volume, included all the legal documentation of Orders (*prikazi*) and current legislation, served as an explanation and supplement to the Law Code (*sudebnik*) and fulfilled their role [24]. In 1649 a new codified legislative act, the Council Code[4], was adopted and served as the basis of Russian legislation for

[2] Khamovny House ('Khaman' - white paper cloth) - state textile manufactories in Moscow in the 17th and early 18th centuries.

[3] Law Code (*sudebnik*) appeared earlier than the same documents in other European countries. An earlier document was the Law Code (*sudebnik*) adopted in 1468 in the Grand Duchy of Lithuania. French king Louis XI (1461–1483) made an attempt, but did not have time to carry out the plan. In 1532 Holy Roman Emperor Charles V (1519–1556) adopted the 'Criminal Charles Constitution' ('Carolina') [22].

[4] The first Western European civil code was developed in Denmark (Danske Lov) in 1683. The code of Sardinia (1723), Bavaria (1756), Prussia (1794), and Austria (1812) followed it. The most famous civil code in Europe, the French code of Napoleon, was adopted in 1803–1804.

the next two centuries [17]. This document did not contain instructions on accounting. However, it described the execution of documentation of Orders (*prikazi*), mainly related to the performance of judicial functions. Accounting registers and books applied by Orders (*prikazi*) and monasteries were also mentioned.

4 Peter the Great Reforms in Public Administration and Accounting

The first years of the 18th century were crucial for Russian culture and mentality. A new civil code was being introduced, administrative divisions were being reorganized, Orders (*prikazi*) were being replaced by *Collegia*, and new government positions were being introduced [25]. The process of replacing the existing system of Orders (*prikazi*) in the Moscow state with 12 *Collegia* was carried out from 1718 to 1724 [6, 26, 27]. 'Revenue management', 'expense management' and 'financial control' were separated. Three *Collegia* were created: the Revenue *Collegia* engaged in operations for calculating and collecting duties; the Expenditures *Collegia* carried out operations on state expenditures; the Revision *Collegia* provided regular control of the accounting books of all state institutions [6]. One of the most important directions of transformation of the public administration system was the systematization and development of legislation. During the reign of Peter the Great 82 decrees related to accounting were adopted, which is almost three times more than in the same previous period (1653–1688) [28]. These documents can be divided into two groups: adopted before the *Collegia* system and adopted during its creation. The first group presented in Table 1.

Table 1. First documents of Peter I regulating accounting practice

Date of approval	Title	Scope of the document
January 8, 1700	(1) The verdict of the City Hall (*Ratusha*)[a] 'On taking to the City Hall (*Ratusha*) from the municipal *zemskih izb* since 205, receipts and expenditure books, as part of the collection from the *zemskih* inhabitants for the maintenance of the *voivodes* and other expenditure'	State institutions
January 1, 1703	(2) The verdict of the City Hall (*Ratusha*) 'On starting bound report books and giving them sealed by *Zemskie* Burgermeisters to customs and *kabak tselovalniks* for daily records of incoming duties and in the wine affair reports of supplies used for making wine and the proceeds obtained from its sale with attach the memorial to the Kursk *Zemskie* Burgermeisters on individual customs and tavern (*kabak*) duties'	State institutions and state-owned and regalia enterprises of Kursk

(*continued*)

Table 1. (*continued*)

Date of approval	Title	Scope of the document
January 9, 1710	(3) Personal decree of Peter I 'On committing in all Orders (*prikazi*) and City Hall (*Ratusha*) statements of receipts, and on sending them to the Governors for drawing up a General statement'	State institutions and City Hall (*Ratusha*)
January 22, 1714	(4) Personal decree 'On revision of all chanceries, Orders (*prikazi*), Mint and salt collection on receipts and expenditure of monetary Treasury in the Near Chancery (*blizhnyaya kancelyariya*)[b], about accounting in each public office where the receipts received and about sending urgent statements to the Near Chancery (*blizhnyaya kancelyariya*)'	State institutions (Orders (*prikazi*) and state-owned enterprises)

Compiled by: [1, 9, 29].

[a]City Hall (*Ratusha*) is the Central state structure. A part of the territorial state structures (Orders (*prikazi*)) and the cities that were removed from subordination to the Orders (*prikazi*) were subordinated to it. It served as the state Treasury.

[b]Near Chancery (*blizhnyaya kancelyariya*) - chancery of the Boyar Duma founded by Peter I for administrative and financial control over the activities of state institutions in Russia; existed during 1699–1719.

Two documents (1 and 4) were devoted to the revision of accounting registers. The appearance of the document dated 1700 (1) was associated with the first transformations of Peter I in financial management, when the financial functions of part of the Orders (*prikazi*) were transferred to Moscow City Hall (*Ratusha*). Document dated 1714 (4) provided the work of the Near Chancery (*blizhnyaya kancelyariya*). This document also defined the rules for accounting and reporting in state institutions and state-owned enterprises. The document dated 1703 (2) described the rules for keeping records in customs and taverns (*kabaks*). Its adoption can be called the first case of rejection from personalized regulations, but its scope was still quite limited; it regulated the rules only in one city – Kursk, while all the other documents had national significance. It prescribed: 1) to keep records in special accounting books issued from the controlling state institution, 2) to make entries daily, 3) to send quarterly reports for verification, etc. Document dated 1710 (3) dealt with the reporting rules for state institutions and prescribed the form of these reports - a table that had to be filled out based on accounting in Orders (*prikazi*) and cities (Table 2).

Table 2. General statement of the Governor

City	*Oklad*	According to the *oklad* (plan)			Arrears		
		1705	1706	1707	1705	1706	1704

Compiled by: [29].

This document changed the practice of reporting in the form of a simple extract from the registers of current accounting in a form that coincides with the accounting books (solid text without graphics), and replaced it with a more analytical data representation. This report used the method of comparing data on accrued and received state revenues, which was previously used for the accounting in state institutions in an implicit form. In this document it was presented in a visual form, as it was established in the practice of the charge and discharge accounting rules [14–16]. At the next stage of Peter the Great state reforms, the development of Russian accounting became even closer to European practice.

The *Collegia* system of state administration was started by a decree dated December 12, 1718 'On the administration in all the *Collegia* and Provinces since 1720 under the new order and on the division of affairs between the *Collegia*'. Then Peter I approved a number of important decrees regulating accounting rules. These documents were developed based on the results of studying and comparing legal acts of European countries, such as Sweden, France, England, and Austria. The main documents of this period were the General Regulations (February 28, 1720), the Regulations of the Admiralty and dockyard (April 5, 1722) and the Order 'On composing regulations in all *Collegia* following the example of the Admiralty' (May 11, 1722).

General Regulations introduced a new system of record keeping called 'collegiate'. The Swedish Charter was used as the basis for it. It was created as a General law regulating the actions of the entire state apparatus. During this period in Russia documentation began to develop as a system: separate genres of business writing were being formed, their forms were being created, and the relationship between clerical structures was being regulated by law. Some documents that were already applied in practice were called differently, others changed their functions or structure, and new documents appeared. There were regulations, charters, instructions, and decrees. Previous memorials, directed from a higher institution to a lower one, were replaced by instructions. There were no direct instructions regarding the accounting practices in the General Regulations, but it included Chapter 49 'On money for small expenditure in the *Collegia*', which prescribed each *Collegia* to draw up an *oklad* (planned) list, where determine the planned amount of office expenses for the year for candles, sealing wax, wax, firewood, and others, and then make a report [29]. In Chapter 40 'On books in offices' six types of books were described. Accounting books were not mentioned, but it was said that other books should be written in special regulations drawn up for each *Collegia* [29].

The innovations of the Regulations of the Admiralty and dockyard for Russian accounting were: 1) the appearance of the terms 'debit' and 'credit'; 2) the approval of the requirement to make book entries based on supporting documents; 3) new rules for the organization of material accounting based on the linear method in two-sided books [1]; 4) an approval of forms of accounting books and their tabular form [28]. It contained many forms of accounting registers, for example, accounting book for the cash receipts and expenditure (Table 3).

Table 3. Form of the book No. 5 from the samples given in the Regulations of the Admiralty and dockyard

Book that will be draw up from the receipts and expenditure book for compiling monthly and annual statements to the *Collegia*									
1 page Debit 1722					Credit				
Months, dates	№ on decree	Pages	Rubles	Kopecks	Months, dates	№ on Decree	Pages	Rubles	Kopecks

Compiled by: [29].

The form of the accounting register (Table 3) was an example of a systematic register of linear entries for cash accounting [4]. Cash accounting was considered among other values; and monetary value was not accepted as a single measure. Materials were recorded in natural units of measurement; and there was no mention of the balance sheet [4]. The accounting system described in the Regulations of the Admiralty and dockyard was built on the principles of single entry bookkeeping with elements of cameral accounting [4]. The Order approved by Peter the Great and dated May 11, 1722 'On the content of accounting books in *Collegia* according to the forms attached to the Regulations of the Admiralty and dockyard' significantly changed Russian accounting practice [29]. Later these rules were extended to state-owned enterprises [28].

The legally imposed requirements did not affect accounting practice, where traditional methods continued to be applied. The excerpt from the decree dated 1735 referred to the forms of books established by the Regulations of the Admiralty and dockyard: 'According to those forms, many books are not kept, and others, although contained, but not according to those forms, and not decently and with great difficulty' [4]. In the archival materials of the Voronezh shipyard for 1723–1741 and 1768–1800, documents of military units that participated in military operations in the Caucasus in the second half of the 18th century accounting books according to the rules of the Regulations of the Admiralty and dockyard were not found [28]. But in the Mining *Collegia* (Berg *Collegia*), the books of cash and supplies were compiled in a two-sided form during 1726–1767; and in the 1720–30s the left and right pages are marked as Debit and Credit, which is consistent with the regulations [28].

The development of Russian accounting through the publication of regulatory documents continued. In 1740 the 'Charter on bankrupts' was approved, which introduced a requirement for private entrepreneurs to make a balance sheet based on accounting books [29]. However, this rule was applied in practice only many years later, when the new version of the 'Charter of bankrupts' dated 1800 was approved. It included an Appendix containing forms of accounting registers with examples of their compiling. The desire of Peter I to use foreign experience become evident in accounting, especially during 1720s. However, the adopted legislative acts were not fully accepted in practice. But they allowed to make an important step in accounting development in terms of introducing tabular forms into accounting registers; after that Russian accountants became more prepared for the perception of European accounting based on double entry.

5 Conclusions

Uniformity of accounting system was explained by the traditions of Russian accounting passed from generation to generation, and supported by local regulations formed by state institutions controlling the activities of economic entities, and in some cases personally by Supreme rulers. Transformations of accounting by Peter I can be divided into two groups: developing the existing accounting practice and transforming it according to European models. The first approach was used by Peter I in the first 15–16 years of the 18th century, and it was implemented by legislating rules for accounting at the regional or national level. That was already the implementation of the second approach, because before Peter the Great all regulations were targeted at specific officials. He made the main innovative changes in the 1720s as part of the reform of state administration, replacing Orders (*prikazi*) with *Collegia*. At this time Peter I developed accounting by adopting national regulations that introduced special accounting terminology and procedures based on double entry and borrowed from Europe. He overcame an old stagnation in the development of Russian accounting, introducing the tabular form of accounting registers that facilitate the analysis of accounting data and simplify their generalization, necessary for the double entry book keeping.

Acknowledgments. The reported study was funded by RFBR, project number 20-010-42004.

References

1. Sokolov, Y.V.: Accounting: from the origins to the present day. Audit, Unity, Moscow (1996). (in Russian)
2. Pronina, S.A.: Reflection of administrative activity in documents of the 13th-17th centuries. Hist. Arch. **8**, 262–268 (2008). (in Russian)
3. Source studies: textbook for universities; under the ed. of A.V. Sirenov (2020). Yurayt publishing house, Moscow. (in Russian)
4. Lvova, D.: The Sources of Accounting Theory of Public Sector Economy. Publishing House of St Petersburg University, St Petersburg (2014). (in Russian)
5. Motyka, W.: The impact of Western Europe on accounting development in Tsarist Russia prior to 1800. Abacus **26**(1), 36–62 (1990). https://doi.org/10.1111/j.1467-6281.1990.tb00231.x
6. Platonova, N.: Peter the Great's government reforms and accounting practice in Russia. Acc. Hist. **14**(4), 437–464 (2009). https://doi.org/10.1177/1032373209342476
7. Sidorova, M.I., Nazarov, D.V.: Account books of the Moscow Print Yard (1622-1700): the origins of cost accounting in Russia. De Computis-Revista Española de Historia de la Contabilidad **16**(2), 188–213 (2019). https://doi.org/10.26784/issn.1886-1881.v16i2.358
8. Sokolov, V., Karelskaia, S., Zuga, E.: Accounting in state-owned companies operated for the production and sale of alcohol in Russia in the fifteenth-seventeenth centuries. Acc. Hist. **24** (3), 425–443 (2019). https://doi.org/10.1177/1032373219846663
9. Shirokiy, V.F.: Balance sheet in legislation, literature and accounting practice in Russia in the 18th century. St Petersburg (1940). (in Russian)

10. Mazdorov, V.: History of accounting development in the USSR (1917-1972). Finance, Moscow (1972). (in Russian)
11. Kuter, M.I., Gurskaya, M.M., Kuznetsov, A.V.: Alexander Galagan: Russian titan of the enlightenment in the history of accounting. Acc. Hist. **24**(2), 293–316 (2019). https://doi.org/10.1177/1032373218761129
12. Lemarchand, Y.: Double entry versus charge and discharge accounting in eighteenth-century France. Acc. Bus. Finan. Hist. **4**(1), 119–145 (1994). https://doi.org/10.1080/09585209400000040
13. Lemarchand, Y.: Introducing double-entry bookkeeping in public finance: a French experiment at the beginning of the 18th century. Acc. Bus. Finan. Hist. **9**(2), 225–254 (1999). https://doi.org/10.1080/095852099330313
14. Jones, M.J.: The role of change agents and imitation in the diffusion of an idea: charge and discharge accounting. Acc. Bus. Res. **38**(5), 355–371 (2008). https://doi.org/10.1080/00014788.2008.9665771
15. Broadbent, J., Guthrie, J.: Public sector to public services: 20 years of 'contextual' accounting research. Acc. Audit. Acc. J. **5**(2), 3–32 (1992). https://doi.org/10.1108/09513570810854383
16. Jacobs, K.: Making sense of social practice: theoretical pluralism in public sector accounting research. Finan. Acc. Manag. **1**, 1–26 (2012)
17. Bogomazov, G.G. (ed.): Economic History of Russia in 2 Parts. Part 1. 9-19th Centuries: Textbook for Universities, 2nd ed. Yurayt Publishing House, Moscow (2020). (in Russian)
18. Muravyeva, L.A.: Socio-economic development of the Moscow state (the second half of the 15th-17th centuries). Int. Acc. **21**(3), 363–374 (2018). (in Russian)
19. Solovyov, S.M.: Readings and Stories on the History of Russia in 2 Parts. Part 2. History in the 17th-18th Centuries. Yurayt Publishing House, Moscow (2020). (in Russian)
20. Kirillova, E.N.: Craft and trade corporations in early modern times. New Mod. Hist. **3**, 80–93 (2009). (in Russian)
21. Pogosyan, N.D.: Establishment and functioning of the Accounting *prikaz* of Russia in the 17th century: historical and legal aspects. Law State: Theory Pract. **7**, 93–102 (2012). (in Russian)
22. Kazakov, R.B.: Sudebnik 1497. Quest. Hist. **3**, 139–145 (2000). (in Russian)
23. Elchaninova, O.Yu., Ospennikov, Yu.V., Romashov, R.A., Yutyaeva, L.E.: System of sources of Russian law 10th-18th centuries. ASGARD, Samara (2014). (in Russian)
24. Sazankova, O.V.: History of civil law. RITS HGUEB, Khabarovsk (2019). (in Russian)
25. Chistyakov, A.V.: The first attempts to systematize Russian legislation and create a Code of laws of the Russian Empire. Curr. Prob. Russian Law **8**(93), 44–53 (2018). (in Russian)
26. Anisimov, E.V.: State transformations and autocracy of Peter the Great in the first quarter of the 18th century. Dmitry Bulanin, Saint Petersburg (1997). (in Russian)
27. Milyukov, P.N.: State Economy of Russia in the First Quarter of the 18th Century and the Reform of Peter the Great, 2nd edn. Saint Petersburg (1905). (in Russian)
28. Sidorova, M.I., Nazarov, D.V.: Regulation of accounting practices by state institutions in Russia in the middle of the 18th century. Audit **9**, 15–21 (2017). (in Russian)
29. Complete collection of laws of the Russian Empire. St Peterburg (1830). (in Russian)

Financial Instruments for Tourism Development: Challenges and Opportunities

Aleksandr Gudkov(✉), Elena Dedkova, and Elena Rozhdestvenskaia

Orel State University, 302020 Orel, Russia
sashaworld777@gmail.com

Abstract. Currently, the development of tourism faces many challenges that make this industry extremely unstable due to political, medical and economic fluctuations. All the ups and downs are usually outside the control of tourism. One of the most important economic problems associated with tourism is the lack of investment and financing. This is a serious problem, considering that tourism is one of the largest employers of labor in the world, and its development has huge benefits for the economy. The purpose of this article is to consider the problems and opportunities in the field of tourism financing. For the purposes of the study, we have achieved a substantial and relevant analysis and research on the main areas of financial support of tourism.

Keywords: Tourism · Finance · Economic growth

1 Introduction

The importance of tourism development for an individual country is due to the fact that it acts as a means of foreign exchange earnings, employment, increasing the contribution to the balance of payments and GDP, economic diversification, income growth and improving the welfare of the nation [23]. Tourism development is particularly important for poor developing countries, where tourism can be a major source of income and generate a large share of their GDP. This is due to the fact that in these countries, many industries are poorly developed and do not bring such income as tourism. This is particularly evident in island countries, where limited territory, small populations, lack of resources, remoteness from countries that generate the main economic activity, exposure to natural disasters, heavy dependence on international trade, etc. make it difficult to develop other industries. The dynamic development of tourism has a multiplier effect on other sectors of the economy, in which investment activity is activated, new jobs are created, and the turnover of funds is expanded [17]. At the same time, tourism is an activity that requires a favorable economic environment for development and attracting significant amounts of financial resources. Therefore, the issue of considering the financial side of the functioning of the tourism industry is important in practical and theoretical terms.

In modern economic conditions, and especially in times of crisis, it is necessary to pay special attention to the analysis of financial support for tourism, as a branch of the

T. Antipova (Ed.): ICCS 2020, LNNS 186, pp. 70–80, 2021.
https://doi.org/10.1007/978-3-030-66093-2_7

national economy that forms various types of socio-economic activities that perform production and sales and coordination and management functions in the economy. There is a need to study the impact of the financial security mechanism (regulatory framework, methods, sources of financial security, financial instruments, incentives, sanctions) and its adaptation to the specifics of the management object. Effective implementation of the mechanism of financial support for tourism is hindered by a number of certain factors. First, there is the problem of organizing financial support due to time constraints, given the outpacing growth of the export tourism potential of developed market economies over developing countries. Secondly, the organization of financial security usually takes place in a context of a shortage of financial resources, high cost of capital raising, as well as in the context of public policy, which in developing countries is usually not marked by a systematic and gradual implementation of powers in the field of financial security regulation.

Many scientists around the world have made a significant contribution to the study of tourism industry financing and tax incentives.

The problems of financing are addressed in the works of C. Chang, H. Hsu & M. McAleer [2], E. Dedkova & A. Gudkov [4, 9, 11], M. Mahmoodi & E. Mahmoodi [13], S. Vujović, D. Vukosavljević & Ž. Bjeljac [21], R. Trynko & Y. Grygorieva [20], I. Carrillo-Hidalgo & J.I. Pulido-Fernández [1]. Aspects of tax incentives for the tourism industry are considered by A. Gudkov et al. [8, 10] and E. Dedkova et al. [5]. A. Shirvani et. al. [19] explain and evaluate market risk before and during an actual distressed market period. M. Gasic et. al. [7], G. Nyikos, G. Soós [16] and O. Manta [14] discussed EU financial instrument that supports member countries in various levels and subsidies of financial intervention (direct payments) or with a financial agency/financial intermediary.

However, the new challenges of the modern world require a review of the priorities of economic problems of tourism development and the improvement of financial instruments to support it. Knowledge of the tools, characteristics, and ways of channeling development cooperation is fundamental if we are to find out whether tourism finance fosters financial inclusion [1].

2 Methodology

Using classic research tools based on analysis and synthesis, induction and deduction, as well as modern methods, we have achieved a substantial and relevant analysis and research in the main areas of financial support for tourism. The theoretical information needed for the study was taken from research articles in the field of tourism financing. Practical data is based on official statistical information.

3 Results

3.1 Financial and Economic Problems of Tourism Development

Tourism is one of the largest highly profitable and most dynamic sectors of the world economy. But at the same time, the development of tourism in developing countries faces many problems that make this industry extremely unstable. The problems of modern tourism are based on political, medical, and economic fluctuations.

In the modern world international tourism is influenced by the following factors:

- demographic (territorial distribution of the population, its size, urbanization);
- socio-economic (the degree of development of productive forces, industrialization and scientific and technical revolution; the nature of the socio-economic system; labor activity, cultural level, educational and professional structure of the population);
- socio-psychological (influence of traditions and the presence of certain behavioral stereotypes; value orientation; religion);
- - environmental and medical-biological (quality of the environment of permanent residence and the state of health of the population);
- natural (various climatic and landscape conditions; exotic flora and fauna; availability of mineral water sources);
- political (expansion of international economic, political, cultural, scientific relations; stability of foreign economic relations and domestic policy; social policy of the state).

The following Table 1 shows the main current economic problems of tourism.

Table 1. Modern economic problems of tourism development.

Infrastructure problems			
Poor transport accessibility	Small number of leisure facilities	An undeveloped health care system	Lack of accommodation facilities
Financial problem			
Insufficient investment and financing	Significant tax burden	Lack of government support programs	Large risks, inadequate system of insurance
Organizational and managerial problems			
Lack of proper management	Lack of control in various areas	Lack of awareness and falsification of data	Lack of an adequate education system
Corruption	Political instability	Gradual loss of national identity	Terrorism, high crime rate

Poor infrastructure is one of the most common economic problems of tourism in developing countries. In order for tourism to flourish in any country, it is necessary to take into account such basic factors as good roads, constant electricity, and drinking water. International tourists are also primarily concerned about accommodation. Countries such as the United States, France, Spain, and Germany are advanced in terms of the development of the tourism industry. In these countries, there are numerous hotels that meet international standards, quality roads, seaports, and airports. Thus, it is obvious that the lack of infrastructure, its non-compliance with international standards, as well as its centralization, for example, only in the country's capital, creates obstacles to the development of tourism.

One of the most important contemporary problems that have significantly affected tourism in recent years and generally affects the entire economy of the country is terrorism and the security problem. On the one hand, the phenomenon of terrorism and its impact on different countries are well studied – and the most obvious conclusion, which came to a variety of researchers, was the idea that the economic systems of rich and successful countries are much easier to cope with the economic consequences of terror than in the case of developing countries. For example, in seven developed Asian countries in 1980–2005, the effect of terrorist attacks was close to zero, while in 35 developing countries, one terrorist attack per million inhabitants caused an economic decline of 1.4% [6]. The explanation is simple: the economies of developing countries are highly dependent on investment from abroad, and fears of threats and instability affect the mood of investors, including in the tourism industry. In addition, in developing countries, governments overwhelmingly use the pretext of fighting terrorism as a justification to tighten the political regime. And of course, terrorism and high crime rates discourage tourists.

The next major economic problem related to tourism is insufficient investment and financing. This is a serious problem, given that tourism is one of the largest employers of labor in the world, and its development has huge benefits for the economy. In developing countries, tourism is underfunded relative to its potential, there are usually no government support systems, there are no tax incentive programs for tourism, and the insurance industry is in its infancy to reduce risks. This has caused tourism development to move very slowly and give priority to the extraction and sale of natural resources. However, this neglect of other important aspects of the economy often contributes only to the economic degradation of the country and the loss of its attractiveness for tourism in the future.

As a rule, problems with financing are related to corruption and lack of adequate quality training. Education in the field of tourism is undoubtedly the main factor contributing to the development of tourism, and its absence will create serious problems in understanding the main processes of its development and impact on the economy of the state. In many developing countries, corruption has penetrated deep into public administration and at the highest levels of government, creating favorable conditions for the theft of public funds and budget allocations, contributing to the stagnation of the economy in general, and the tourism industry in particular. Corruption is also a system-forming factor for the development of crime. Sometimes the crime rate in poor countries is so high that other states warn their citizens against traveling to these countries, which is a devastating blow to any possibility of tourism developing.

Recently, the level of development of the health care system has become a particularly sensitive factor in maintaining a good tourist image of the country. The spread of Covid-19 virus in 2020 revealed the inability of many countries to fight not only for the lives of tourists, but also their own residents in such difficult unexpected conditions. Health is a serious problem for people of all classes, and if travelers are not convinced of the availability of good medical care in the host country, if necessary, they will refuse to travel to that country. Safety in tourism will always be a top priority, so solutions that ensure safety and peace of mind should be the government's first steps in the right direction.

The Fig. 1 presents a system for developing the export potential of tourism based on financial, investment, and resource support.

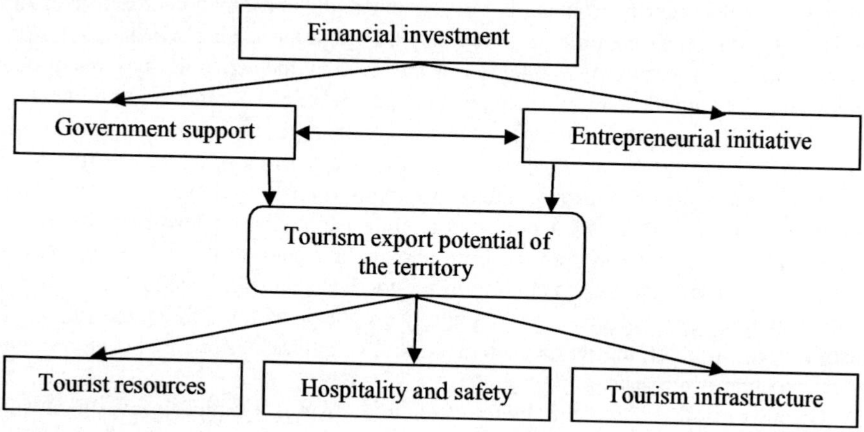

Fig. 1. The main elements of developing and supporting the export potential of tourism.

3.2 Financial Support of Tourism Industry

Statistics provide an accurate description that today tourism is one of the most profitable economic sectors in the world, which in modern conditions is constantly and dynamically developing, contributing to the solution of a whole range of vital socio-economic problems [15].

The tourism industry is becoming increasingly important for the development of the economy and social sphere of any country in the world. The priority direction for the development of inbound and domestic tourism is an important factor in improving the quality of life of people, increasing additional jobs, replenishing the state's currency reserves, and increasing the revenue part of the state budget.

However, to stimulate tourism, it is necessary to provide stable and timely financial incentives to the industry, including the implementation of an adequate financial policy. Historically, fiscal policy was quite often used as a tool to ensure financial stability and boost economic growth. Applying a wide range of fiscal instruments and levers, this policy provides a certain regulation of business activity of economic agents, investment and consumer demand, and the household incomes as well [3].

Financial support of the tourism industry has a number of features and specific features associated primarily with the mechanism of its functioning. Basically, this industry needs financing in related areas of activity (hotels and restaurants, various types of transport and its infrastructure, entertainment facilities, folk crafts, etc.), and on the other hand, the social effect is that the upgraded infrastructure is used not only by tourists but also by the local population.

Researchers of the tourism industry quite highly assess the effectiveness of its financing. All of them note that this is an industry that can generate significant profits at relatively low costs, since the capital turnover in this area is 4 times higher than the average for other sectors of the world economy, and the cost of creating one job in the tourism infrastructure is 20 times less than in industry. Therefore, the solution of financial issues for the tourism industry is of national importance.

The state should play a leading role in creating an enabling environment for tourism development in developing countries. Pandemic 2020 once again proved that, despite the presence of private financial initiatives in crisis, much greater efforts are required to reverse the negative trend.

The Fig. 2 shows the ratio of the number of trips of foreign citizens to the Russian Federation and the amount of budget allocations of the Federal budget for implementation of the tourism development program.

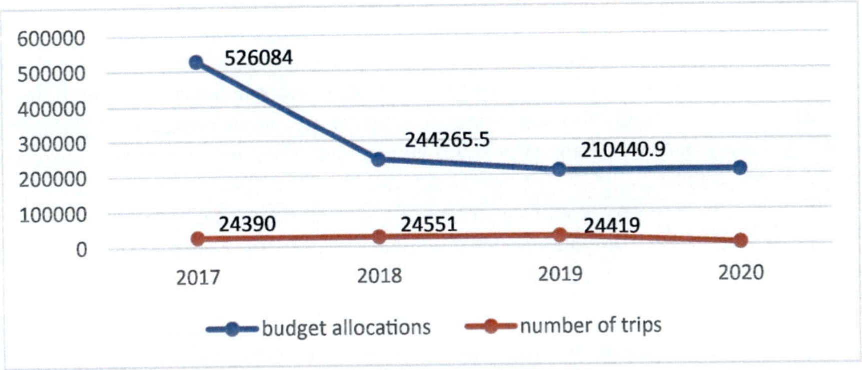

Fig. 2. The ratio of the number of trips of foreign citizens (in thousands of people) to the Russian Federation and the amount of budget allocations (in thousands of rubles) according to data from the state program of the Russian Federation "Development of culture and tourism' for 2013–2020 and statistical indicators of Rostourism.

Figure 2 shows that the decline in funding for the tourism development program aimed at attracting foreign citizens to the Russian Federation has led to a complete stop in the growth of the number of foreign arrivals. The lack of growth in foreign tourists is also explained by the fact that since the crisis of 2014, the business initiative has not yet regained its previous values.

As for public financing of the tourism industry, its form and scale depending on the place and role of tourism in the national economy. The main directions of tourism financing policy are both direct and indirect sources of financing. Among the direct sources, it is necessary to allocate the enterprises ' own resources, budgetary and extra-budgetary sources of financing [21]. Indirect sources are attracted capital, mainly borrowed, at the expense of private and foreign capital [13], as well as funds of public organizations, tax preferences, etc. In the conditions of recovery from the financial and economic crisis and limited opportunities to provide direct financial support, the role of indirect forms of financial support should increase (in particular, reducing the rates of certain types of taxes and fees [5, 10], establishing benefits, etc.).

World experience provides the following types of state assistance to tourist enterprises:

- subsidies to help solve the monetary problem, especially at the first stage of implementing tourism projects;
- soft (preferential) to compensate for the gap between the fixed and commercial rates, monitoring the implementation of the project for which the loan is issued is carried out throughout the entire period of its implementation;
- interest bonuses and guarantees on loans and subsidies-the state or a special body guarantees loans provided by commercial banks for the development of tourism. At the same time, it is expected to assess the project's ability to meet the conditions set for granting loans, as well as to assess the risk of using guarantees;
- tax benefits that are provided before and after a certain tourist project begins to generate revenue.

However, it should be noted that by financing tourism and creating conditions for its development, detailed planning should be balanced, since an increase in the flow of tourists can have negative consequences (rising prices, environmental degradation, growing dissatisfaction with the population, etc.) this is observed in some regions, such as Catalonia (Spain).

So, given the importance and specificity of the functioning and development of the tourism industry, the solution of financial issues should be based on a systematic approach and take into account the following aspects:

- rational and effective use of tourism resources based on sustainable development in order to form a national tourism product [17];
- introduction of a systematic approach to the regulation of socio-economic and financial relations in the field of tourism, since tourism covers various sectors of the economy (according to research by scientists – about 30), so the concept of tourism development with a clear regulatory and socio-economic mechanism should be adopted [11];
- the functioning of the tourism industry is impossible without the availability of modern infrastructure, so the main direction of financing involves attracting funds for the development of tourism infrastructure [18];
- promotion of domestic and social tourism. A certain share of revenues from the tourism industry should be allocated to support domestic tourism and, above all, social tourism, as well as to restore tourism resources and infrastructure [12].

Financial instruments are increasingly used in tourism to stimulate its export potential, as domestic tourism has great limitations for growth and development. With an invisible export orientation, as well as providing direct and indirect employment, the tourism sector has become a significant part of the economy and finance [23]. Financial support allows to reduce risks and get tax benefits, but requires additional investment and the collection of a large amount of documentation. Therefore, many participants in the tourist market often prefer to operate without state support due to currency fluctuations, strict requirements for the provision of state benefits and subsidies, on the one hand, and high competition in the world market, on the other hand.

The following Table 2 shows the main proposed ways to improve financial instruments to support tourism exports.

Table 2. The main ways of improvement of financial instruments to support tourism exports.

Type	Description
Credit availability	The availability of export credits means that they can be obtained with minimal costs. For example, the methods can be used to simplify the procedure for obtaining, reduce the time for reviewing the application, and limit the period for creating a commercial offer by the bank
Export insurance development	The availability of Internet services for reviewing applications, expanding the range of services provided, improving the skills of employees of insurance companies, as well as state control of compensation payments-these measures will make the insurance market more accessible, reliable, and convenient for all participants in legal relations
Simplification of the taxation and tax accounting system	In order to operate within the framework of tax legislation, in many countries it is necessary to keep separate records of goods and services subject to different tax rates. To simplify it, integrated electronic commodity turnover systems are being created with customs and tax services, and simplified taxation systems are being introduced for the tourism industry
Reduction of currency risks	Reducing currency risks is very important for tourism enterprises. It is important to stabilize the national currency, ensure the independence of the domestic market from intervention, develop industry and avoid commodity exports, reduce capital outflows, and increase the country's balance of payments
Fuel subsidies	Promotion of inbound tourism can be carried out by providing state fuel subsidies to air carriers that operate tourist flights to the necessary airports

Thus, financial support tools are primarily designed to ensure the price attractiveness and competitiveness of tourist destinations, attract the interest of tourists from various countries, facilitate the activities of national companies engaged in the hospitality sector, which will eventually stimulate the development of the national economy, create new jobs and improve the well-being of citizens.

In order for the existing financial instruments to be effective, it is necessary to have a comprehensive interaction between exporters and importers, financial institutions and the state [8, 22], and to create an effective foreign economic policy [4, 9], the performance of insurance companies' obligations to the insured, availability of highly qualified personnel. Only then will the created financial support measures fully work, allowing the tourism sector to develop effectively.

4 Conclusions

The tourism sector is one of the types of business that is developing quite dynamically and is capable of rapid recovery even after significant crises, including the Covid-19 pandemic. Interest in tourism is explained by a number of factors: despite competition, large, medium and small firms successfully interact in the tourism market, the tourism business is characterized by a rapid turnover of capital, tourism is the largest industry in terms of the number of people involved and the income received. Despite the fact that international tourism has a number of negative consequences: violation of the way of life of the local population and traditions, inflation, destruction and pollution of the environment, the effective functioning of the tourism industry makes it possible to mitigate these shortcomings by partially directing financial flows from tourism to compensate for its negative consequences.

The problem of financial support for the tourism industry is particularly relevant, since the development of tourism stimulates other sectors of the economy and contributes to strengthening the authority of the state at the international level. In times of crisis, the following financial incentives are most often used: direct financial support for industry entities and tax preferences.

Financial support should solve the following main tasks of the tourism industry:

- achieving maximum efficiency in the use of available financial resources-maximizing the volume of GDP produced based on the choice of a rational financial security model;
- establishment of optimal proportions of distribution and redistribution of the produced GDP in order to fully meet the needs of citizens, enterprises of the tourism industry, the state;
- comprehensive assistance in attracting all temporarily available funds and income received through financial market institutions for the needs of financial support for non-production sectors, in particular the tourism industry.

Thus, the main goal of financial support for the tourism industry is to create a favorable conditions for the allocation, distribution, and attraction of financial resources for the effective and balanced development of tourism. Given that the financial support of the tourism industry has a number of features and specific features associated primarily with the mechanism of its functioning, the solution of these problems requires a comprehensive and prolonged approach.

References

1. Carrillo-Hidalgo, I., Pulido-Fernández, J.I.: The role of the world bank in the inclusive financing of tourism as an instrument of sustainable development. Sustainability **12**, 285 (2020)
2. Chang, C., Hsu, H., McAleer, M.: A tourism financial conditions index for tourism finance. Tinbergen Institute Discussion Papers 17-071/III (2017)
3. Chugunov, I., Pasichnyi, M.: Fiscal stimuli and consolidation in emerging market economies. Invest. Manag. Finan. Innov. **15**(4), 113–122 (2018)
4. Dedkova, E., Gudkov, A.: Tourism export potential: problems of competitiveness and financial support. In: Antipova, T. (ed.) Integrated Science in Digital Age. ICIS 2019. Lecture Notes in Networks and Systems, vol 78. Springer, Cham (2020). https://doi.org/10.1007/978-3-030-22493-6_17
5. Dedkova, E., Gudkov, A., Dudina, K.: Perspectives for non-primary export development in Russia and measures of its tax incentives. Prob. Perspect. Manag. **16**(2), 78–89 (2018)
6. Gaibulloev, K., Sandler, T.: The impact of terrorism and conflicts on growth in Asia. Econ. Polit. **21**(3), 359–383 (2009)
7. Gasic, M., Oklobdžija, S., Perić, G., Ilić, D.: Financial instruments of the European union for the development of tourism in rural areas of Serbia. Manag. Educ. **11**(2), 19–25 (2015)
8. Gudkov, A., Dedkova, E.: Development and financial support of tourism exports in the digital economy. In: JDS, vol. 2, no. 1, pp. 54–66 (2020). https://doi.org/10.33847/2686-8296.2.1_5
9. Gudkov, A., Dedkova, E.: What does ICT mean for tourism export development? In: Antipova, T., Rocha, Á. (eds.) Digital Science 2019. DSIC 2019. Advances in Intelligent Systems and Computing, vol 1114. Springer, Cham (2020). https://doi.org/10.1007/978-3-030-37737-3_15
10. Gudkov, A., Dedkova, E., Dudina, K.: Tax incentives as a factor of effective development of domestic tourism industry in Russia. Prob. Perspect. Manag. **15**(2), 90–101 (2017)
11. Gudkov, A., Dedkova, E., Dudina, K.: The main trends in the Russian tourism and hospitality market from the point of view of Russian travel agencies. Worldwide Hosp. Tour. Themes **10**(4), 412–420 (2018)
12. Kukharenko, O.G., Gizyatova, A.Sh.: Sources of investment and financing of social tourism. IOP Conf. Ser.: Earth Environ. Sci. **204**, 012022 (2018)
13. Mahmoodi, M., Mahmoodi, E.: Foreign direct investment, exports and economic growth: evidence from two panels of developing countries. Econ. Res. **29**(1), 938–949 (2016)
14. Manta, O.: Financial instruments for tourism and agrotourism in Romania. In: Caring and Sharing: The Cultural Heritage Environment as an Agent for Change, pp. 345–359 (2019).
15. McAleer, M.: The fundamental equation in tourism finance. J. Risk Finan. Manag. **8**, 369–374 (2015)
16. Nyikos, G., Soós, G.: Financial instruments now and after 2020. In: A World of Flows: Labour Mobility, Capital and Knowledge in an Age of Global Reversal and Regional Revival, pp. 150–151 (2018).
17. Sheresheva, M., Polukhina, A., Oborin, M.: Marketing issues of sustainable tourism development in Russian regions. J. Tour. Herit. Serv. Mark. **6**(1), 33–38 (2020)
18. Sheresheva, M.Y.: The Russian tourism and hospitality market: new challenges and destinations. Worldwide Hosp. Tour. Themes **10**(4), 400–411 (2018)
19. Shirvani, A., Stoyanov, S.V., Rachev, S.T., Fabozzi, F.J.: A new set of financial instruments (2019). https://ssrn.com/abstract=3486655

20. Trynko, R., Grygorieva, Y.: Sources of financing tourism: challenges and solutions. Bull. Taras Shevchenko Natl. Univ. Kyiv. Econ. **11**(176), 35–39 (2015)
21. Vujović, S., Vukosavljević, D., Beljac, Ž.: Financing of tourist activity. Proc. Geogr. Inst. "Jovan Cvijic' **64**(2), 207–214 (2014)
22. Zameer, A.: Significance of relation between retailing and tourism in designing credit derivative products for financing tourism and retailing infrastructure. In: Conference on Tourism in India – Challenges Ahead, pp. 279–287 (2012)
23. Zeren, F., Koç, M., Konuk, F.: Interaction between finance, tourism and advertising: evidence from Turkey. Tour. Hosp. Manag. **20**(2), 185–193 (2014)

Intellectual Capital of a Company in the Financial Statements: The Reasons of Information Deficit and the Ways of it Overcoming

Yana Ustinova(✉)

Novosibirsk State University of Economics and Management,
630099 Novosibirsk, Russia
ustinova_pr@mail.ru

Abstract. The paper discusses the concepts of "intellectual capital" and "intangible assets", substantiates the need to disclose information on the intellectual capital of the company in addition to traditional financial statements. Specific features of the elements of intellectual capital as objects of accounting and the associated difficulties in their accounting are investigated. The limitations of the current concept of accounting for intangible assets, the factors that determined them, the methods used to overcome them, and the reasons for their insufficient effectiveness are analyzed. It contains an analysis of the work of scientists studying the problems to be solved in the report on intellectual capital, its composition and structure, as well as the impact of the report on statement users' perceptions of the company. The content and structure of information on intellectual capital presented in the public financial statements of the largest Russian companies, as well as the information needs of statement users, are examined. The study shows the need to develop the theory and methodology of accounting in order to expand the informational restrictions on the reflection of intellectual capital in financial statements (including through the management report recommended by the IASB). Recommendations on overcoming the identified information deficit are formulated.

Keywords: Information limitations of traditional financial statements · Information needs of users of financial statements · Intellectual capital

1 Introduction

In the modern economy, there is a noticeable shift in emphasis in terms of economic importance from physical and financial capital to intellectual capital. A few decades ago, the economic potential of companies was determined by the valuation of their fixed assets, production and inventory, the availability of cash, debts. At present, the assessment of the economic potential for the availability of intellectual capital is slowly but surely coming to the fore, since it is it that directly affects the efficiency of use of all other types of capital of companies. The globalization of the economy, the intensification of competition, the development of engineering and technology, the widespread

T. Antipova (Ed.): ICCS 2020, LNNS 186, pp. 81–90, 2021.
https://doi.org/10.1007/978-3-030-66093-2_8

use of intellectual resources stimulate the interest of stakeholders in information about its intellectual capital.

But an analysis of the multiple informational limitations of financial reporting in relation to intellectual capital forces us to recognize the inadequacy of traditional financial statements to comprehensively reflect the intangible, intellectual resources of a company. As a result, users of financial statements, including investors and creditors, are unable to adequately assess the financial performance of the company, its solvency, the profitability of its assets, investment attractiveness and prospects, which affects the company's capitalization.

At the same time, the growth of intellectual capital is closely related to the growth of capitalization of companies. In particular, according to the results of V. Dzenopoljac (2018), obtained on the basis of analysis of reporting data of oil and gas companies included in the 2016 Forbs global 2000 list for the period from 2000 to 2015, an increase in the estimated value of intellectual capital by 1 percentage point leads to an increase in capitalization companies for 113–157 million dollars. At the same time, 35% of the formed sample of companies did not disclose information on intellectual capital in general, and 65% of companies disclosed it only partially [1, p. 14–15].

The insufficiency of information disclosed in traditional financial statements requires the submission of additional information on intellectual capital to users of financial statements. Its volume, structure, presentation form are the subject of serious scientific discussions.

The work aimed on a comprehensive study of the reasons of the information deficit regarding intellectual capital in traditional financial statements and on analysis of the possible ways to overcome it, their strengths and weaknesses.

2 Literature Review

A detailed analysis and correlation of the concepts of "intellectual capital" and "intangible assets" is widely represented in the works of Russian and foreign authors. At the same time, some researchers consider these concepts as identical, and some researchers when determining the concept of "intangible assets" proceed from the provisions of accounting standards.

In particular, the extremely broad concept of intangible assets goes back to the works of K. Sveiby (1986) [2, p. 18], T. Stewart (1997) [3, p. 23–47], L. Edvinsson and M. Malone (1997) [4, p. 37], R. Reilly and R. Schweiss (1998) [5, p. 148], B. Lev (2001) [6, p. 5–57], which consider the concepts of intangible assets and intellectual capital interchangeably. The definition of intangible assets in the narrow sense is enshrined in accounting standards (Meritum project (2002) [7, p. 62], A. Wyatt and M. Abernethy (2003) [8, p. 7, 20, 25], D. Starovic and B. Marr (2005) [9, p. 6], L.S. Moerman and S. van der Laan (2006) [10, p. 4–6]), I. Abeysekera (2007) [11, p. 10, 21].

The lack of information on intellectual capital in financial statements has attracted the attention of many researchers. In particular, the existence of the gap between the information needs of stakeholders and information content of traditional financial statement researched in the works by A. Wyatt and M. Abernethy (2003) [8, p. 10–20], D. Starovic and B. Marr (2005) [9, p. 7–15], L.S. Moerman and S. van der Laan (2006) [10, p. 6–8], I. Abeysekera (2007) [11, p. 3–4], D. Andriessen and R. Tissen [12], F. Castilla-Polo and M.C. Ruiz-Rodriguez (2018) [13, p. 3–6].

The influence of the lack of adequate information on intellectual capital in the financial statements on the assessment of the financial picture of the company was studied by L. Edwinsson and M. Malone (1997) [4, p. 38–40], B. Lev and P. Zarowin (2003) [14, 358–360], I. Lopes (2011) [15, p. 102–104], S. Chander and W. Mehra (2011) [16, p. 5–8].

An analysis of the reasons for the lack of information on intellectual capital in traditional financial reporting was performed in (P.H. Sullivan Jr. and P.H. Sullivan Sr. (2000) [17, p. 340], E. Oliveras and O. Amat (2003) [18, p. 5], A. Wyatt and M. Abernethy (2003) [8, p. 8], D. Cormier and M.J. Ledoux (2010) [19, p. 3], J.-Ch. Wu and Ch. Lin (2013) [20, p. 343], T. Karius (2016) [21, p. 19].

Specific features of intellectual capital as an object of accounting, complicating the formation and disclosure of accounting information, became the subject of research in the works of M. Greenwell and G. Tibbits (1992) [22, p. 3, 19], E. Webster (2002) [23, p. 4–8], A. Wyatt and M. Abernethy (2003) [8, p. 8], B. Lev (2005) [24, p. 302–303], L.S. Moerman and S. van der Laan (2006) [10, p. 248]), L. Hunter, E. Webster, A. Wyatt (2009) [25, p. 7], K. Artsberg and N. Mehtiyeva (2010) [26, p. 24], R. Heinrich (2011) [27, p. 38], R. Petrova (2011) [28, p. 64], T. Karius (2016) [21, p. 4, 9, 16].

An analysis of the works of researchers on the study of overcoming the information gap in terms of disclosing intellectual capital in financial statements so as the study of the volume, structure, and form of presentation additional information on intellectual capital made described in author previous work (Y. Ustinova (2019) [29, p. 105]).

3 Data and Methodology

The study was based on the results of foreign and Russian researcher works aimed at developing a concept and tools for reflecting information on intellectual capital in financial statements that would allow users of financial statements to form an adequate idea of the impact of this object on the future financial performance of the company.

In addition, an analysis of the data of the consolidated financial statements for 2015–2018 was carried out (compiled in accordance with IFRS, including explanations), of the largest Russian companies (included in the rating 2000 Global from the journal website http://www.forbes.com), in terms of the presentation and disclosure of information on intellectual capital.

As part of the study, we used materials from surveys of the company's administration, financial analysts, lenders and investors regarding information on intellectual capital that they would like to see in the reporting data (traditional financial statements (explanations) or additional reports).

4 Results

4.1 Intellectual Capital and Intangible Assets

So, the concept of intellectual capital and its structure are the subject of scientific discussions. There is a need to specify them for research purposes.

Intellectual capital is the totality of all the company's intellectual resources (associated with the corresponding assets and liabilities), as well as a system of economic relations for the use of these resources in order to create business value. By "intellectual resources" of a company are understood the results of intellectual activity and other intangible resources that ensure the individualization of the product, organization and conduct of the business and the company itself in relations with the external environment.

However, in financial statements, intellectual capital is disclosed through the category of intangible assets. In the broad sense, intangible assets and intellectual capital are considered as synonyms. In a narrow sense, intangible assets include a part of intellectual capital that meets the recognition criteria provided for by the accounting standard. The formation of financial statements in accordance with accounting standards restrains intangible assets within a narrow framework, extremely impoverishing information content and reducing the usefulness of financial statements for users.

4.2 Limitations of the Modern Concept of Accounting for Intangible Assets

The current concept of accounting for intellectual capital, both Russian and international, considers it exclusively through the prism of the category of intangible asset. The result of the formation in the financial statements of data on intangible assets in accordance with the requirements of accounting standards are multiple informational restrictions.

Information restrictions are comprehensive, ranging from the spectrum of facts of economic life in relation to intellectual resources reflected in accounting and reporting, and ending with cost estimates of these facts. As examples of the described limitations, mention may be made of:

- refusal to include an object into the intangible assets, the possibility of deriving future economic benefits from it uses is not obvious at the time of recognition;
- reflection in the structure of intangible assets of an object at actual costs from the moment the development stage is recognized (in fact, only the costs of patenting);
- lack of reflection of the burdens accepted by the copyright holder when issuing a certain number of licenses to use the facility.

Moreover, the lack of uniformity is observed also in the disclosure of information even on basic issues (structure, presentation and disclosure rules, etc.), as well as a clear deficit of disclosed information, complicates the perception of information on intellectual capital by users of financial statements.

According to the consolidated financial statements for 2015–2018 (compiled in accordance with IFRS) of the largest Russian companies (included in the rating 2000

Global from the journal website http://www.forbes.com in May 2015), information on intellectual capital is disclosed as follows (see Table 1).

As can be seen from Table 1, the following trends are observed in the structure of disclosure of information about intellectual capital:

- 33.3–66.7% of cases of disclosure - information on the intellectual property;
- 40.7–25.9% of cases of disclosure - information about goodwill;
- 55.6–92.6% of cases of disclosure - information about other elements of intellectual capital, including brand, reputation, staff qualifications, corporate culture and management policy;
- 74.1–93.6% of cases of information disclosure in the discursive form, in addition to numerical indicators;
- 6.4–25.9% of companies did not disclose relevant information at all.

At the same time, the quality and content of information disclosure require separate consideration. Numerical information is disclosed only for aggregated indicators (without further elaboration). The disclosed discursive information contains only brief explanations of the data included in the aggregated indicators. The procedure for the formation of these indicators, including the uncertainties associated with them, plans for their further use and prospects for the development of companies do not find adequate disclosure.

Thus, the analysis of the structure and content of the information on intellectual capital disclosed by the largest Russian companies in the consolidated public financial statements allows us to conclude that it is insufficient for users to make adequate economic decisions. So, we can state the "information deficit" in the financial statements in relation to intellectual capital. It explains by the specific features of intellectual capital as an object of accounting.

4.3 Specific Features of Intellectual Capital as an Object of Accounting

The specific features of intellectual capital as an object of accounting and part of intangible assets can be explained by a combination of three factors:

1) Intangibility. Intellectual capital does not have a hard limit in space; the fact of its presence and condition is not subject to physical control. The criterion of control over an object seems ambiguous, because, unlike the objects of the material world, an intangible object can belong to several persons at the same time.
   The interpretation in practice of the principle of priority of economic content over the legal form as oblivion of the legal nature of the object only strengthens multiple accounting problems, since the use of economic and accounting potential is possible only in conjunction with the legal one. The intangibility of intellectual capital is closely associated with a higher degree of uncertainty associated with them and a lower degree of predictability. The high degree of uncertainty associated with intellectual property is also predetermined by the weakness of property rights to its objects, the lack of full control over the economic benefits of their use, which in practice is significantly reduced by the actions of competitors, and the variability of external environment (including tastes of consumers).

Table 1. Information on the disclosure of intellectual capital in public financial statements

Company	2015				2016				2017				2018			
	Numerical			Discursive	Numerical			Discursive	Numerical			Discursive	Numerical			Discursive
	Intellectual property	Goodwill	Other		Intellectual property	Goodwill	Other		Intellectual property	Goodwill	Other		Intellectual property	Goodwill	Other	
1. Gazprom	+	+	+	+	+	+	+	+	+	+	+	+	+	+	+	+
2. Rosneft			+	+			+	+			+	+			+	+
3. LukOil		+		+		+		+		+		+	+		+	+
4. Sberbank			+	+			+	+			+	+			+	+
5. Surgutneftegas		+		+		+		+		+		+		+	+	+
6. Transneft		+		+		+		+		+		+		+		+
7. Norilsk Nickel			+	+			+	+			+	+	+		+	+
8. VTB Bank			+	+			+	+			+	+			+	+
9. Magnit	+	+	+	+	+	+	+	+	+	+	+	+	+		+	+
10. Tatneft											+	+			+	+
11. Novatek			+	+			+	+			+	+			+	+
12. IDGC Holding													+		+	
13. Novolipetsk Steel													+	+	+	+
14. MegaFon	+	+	+	+	+	+	+	+	+	+	+	+	+		+	+
15. Sistema													+			
16. RusHydro													+		+	+
17. UC Rusal	+			+	+			+	+			+	+	+	+	+
18. Rostelecom	+			+	+			+	+			+	+		+	+
19. Severstal	+	+	+	+	+	+	+	+	+	+	+	+	+		+	+
20. X5 Retail Group		+		+		+		+		+		+	+		+	+
21. Inter Rao									+			+			+	+
22. Moscow Exchange	+	+	+		+	+	+		+	+	+		+	+	+	
23. United Aircraft Corporation		+	+	+		+	+	+		+	+	+	+		+	+
24. Nomos Bank			+	+			+	+			+	+			+	+
25. Mechel	+		+	+	+		+	+	+	+	+	+	+		+	+
26. Mail.ru Group Ltd. Sponsored GDR RegS	+	+	+	+	+	+	+	+	+	+	+	+	+	+	+	+
27. Alrosa			+	+			+	+			+	+	+		+	+
Total,% of the total number of companies	33.3	40.7	55.6	74.1	33.3	40.7	55.6	74.1	37.0	44.4	59.3	81.5	66.7	25.9	92.6	93.6

The intangibility of intangible assets also implies a close relationship with tangible assets in the process of use for the purpose of generating income, which, in turn, complicates the assessment of the potential for their future profitability and increases the degree of uncertainty associated with them.

2) Dualism. Duality manifests itself, first of all, in the very nature of intellectual capital.

   Part of the intellectual capital does not fall under the scope of legal protection instruments and guarantees, and part, primarily intellectual property, is an object of civil legal relations in respect of which a company may own a different scope of rights and different the degree of control over the extraction of income from the use of the object, and therefore their analysis can be very informative for the company's stakeholders.

   Dualism also reflects an internal contradiction in the valuation of intangible assets. On the one hand, there are no restrictions on the extent of use with respect to intangible assets, which is associated with the weakness of property rights and network effects (increase in value with an increase in the number of users). Moreover, intangible assets are goods (through the assignment of patents, trade in licenses and the results of research and development, etc.). But, on the other hand, most intangible assets are specific to their company. They may not cost anything on their own, but they work in the system and acquire value due to the synergistic effect. This dualism, taking into account the sequence of elements of the accounting procedure, entails the multi-aspect nature of the problem of accounting disclosure of information on intellectual capital.

3) Focus on the future. If traditional accounting is focused on facts that have happened in the past, then with regard to accounting for intellectual capital for users of financial statements, of interest are, first of all, those facts that are directly related to the future. The rules of financial accounting should not ignore the principle of conservatism, but should include methods for the economic evaluation of future income from efficient use of intellectual capital. The modern stakeholder is less and less interested in the traditional contents of the balance sheet and income statement. He needs information about the future of the company, the possibilities of its development, the creation of future value, the preservation and development of competitive advantages, which requires the provision of detailed information about the intangible assets of the company.

   These features of intellectual capital as an accounting object greatly complicate the methodology for preparing and disclosing accounting information and at the same time minimize the possibility of using traditional accounting tools that have become generally accepted within the framework of the industrial accounting paradigm. All this determines the informational limitations of the modern concept of accounting for intellectual capital.

4.4 The Ways for Overcoming of the Information Deficit Regarding Intellectual Capital in Traditional Financial Statements

Given the steadily increasing importance of intellectual capital in the global economy, this problem requires its immediate and effective solution. Key requirements for this decision: adaptability to the growing synergistic effect of intellectual capital in the company's business, the ability to take into account the individual characteristics of the company and focus on the numerous, diverse and dynamically changing information needs of its stakeholders.

Today, researchers are discussing two main ways to overcome the information deficit regarding intellectual capital in financial statements:

1) Supplementing traditional financial statements (while maintaining the accepted concept of its preparation) with a report on intangible assets (a report on intellectual capital). The obvious advantage of this solution is the stability of accounting regulation, which is not subject to revision. However, additional disclosure does not compensate for gaps in traditional financial statements, reduce its value for decision-making and trust in it, and does not guarantee the satisfaction of the information needs of users of statements.
2) Modification (expansion, revision of the concept of preparation) of traditional financial statements. This option is aimed at a long-term perspective, involves the development of the theory and methodology of accounting and reporting, the result of which should be the expansion of the information boundaries of traditional reporting in general and intangible assets in particular. It includes the revision of accounting regulation concepts and the development of accounting standards, the development of the concept of professional judgment of an accountant.

It is not surprising that in practice the first option, which is focused on the short term, preserving the prevailing accounting practices and accounting standards through the introduction of an additional information block, gained the most recognition.

To date, the IASB has decided to conduct a series of studies, followed by a discussion of the option of providing investors with information about intangible assets through a management report. However, based on the tasks set for the researchers by the IASB, there is reason to believe that the new management report will most likely not overcome the problematic situation that is taking place.

The author's analysis of the main trends in the formation of intellectual capital reports in practice, along with an analysis of the information needs of users, made it possible to establish the list of items which disclosure would be optimal from the position of financial statements users (Y. Ustinova (2019) [29, p. 111]).

5 Conclusion

So, in the traditional financial statements there is a deficit of information regarding intellectual capital both in terms of the range of recorded transactions and in terms of their valuation. Financial statements that do not reflect reliable and faithfully prepared information about intangible assets lose their status as the most important source of information when making management decisions.

At the same time, additional disclosures the information about the intellectual capital in special reports in some cases are used as a tool to bridge the information gap between the information content of financial statements and the information needs of its users. But the preparation of additional reports on intangible assets, in addition to traditional financial statements, does not solve the problem of disclosing information that users need, does not ensure its comparability and completeness of coverage.

The considered specific features of intellectual property as an object of accounting not only determine the informational limitations of the modern concept of its accounting, but also require the development of accounting theory and methodology in order to expand these restrictions. At the same time, it is necessary to analyze the information needs of stakeholders, factors explaining the boundaries of informational content of reporting, the possibilities of expanding these boundaries by presenting and disclosing relevant data. The unique nature of intangible assets as an accounting item implies a certain degree of freedom in making accounting decisions in preparing financial statements, with the mandatory disclosure of the approaches used, the circumstances taken into account and the associated uncertainties in the explanations.

In general, it should be noted that overcoming the information deficit of traditional financial statements in terms of intellectual capital becomes a fundamentally important issue when designing a company's development strategy, maintaining and enhancing its competitive advantages, including within the industry, in assessing its profitability and investment attractiveness. At the same time, the study shows the actual gap between the information on intellectual capital disclosed in public traditional financial statements and the information needs of financial statements users. Thus, a revision of the concept of preparing reporting information is required, the development of new theoretical and methodological provisions that would provide the opportunity to go beyond the informativeness of the industrial accounting paradigm and users to build reports based on it based on their own models of economic decision-making.

References

1. Dzenopoljac, V., Muhammed, Sh., Janosevic, S.: Intangibles and performance in oil and gas industry. Management Decision (2018). https://doi.org/10.1108/MD-11-2017-1139. Accessed 18 May 2019
2. Sveiby, K.E.: Knowledge companies (1986). http://www.sveiby.com/Portals/0/books/knowledgecompanies.html. Accessed 20 May 2018
3. Stewart, T.A.: Intellectual Capital: The New Wealth of Organizations, p. 125. Doubleday/Currency, New York (1997)
4. Edvinsson, L., Malone, M.S.: Intellectual Capital: The Proven Way to Establish Your Company's Real Value by Measuring its Hidden Brainpower, p. 430. Platkus, London (1997)
5. Reilly, R., Schweiss, R.: Otsenka nematerial'nykh aktivov [Valuation of intangible assets.]. Quinto Consulting, 796 p. (1998, 2005). (in Russian)
6. Lev, B.: Intangibles: Management, Measurement and Reporting, 240 p. Brookings Institute Press, Washington DC (2001)
7. Meritum project. Guidelines for managing and reporting on intangibles. Intellectual Capital Report (Vodafone Foundation), 130 p. (2002)
8. Wyatt, A., Abernethy, M.A.: Framework for Measurement and Reporting on Intangible Assets. Working Paper № 12/03. Intellectual Property Research Institute of Australia. Melbourne, 36 p. (2003)

9. Starovic, D., Marr, B.: Understanding Corporate Value: Managing and Reporting Intellectual Capital. Chartered Institute of Management Accountants. Cranfield University. School of Management, 28 p. (2005)
10. Moerman, L.C., Van der Laan, S.: Accounting for intellectual property: inconsistencies and challenges. J. Intellect. Prop. **11**(4), 243–248 (2006)
11. Abeysekera, I.: Intellectual Capital Accounting: Practices in a Developing Country, 224 p. Routledge (2007)
12. Andriessen, D., Tissen, R.: The future of financial reporting (in the field of intangible assets) (2017). https://gaap.ru/articles/budushchee_finansovoy_otchetnosti_v_sferenematerialnykh_aktivov. Accessed 20 May 2018
13. Castilla-Polo, F., Ruiz-Rodriguez, M.C.: Intangible assets disclosures in the olive oil differentiation strategy: a theoretical review. Agricul. Res. Technol. Open Access J. **14**(1), 1–8 (2018)
14. Lev, B., Zarowin, P.: The boundaries of financial reporting and how to extend them. J. Account. Res. **37**(3), 353–386 (1999)
15. Lopes, I.T.: The boundaries of intellectual property valuation: cost, market, income based approaches and innovation turnover. Intellect. Econ. **1**(9), 99–116 (2011)
16. Chander, S., Mehra, V.: A study on intangible assets disclosure: an evidence from indian companies. Intang. Cap. **7**(1), 1–30 (2011)
17. Sullivan Jr., P.H., Sullivan Sr., P.H.: Valuing intangible companies. An intellectual capital approach. J. Intellect. Cap. **1**(4), 328–340 (2000)
18. Oliveras, E., Amat, O.: Ethics and Creative Accounting: Some Empirical Evidence on Accounting for Intangibles in Spain. Economics Working Paper. № 732. Department of Economics and Business. Universitat Pompeu Fabra, 21 p. (2003)
19. Cormier, D., Ledoux, M.-J.: The Influence of Voluntary Disclosure about Intangible Assets Reported in French Financial Statement: The Role Played by IFRS. Corporate Reporting Chair, October, 36 p. (2010)
20. Wu, J.-Ch., Lin, Ch.: A balance sheet for knowledge evaluation and reporting. Management, Knowledge and Learning, International Conference, Zadav, Croatia, pp. 341–348 (2013)
21. Karius, T.: Intellectual Property and Intangible Assets: Alternative Valuation and Financing Approaches for the Knowledge Economy in Luxemburg. Research Report. EIKV-Schriftenreihe Zum Wissens und Wertemanagement, № 3, 92 p. (2016)
22. Greenwell, M.M., Tibbits, G.E.: Accounting for Identifiable Intangible Assets: No One Right Way. Accounting and Finance Working Paper 92/37, School of Accounting & Finance, University of Wollongong, June, 20 p. (1992)
23. Webster, E.: Intangible and Intellectual Capital: A Review of the Literature. Melbourne Institute of Applied Economic and Social Research, and Intellectual Property Research Institute of Australia, The University of Melbourne, 73 p. (2002)
24. Lev, B.: Intangible assets: concepts and measurements. Encycl. Soc. Measur. **2**, 299–305 (2005)
25. Hunter, L.C., Webster, E., Wyatt, A.: Identifying Corporate Expenditures on Intangibles Using GAAP. Melbourne Institute Working Paper Series, Working Paper No. 12/09, 15 p. (2009)
26. Artsberg, K., Mehtiyeva, N.: A literature review on Intangible assets. Critical questions for Standard setters. Working Paper. School of Economics and Management, June, 34 p. (2010)
27. Heinrich, R.: Valuation in Intellectual Property Accounting. UNECE Team of Specialists on Intellectual Property. Bishkek, 11 October, 43 p. (2011)
28. Petrova, R.: Accounting treatment of Intellectual Property in the pharmaceutical industry. Trakia J. Sci. **9**(4), 63–68 (2011)
29. Ustinova, Y.: Intellectual capital of a company: presentation and disclosure of information. In: Antipova, T., Rocha, Á. (eds.) DSIC 2019. AISC, vol. 1114, pp. 104–113. Springer, Cham (2020). https://doi.org/10.1007/978-3-030-37737-3_10

Model for Assessment of the Quality of Financial Management in Solving the Problem of Agency Conflicts

Nikolai Tolstov(✉)

Kuban State University, 149, Stavropolskaya Street,
Krasnodar 350040, Russian Federation
ntolstov@yahoo.com

Abstract. The study examines the problem of agency conflicts and their impact on the value of a business and the welfare of shareholders. This problem may lead to a decrease in the business capitalization, become the basis for unfriendly acquisitions by competitors and foreign capital, may become the reason for an increase in the cost of debt financing, thus affecting the overall financial stability and solvency of the company. The study offers an updated model for evaluating the quality of financial management of public joint-stock companies, based on EMR, TSR, FARR, SGR, EQ/TA, CR and ND/EBITDA indicators as the tool for eliminating and preventing agency conflicts that affect the business value; its efficiency is demonstrated basing on the data obtained from the largest corporations in Russia. Evaluation of the model effectiveness shows a stable direct correlation between the scores obtained using the model and the future indicators of a business value, which testifies to the appropriateness of using the model in the context of the Russian economic environment.

Keywords: Agency conflicts · Value based management · Management quality evaluation

1 The Problem of Agency Conflicts in Russian Corporations

The current stage of development of the Russian economy is characterized by significant strengthening of the role of the corporate sector, which, to various extents, affects almost all areas of the society activities and generates a significant portion of revenues to the state budget. At the same time, topicality and significance of problems associated with corporate management has increased, making not only competitiveness of corporations, but also the degree of satisfaction of public needs and interests dependable on the effectiveness of solving them. Agency conflicts between shareholders and the company managers have become one of the most common problems of corporate management, which consist in discrepancy between the decisions made by the top management and interests of the corporate shareholders.

According to the topical concept of value-based management (VBM) [1] the main goal of a business is to maximize wealth of its owners, therefore all management decisions taken in a corporation in the sphere of financial management should be aimed

T. Antipova (Ed.): ICCS 2020, LNNS 186, pp. 91–100, 2021.
https://doi.org/10.1007/978-3-030-66093-2_9

at maximizing the value of a business which depends not only on current financial results but also on the development prospects included into the business model [2, 3]. However, the top management may have personal interest in certain events of financial and economic activities of a corporation [4], or they may not be able to adequately and timely respond to the processes taking place in the company, causing decrease in the value of the business, which contradicts interests of the shareholders and becomes basis for agency conflicts which weaken the financial management system formed in the company [5].

Topicality of the agency relationship problem has increased with the development of financial markets. In the context of rapid digitalization of economic relations, agency conflicts may cause significant undervaluation of a business, become the basis for unfriendly acquisitions by competitors and foreign capital, may be the reason for an increase in the cost of debt financing, thus affecting the overall financial stability and solvency of the company, thereby increasing the significance of the problem and causing the need to find the solution.

It should also be noted that active study of the problem of agency relationships and its impact on the cost of businesses in Western scientific circles (the United States in the first place) began more than 80 years ago on the basis of long-term practices of managing joint-stock companies in a market economy environment. Western economists have developed and implemented models for assessing effectiveness of financial management and value of a business, mechanisms of corporate management and control, effective in their economic and institutional conditions, which made it possible for them to reduce the importance of the problem under consideration.

At the same time, the problem of agency conflicts in Russia first attracted attention of the specialists less than 20 years ago as the result of large-scale privatization and spontaneous formation of the private corporate sector. Most often, scientific works of domestic economists aim at reviewing the nature and manifestation of agency conflicts, while modified models and tools developed by Western economists, the effectiveness of which is empirically evaluated solely in terms of functioning of the Western financial markets and institutions, are offered as instruments of their prevention. At the same time, due attention is not paid to peculiarities of the national model of corporate management in Russia, which significantly reduces the effectiveness of Western models and tools in domestic economic and institutional environment [6, 7].

In our previous research, we considered the possibility of using financial management tools in solving agency conflicts. We analyzed the most popular Western indicators used to estimate the company's value, developed a model for evaluating the quality of financial management of a Public Corporation, taking into account specific character of the national model of corporate management in Russia, and evaluated its effectiveness basing on the data obtained from the largest corporations in Russia [8], which proved the existence of a stable correlation between the data obtained in the model and future value indicators of a business.

Further research has shown that the results can be improved by revising the list of indicators used in the model and adjusting the scale of standard criterion values used to form final assessments of the quality of the corporate financial management. Apart from that, expanding the observation scope not only in terms of increasing the number of analyzed companies, but also extending the time interval is of a particular value which will allow us to assess effectiveness of the model in different reporting periods and analyze dynamics of changes in the quality of financial management in the largest corporations in Russia.

In this regard, we consider it necessary to propose a new list of indicators used in the model and evaluate the effectiveness of the updated model for evaluating the quality of financial management of public corporations in dynamics, which will allow us to draw conclusions about the possibility and appropriateness of using the model in Russian economic environment on the basis of a broader empirical base.

2 Improving the Model for Evaluating the Quality of Financial Management of Public Companies

Previous research demonstrated appropriateness of using complex interdependent indicators, forming the final score on the basis of the coefficients of priority, in settling agency conflicts, which on the one hand encourages the top management to improve efficiency of the most vulnerable areas of the corporation activity, and on the other hand makes attempts to influence key performance indicators ineffective due to relatively low impact of each individual criterion on the final evaluation. In this regard, we previously suggested using a complex of 7 indicators, which included EMR, TSR, SVA, Sales Growth Rate, Equity to Total Assets, Return on Sales and Current Ratio [9]. In this study, we propose to replace two previously selected indicators in the model, which, in our opinion, will strengthen the balance of the model and increase its effectiveness in the Russian economy.

Firstly, we suggest replacing SVA indicator with the Fixed Assets Renewal Ratio. This is due to the fact that the SVA indicator is quite difficult both to calculate and understand, which makes its calculation polyvariant and gives the management an opportunity to manipulate the evaluation results. Besides, changing of the indicator values covers a rather large number of factors (free cash flow to firm, discount rate, terminal value, cost of capital, etc.), this indicator does not allow you to fully balance TSR indicator as to the possibility of manipulating it through excessively high dividend payments with prejudice to the promising investment programs, as a result. In its turn, the Fixed Assets Renewal Ratio is directly related to the CAPEX item in the cash flow statement and gives a more accurate idea of the net profit reinvestment volume.

Secondly, we considered it reasonable to remove the Return on Sales indicator from the model. Initially it was assumed that this indicator will allow evaluating financial performance of the company under analysis versus other comparable enterprises in the industry; and its dynamics will show how stable the financial results are. However, the study demonstrated that the company's profit indicators in the profit and loss account are often influenced by the accounting policies of each individual company, which makes it impossible to compare them correctly. In addition, a net loss reflected in the profit and loss account for a certain period does not always mean that the company has not created value for shareholders over a specified period of time, and vice versa. In this regard, we suggest removing the Return on Sales indicator from the model.

Thirdly, we consider it appropriate to include Net Debt/EBITDA into the model. This indicator is extremely popular among investors and brokerage firms. The peculiarity of this multiplier is that it compares the borrowed funds with the revenue generated by the company defining the period during which the enterprise will be able to pay back its debts provided that the amount of debt and the volume of profit are preserved, and not with other sources of financing as is the case with the financial independence and financing ratios.

Thus, the model for assessing the quality of financial management of public joint-stock companies includes 7 indicators, namely: EMR, TSR, FARR, Sales Growth Rate, Equity to Total Assets, Current Ratio and ND/EBITDA. As a result, the model takes the form shown in Fig. 1.

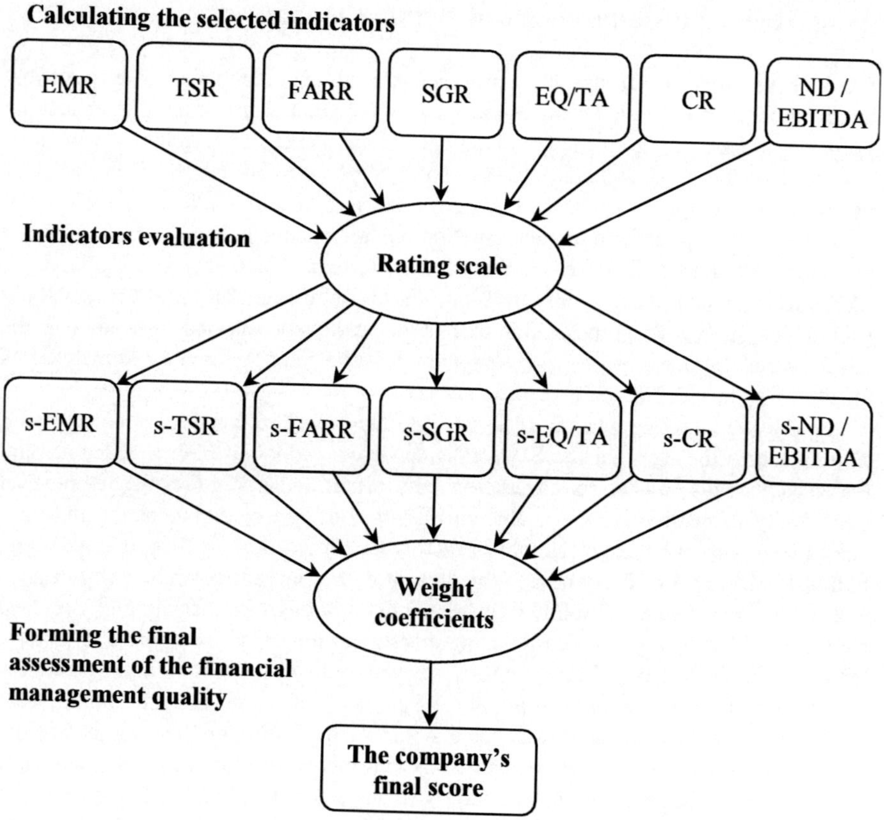

Fig. 1. A model for assessing the quality of financial management

To illustrate the effectiveness of the model, we selected the largest Russian corporations operating in the fields of metallurgy, telecommunications and oil and gas which have shares freely traded on the Moscow Exchange. The results of the first calculation stage are shown in Table 1.

Table 1. Values of indicators selected to be included into the model for assessment of the quality of financial management based on the results of the year 2018

Corporation	EMR	TSR	FARR	SGR	EQ/TA	CR	Net Debt/EBITDA
VSMPO-AVISMA	−11,6%	17,9%	4,6%	6,5%	0,54	3,6	1,5
Mechel	−37,2%	−31,2%	5,2%	6,4%	−0,73	0,2	6,3
MISW	−33,1%	22,4%	17,8%	17,3%	0,70	2,0	−0,1
NLMK	−25,9%	28,4%	13,9%	22,0%	0,59	2,1	0,3
NorNickel	−4,7%	23,9%	15,6%	15,3%	0,23	1,9	1,6
Polyus	4,4%	15,8%	22,8%	6,4%	0,07	1,9	3,3
Severstal	−20,3%	13,3%	20,4%	17,0%	0,49	1,8	0,4
MGTS	13,0%	62,3%	14,9%	−0,7%	0,73	2,6	−0,9
MegaFon	10,8%	8,8%	28,5%	3,0%	0,14	0,8	2,1
MTS	7,1%	6,5%	29,5%	5,0%	0,08	0,9	1,2
Rostelecom	−12,4%	0,2%	20,1%	3,8%	0,39	0,6	2,0
Tattelecom	−12,5%	15,3%	10,5%	5,9%	0,50	1,2	0,9
Bashneft	−55,1%	−18,2%	12,3%	20,4%	0,60	2,7	0,3
Gazprom	−48,8%	5,1%	14,1%	16,0%	0,66	1,7	1,2
Gazprom Neft	−37,4%	32,5%	17,4%	26,9%	0,57	1,2	0,7
LUKOIL	−20,8%	25,5%	12,1%	24,0%	0,71	1,6	0,0
NOVATEK	−31,7%	21,5%	24,0%	24,8%	0,73	2,7	0,2
Rosneft	−37,8%	8,2%	13,5%	28,5%	0,36	1,1	1,8
Surgutneftegas	−32,2%	−1,1%	11,8%	23,5%	0,86	3,6	−1,1
TATNEFT	−25,6%	40,4%	13,6%	25,3%	0,65	0,9	−0,5

Further, at the second stage, we will need to evaluate the obtained results. For this purpose, we have developed a scale of standard criterion values (Table 2), which allows us to rank the obtained results and assign a strictly defined score to them. The scale contains both generally accepted standard values of indicators and the ranges formed using the Three Sigma Rule.

Table 2. Scale of standard criterion values of indicators

Indicator	1 score	4 scores	7 scores	10 scores
EMR	$Xi < -30\%$ $30\% < Xi$	$-30\% \leq Xi < -20\%$ $20\% < Xi \leq 30\%$	$-20\% \leq Xi < -10\%$ $10\% < Xi \leq 20\%$	$-10\% \leq Xi \leq 10\%$
TSR (+)	$Xi < 0$	$0 \leq Xi < \dot{X} - 5\%$	$\dot{X} - 5\% \leq Xi \leq \dot{X} + 5\%$	$\dot{X} + 5\% < Xi$
TSR (+/−)	$Xi < \dot{X} - 5\%$	$\dot{X} - 5\% \leq Xi < 0$	$0 \leq Xi < \dot{X} + 5\%$	$\dot{X} + 5\% < Xi$
TSR (−)	$Xi < \dot{X} - 5\%$	$\dot{X} - 5\% \leq Xi \leq \dot{X} + 5\%$	$\dot{X} + 5\% < Xi \leq 0$	$0 < Xi$
FARR	$0 \leq Xi < \dot{X} - 9\%$	$\dot{X} - 9\% \leq Xi < \dot{X} - 3\%$	$\dot{X} - 3\% \leq Xi \leq \dot{X} + 3\%$	$\dot{X} + 3\% < Xi$
SGR (+)	$Xi < 0$	$0 \leq Xi < \dot{X} - 2\%$	$\dot{X} - 2\% \leq Xi \leq \dot{X} + 2\%$	$\dot{X} + 2\% \leq Xi$
SGR (+/−)	$Xi < \dot{X} - 2\%$	$\dot{X} - 2\% \leq Xi < 0$	$0 \leq Xi \leq \dot{X} + 2\%$	$\dot{X} + 2\% \leq Xi$
SGR (−)	$Xi < \dot{X} - 2\%$	$\dot{X} - 2\% \leq Xi \leq \dot{X} + 2\%$	$\dot{X} + 2\% < Xi < 0$	$0 \leq Xi$
EQ/TA	$Xi < 0{,}4$	$0{,}4 \leq Xi < 0{,}5$	$0{,}5 \leq Xi < 0{,}6$ $0{,}7 < Xi \leq 1$	$0{,}6 \leq Xi \leq 0{,}7$
CR	$Xi < 1$	$1 \leq Xi < 1{,}4$	$1{,}4 \leq Xi < 1{,}7$ $2{,}3 < Xi$	$1{,}7 \leq Xi \leq 2{,}3$
ND/EBITDA	$3 < Xi$	$2 < Xi \leq 3$	$1 < Xi \leq 2$	$Xi \leq 1$

We will evaluate the results of the first stage using the evaluation scale of standard criterion values. The results of the second stage are presented in Table 3.

Table 3. The results of evaluating the indicators of the companies under the study for the year 2018

Corporation	EMR	TSR	FARR	SGR	EQ/TA	CR	Net Debt/EBITDA
VSMPO-AVISMA	7	7	1	4	7	7	7
Mechel	1	1	1	4	1	1	1
MISW	1	10	7	7	10	10	10
NLMK	4	10	7	10	7	10	10
NorNickel	10	10	7	7	1	10	7
Polyus	10	7	10	4	1	10	1
Severstal	4	7	10	7	4	10	10
MGTS	7	10	1	1	7	7	10
MegaFon	7	7	7	7	1	1	4
MTS	10	7	10	7	1	1	7
Rostelecom	7	4	4	7	1	1	4
Tattelecom	7	10	1	7	7	4	10
Bashneft	1	1	7	4	10	7	10
Gazprom	1	4	7	4	10	10	7
Gazprom Neft	1	10	10	10	7	4	10
LUKOIL	4	10	7	7	7	7	10
NOVATEK	1	10	10	7	7	7	10
Rosneft	1	4	7	10	1	4	7
Surgutneftegas	1	1	7	7	7	7	10
TATNEFT	4	10	7	10	10	1	10

At the third stage, we will have to form final assessments of the quality of financial management for each company using static weighting coefficients presented in Table 4. Justification of the values of the weighting coefficients is a rather voluminous and, in some respects, debatable issue not possible to be considered in this study. It should be understood that these weighting coefficients do not change depending on the company under assessment, however, they may change over time depending on the macroeconomic environment. Further in this study, when evaluating the effectiveness of the model for all the considered periods, the coefficients presented in the table will be taken as constants.

Table 4. Values of weighting coefficients used to form the final score

Показатель	EMR	Net Debt/EBITDA	TSR	FARR	SGR	EQ/TA	CR
Weight coefficient	35%	15%	10%	10%	10%	10%	10%

Using these weighting coefficients, based on the previously obtained scores, we will form final assessments of the quality of financial management for the corporations under study. The results of calculations are given in Table 5.

Table 5. Assessment results of the corporations under study for the year 2018

Corporation	EMR	TSR	FARR	SGR	EQ/TA	CR	Net Debt/EBITDA	Final score
VSMPO-AVISMA	2,5	0,7	0,1	0,4	0,7	0,7	1,1	6,1
Mechel	0,4	0,1	0,1	0,4	0,1	0,1	0,2	1,3
MISW	0,4	1,0	0,7	0,7	1,0	1,0	1,5	6,3
NLMK	1,4	1,0	0,7	1,0	0,7	1,0	1,5	7,3
NorNickel	3,5	1,0	0,7	0,7	0,1	1,0	1,1	8,1
Polyus	3,5	0,7	1,0	0,4	0,1	1,0	0,2	6,9
Severstal	1,4	0,7	1,0	0,7	0,4	1,0	1,5	6,7
MGTS	2,5	1,0	0,1	0,1	0,7	0,7	1,5	6,6
MegaFon	2,5	0,7	0,7	0,7	0,1	0,1	0,6	5,4
MTS	3,5	0,7	1,0	0,7	0,1	0,1	1,1	7,2
Rostelecom	2,5	0,4	0,4	0,7	0,1	0,1	0,6	4,8
Tattelecom	2,5	1,0	0,1	0,7	0,7	0,4	1,5	6,9
Bashneft	0,4	0,1	0,7	0,4	1,0	0,7	1,5	4,8
Gazprom	0,4	0,4	0,7	0,4	1,0	1,0	1,1	4,9
Gazprom Neft	0,4	1,0	1,0	1,0	0,7	0,4	1,5	6,0
LUKOIL	1,4	1,0	0,7	0,7	0,7	0,7	1,5	6,7
NOVATEK	0,4	1,0	1,0	0,7	0,7	0,7	1,5	6,0
Rosneft	0,4	0,4	0,7	1,0	0,1	0,4	1,1	4,0
Surgutneftegas	0,4	0,1	0,7	0,7	0,7	0,7	1,5	4,8
TATNEFT	1,4	1,0	0,7	1,0	1,0	0,1	1,5	6,7

As we can see, according to the results of 2018, the companies received significantly different ratings. It is worth noting that at this stage, basing on the information presented in the table, it is possible to determine the most vulnerable areas of the existing corporate business model compared with other representatives of the industry in question, which allows you to identify the areas where agency conflicts may occur in order to eliminate them in the future.

3 Evaluating Effectiveness of the Model in the Context of the Russian Economic Environment

Despite the fact that the obtained estimates characterize companies at a strictly defined moment of time, it should be taken into account that large corporations have a low level of the assets structure and capital mobility and are often unable to radically change their financial and economic structure within a short period of time. This, to some extent, gives us the opportunity to compare the obtained estimates of corporations with the return on their equity instruments. With this end in view, we similarly evaluated the companies under study, basing on the results of 2016 and 2017. We also calculated the MC and TSR indicators using the actual results of 2017–2019. The data is presented in Table 6.

Table 6. Initial data for correlation analysis of the effectiveness of the proposed model

Corporation	Score 2016			Score 2017			Score 2018		
	Score	MC 2017	TSR 2017	Score	MC 2018	TSR 2018	Score	MC 2019	TSR 2019
VSMPO-AVISMA	7,5	21,8%	36,7%	6,3	-7,9%	1,9%	6,1	30,6%	41,8%
Mechel	2,2	-17,3%	-16,4%	2,5	-46,0%	-44,1%	1,3	-14,0%	-10,1%
MISW	9,1	25,8%	35,5%	7,6	2,9%	15,5%	6,3	-2,6%	5,7%
NLMK	8,2	28,0%	39,0%	8,8	6,9%	20,1%	7,3	-8,7%	5,8%
NorNickel	7,6	7,2%	13,8%	4,9	20,2%	32,9%	8,1	46,5%	64,0%
Polyus	7,3	3,7%	9,5%	7,5	17,5%	23,6%	6,9	31,6%	37,2%
Severstal	6,1	-5,8%	5,9%	6,1	6,3%	23,9%	6,7	-0,6%	9,4%
MGTS	8,2	64,8%	93,4%	7,6	25,4%	42,6%	6,6	14,9%	28,7%
MegaFon	6,3	-11,5%	-1,8%	5,5	24,7%	24,7%	5,4	1,7%	1,7%
MTS	6,1	6,6%	16,6%	6,1	-13,8%	-4,3%	7,2	34,4%	46,5%
Rostelecom	5,8	-23,3%	-16,8%	4,5	14,2%	22,2%	4,8	7,5%	14,4%
Tattelecom	7,5	-5,0%	1,1%	6,4	21,0%	34,3%	6,9	32,0%	43,1%
Bashneft	4,5	-34,0%	-29,4%	7,8	-13,0%	-5,6%	4,8	2,2%	10,7%
Gazprom	5,2	-15,6%	-10,4%	4,3	17,7%	23,8%	4,9	66,9%	77,8%
Gazprom Neft	8,1	14,1%	23,7%	6,3	42,1%	53,2%	6,0	21,1%	28,6%
LUKOIL	8,2	-3,3%	2,6%	8,8	49,9%	56,6%	6,7	23,5%	30,4%
NOVATEK	7,8	-14,4%	-12,7%	9,1	66,9%	69,4%	6,0	11,6%	14,3%
Rosneft	5,4	-27,6%	-25,2%	6,7	47,8%	55,1%	4,0	4,4%	10,5%
Surgutneftegas	6,4	-10,3%	-8,4%	5,1	4,2%	6,9%	4,8	65,7%	72,2%
TATNEFT	8,1	13,7%	25,9%	8,8	53,5%	62,5%	6,7	4,8%	17,8%

This data allows us to compare the scores received in the reporting year with the future indicators of the value of a business. To determine the level of effectiveness of the presented model, we will perform a correlation analysis between the obtained actual scores and the values of the following reporting period, i.e. score 2016 – MC/TSR 2017, score 2017 – MC/TSR 2018 and score 2018-MC/TSR 2019. The results of the correlation analysis are presented in Table 7.

It should be pointed out that in the correlation analysis 6 observations were excluded from the 60 presented observations (marked in red in Table 6). This is due to the fact that during the period under review certain force majeure occurred in these companies, which significantly reduced or increased their capitalization within a short period of time. For example, dramatic drop in the value of MTS shares in 2018 was caused by the US Securities and Exchange Regulatory Commission investigation into its activities in Uzbekistan, which resulted in the necessity for the company to accumulate a reserve of 59 billion Rubles for the payment of a fine at the end of November [10].

Table 7. Correlation analysis of the relationship between the scores and corporate value indicators

Year	Metallurgy		Telecommunications		Oil and gas		Total	
	MC	TSR	MC	TSR	MC	TSR	MC	TSR
2016	0,87	0,87	0,81	0,80	0,84	0,80	0,67	0,65
2017	0,67	0,68	0,76	0,98	0,88	0,86	0,66	0,70
2018	0,71	0,75	0,87	0,94	0,57	0,71	0,72	0,76

The data presented in the table indicate that there is a direct stable relationship between the scores obtained in the course of testing the model and future indicators of the value of a business during the three periods under review. The analysis testifies to the fact that the higher the final scores the companies received, basing on the evaluation results, the greater the increase (the smaller the decrease) in value they provided for the shareholders according to the results of the following reporting period, which confirms the effectiveness of the used model in the context of the national corporate management model in Russia and the current market environment. It should also be noted that with this set of indicators, correlation is more stable than in the previous studies, which indicates the correct direction of development of this approach to solving the problem of agency conflicts.

4 Conclusion

The problem of agency conflicts with regard to their impact on the value of a business is a fairly wide, but still little-studied area of research in the Russian corporate sector environment. In this article we presented an updated version of the model for assessing the quality of financial management of public corporations and demonstrated its efficiency basing on the activity of 20 largest Russian corporations operating in the fields of metallurgy, telecommunications and oil and gas. Evaluation of the model effectiveness shows that the updated list of indicators and weighting coefficients presented in this article demonstrates a more stable correlation with the future indicators of the value of a business than the previously presented version of the model, which testifies to the correctness of the changes made and appropriateness of further research in this area.

As for the practical application, it is worth noting that the model is mainly aimed at strengthening the shareholders control over the corporation and identifying opportunistic behavior of its management [11] as well as low-performance management decisions. At the same time, the presented model can be used by the management in order to identify the weakest areas of the existing business model, adjusting which will result in improvement in the financial and economic situation of the company and the corresponding strengthening of the position of its top management.

At the same time, we assume that further research can be continued towards adjusting the list of indicators involved in the model and their number. It is also worth paying attention to changes in the macroeconomic business environment, which alter the priority of the indicators involved and require a corresponding change in the weighting coefficients. We also consider further extension of the empirical base of the study to be of importance. Company estimates for the year 2019 and actual stock prices for the year 2020 will be of particular interest due to the significant decline in economic activity caused by the pandemic, the analysis of which will allow us to evaluate the proposed model in the context of the global crisis.

References

1. Aleksandrova, A.V., Gorohova, A.E., Sekerin, S.V.: The evolution of views on corporate governance. Izvestiya MGTU «MAMI» **4**(18), 141–147 (2013)
2. Jensen, M.: Value maximization, stakeholder theory, and the corporate objective function. J. Appl. Corp. Finan. **14**(3), 8–21 (2001)
3. Kovalev, V.V., Kovalev, V.V.: Profit: the logic of definition and identification. St Petersburg Univ. J. Econ. Stud. **2**, 106–117 (2005)
4. Murphy, K., Jensen, M.: CEO bonus plans: and how to fix them. Harvard Business School NOM Unit Working Paper, no 12-022 (2011)
5. Jensen, M., Meckling, W.: Theory of the firm. Managerial behavior, agency costs and ownership structure. J. Finan. Econ. **3**(4), 305–360 (1976)
6. Lvova, N.A.: The functioning of public companies in the Russian Federation. Econ. Manag. **7**(105), 54–58 (2014)
7. Lvova, N.A.: Financial diagnostics of public companies: choice of methodological alternatives for the emerging financial market. Econ. Manag. Theory Pract. **1**(5), 53–60 (2019)
8. Shevchenko, I., Puchkina, E., Tolstov, N.: Shareholders' and managers' interests: collisions in Russian corporations. HSE Econ. J. **1**, 118–142 (2019)
9. Tolstov, N., Shevchenko, I.: The problem of agency conflicts in Russian corporations and ways to overcome it. In: Lecture Notes in Networks and Systems, vol. 136, pp. 135–143 (2020)
10. MTS will pay almost a billion dollars to the USA on charges of bribes in Uzbekistan. BBC News (2019). https://www.bbc.com/russian/news-47478260. Accessed 14 July 2020
11. Williamson, O.E.: The economic institutions of capitalism: firms, markets, relational contracting. The Free Press, New York (1985)

Role of Accounting Policy in Fair Statement's Representation

Elena Dombrovskaya(✉)

Financial University under the Government of the Russian Federation, Leningradsky pr., 49, Moscow 125993, Russian Federation
den242@mail.ru

Abstract. Financial reporting has always been one of the most important characteristics of a company. Financial reporting serves as a source of information for managers to make critical managerial and strategic decisions. Adherents of fundamental analysis of financial markets base their recommendations on purchasing securities, inter alia, on the analysis of the financial reporting of the company of interest to an investor. The state not only charges a tax on the company but also can subsidize it or restrict its activities based on financial reporting. These and many other factors explain the critical importance of the accuracy of financial reporting. The accounting policy determines the choice of a method for reflecting business transactions allowed by law, which affects the financial reporting. That is, the company's accounting policy, as an instrument for preparing its financial reporting, has a significant effect on the accuracy of financial reporting. The paper is devoted to the study of the accounting policy effect on the accuracy of financial reporting as the most important qualitative characteristic of such reporting information.

Keywords: Accounting policy · Financial reporting · Accuracy

1 Introduction

The accuracy of financial reporting has always been one of its most important characteristics in international business. At first glance, it seems to be easy enough to define the concept of "accuracy" with respect to financial reporting: the operations reflected in accounting must be true. However, this is only one of many aspects of the financial statement characteristics. The concept of accuracy in accounting has evolved over many years along with the development of accounting science.

An accounting policy, subject to its effective integration into the company's internal audit system, can significantly help to provide accurate reporting information both in terms of the contents of financial reporting and the compliance of accounting data, and its presentation in reporting, with current legal standards.

The aim of this paper attempts to identify areas of influence of accounting policies on the reliability of financial statements. The study was performed using materials from an oil production company. The accounting policy is presented as a tool to minimize the risks of presenting false financial statements.

T. Antipova (Ed.): ICCS 2020, LNNS 186, pp. 101–110, 2021.
https://doi.org/10.1007/978-3-030-66093-2_10

2 Literature Review and Current State of the Issue

Having analyzed the prevailing accounting practice in various countries, we can conclude that, even before the spread of the IFRS, many of them widely used the concept of "true and fair view." Australian researchers G. Dean and F. Clarke argue that such a concept was, one might say, the cornerstone for accounting principles that are applied in Australia, the UK, and other Anglo-Saxon countries [1]. Moreover, there has been the so-called adjustment for true and fair reflection in Australia for a long time, allowing the companies to deviate from the national standards if reflecting a business transaction in accordance therewith might impair the accuracy of financial reporting. Currently, Australian Accounting Standards allow, in very rare cases, to neglect their use, but the company must substantiate this need in detail [2].

On the other hand, Australian theorists D. Alexander and E. Eberhartinger note in their study that, even though the goal of financial accounting and reporting is a true and fair reflection, no such adjustment was applied in Germany [3]. The authors further describe the main differences between the Anglo-Saxon and continental systems. Thus, in the UK and other Anglo-Saxon countries, financial reporting is fundamentally economic in their nature, while the continental model—where much attention is paid to tax issues and creditors protection—applies a more "prudent" approach from a legal point of view. To prove the absence of the need to include such an adjustment into their national accounting standards, the authors refer to the fact that the differences in accounting methods, the existence of which may cause the need to deviate from current norms, are too technical and cannot, by their nature, contradict the concept of true and fair reflection. As for the absence of adjustment for the prevalent principle of such reflection in German standards, the authors explain it by the specifics of the German law and by the fact that this concept does not affect fundamental issues of recognition or evaluation, involving only the issues of disclosure of additional information in comments to reporting and their presentation in general.

When analyzing the concept of the accuracy of financial reporting, it is impossible to ignore its audit. Romanian theorists M. Dumitru and B. Florin, in their study of audited reporting in terms of accuracy, conclude that it is the auditor who plays a key role in ensuring the true and fair reflection of financial transactions in reporting. According to the researchers, the main strengths of auditors are their exceptional experience, knowledge, and an accumulated system of auditing methods. A curious outcome of their study is the fact that the main (58% of cases) ground for issuing a negative auditor's opinion is "too much discrepancy with the principles of true and fair reflection in financial reporting". At the same time, the risk of making a wrong economic decision based on erroneous reporting is a sufficient reason for issuing such an opinion in only 11% of cases [4].

Although many foreign researchers are debating about the nature of true and fair reflection, no official definition of this concept is currently enshrined in international regulatory documents. Moreover, many scientists believe that the requirements of accuracy and relevance may be antithetical. A. Burleau, who explores this issue, identifies three periods in the history of accounting, depending on its most significant areas: the asset inventory period, the profit or loss evaluation period, and the period when the assessment (diagnosis) of the company's general financial situation becomes important. It was in the third period that fell on the development of global financial markets and international (and European) financial reporting standards, that there was a shift of priority towards true and fair reflection. Burleau substantiates his point of view by the fact that relevance is understood in the IFRS as the financial statement compliance with the interests of its users, and this does not entail the requirement of its accurate presentation. On the other hand, P. R. Bahnson and P.B.W. Miller note that both of the above characteristics are fundamental for financial reporting. However, neither accuracy, nor comparability, nor comprehensibility of reporting (other characteristics emphasized by international standards) would make sense in case of reflecting irrelevant operations [5].

Another curious suggestion is that of Canadian researchers W. Smieliauskas, J. Amernik, and R. Craig to replace the term of "fair and true reflection" with the term of "acceptable risk of material misstatement" when issuing an auditor's opinion. The scientists believe that such a phrase describes more accurately the essence of their activity outcome, and it is more applicable to the characteristics of the information provided in reporting [6].

Amendments to the Framework for the Preparation and Presentation of Financial reporting, made by the IFRS Board in 2018, returned prudence to the list of financial reporting characteristics. At the same time, the need for an asymmetric approach to different objects is rejected, although the existence of asymmetric requirements in individual standards is recognized to select the most relevant information. Commenting on the decision made by the IFRS Board, Norwegian scientist C. Pelger noted that it would be difficult to combine diligence and asymmetric prudence in practice. According to the IFRS Board, asymmetric prudence can sometimes be used to ensure the usefulness of decisions made. However, C. Pelger believes that its significance as a qualitative characteristic of financial reporting is highly doubtful [7].

Of particular interest is the dialectics of the requirements for ensuring the accuracy of financial reporting in the context of developing digital technology. O. Efimova and O. Rozhnova believe that there is currently a demand among the business community for transparent reporting with analytical and predictive value. This resulted in significant changes in the basic financial reporting system and contributed to the development of nonfinancial reporting. Digital technology provides technical capabilities for creating and structuring of a large amount of reporting information [8].

According to author, an increased amount of data processed makes it difficult to ensure the accuracy of financial statement. To prevent any distortion of information and to provide an objective basis for its audit, the statement preparation principles should be stricter. High certainty in the estimates should be more important than the information relevance. Extremely important is returning the concept of contents prevailing over the form into the Framework for the Preparation and Presentation of Financial reporting, which was excluded in 2010. Adherence to all these approaches aims at one goal: to present the information in financial reporting accurately [9].

The analysis of the genesis of ideas about the accuracy of financial reporting in the Russian and international academic practice allows us to conclude that, although the accuracy of financial reporting is defined in the current Russian law as its compliance with regulatory requirements, scientists agree in many respects. Thus, both Russian and international researchers note that the human factor plays an important role in the accuracy of financial statement. At the same time, differences in the approach to defining the concept of "accuracy of financial reporting" in Russian and foreign practices largely determine the directions of scientific research on this topic. International research practice pays more attention to studying the influence of accountant's professional judgment, whereas Russian authors often focus on research and resolving conflicts in the legal regulation of accounting as one of the factors determining its accuracy.

Russian researchers D. Lugovsky and Kuter identify three areas that are determined individually by each company and have a direct impact on its financial reporting [10]. These areas are the choice of methods for accounting organization, the methods of accounting, and the method of evaluating its items. It is the first two areas that are covered by the accounting policy adopted by a company for its accounting purposes. However, even in cases where regulations in the field of accounting are imperative, i.e., provide for the only possible way to account for one object or several objects, the accountant's professional judgment allows applying such accounting methods properly. The provisions of the above regulations may also be optional; that is, they may provide for several ways to reflect some operations or accounts—e.g., the choice of methods for depreciation or write-off of inventory. In this case, the accountant's professional judgment allows him or her to evaluate the suggested options in terms of appropriateness of their use for a particular company and, on this basis, to include some method or technique into the accounting policy. Therefore, we can conclude that accounting policies directly affect the accounting information accuracy in terms of a formal approach, as it is a direct way of expressing the accountant's professional judgment.

Czech researchers J. Kapič and M. Basič argue that the most important role in the compliance with accounting rules adopted in accounting policy is played by a well-developed system of internal audit: in the absence of a well-organized auditing system, following the most rigorous standards makes no sense [11]. The authors also believe that the system of internal audit should evolve along with new challenges faced by the company, its accounting system, and emerging risks. Thereat, we should distinguish internal audit at the stage of developing the company's accounting policy when, after preparing this document by the chief accountant or another authorized person, it can be amended and improved, and the company's internal audit system as such—aiming at the compliance by the company's employees with the accounting methods and techniques set forth by the accounting policy. Moreover, some authors, such as E.V. Shlifer, believe that the company's accounting policy by itself may be an element of its internal auditing system, since the requirements for developing an accounting policy, as listed above, meet the criteria of the internal audit system for achieving the goals of such system [12].

It is possible to carry to number of researches of the early reporting [13, 14].

Accounting policies directly affect the accuracy of accounting information, since an accounting policy I a summary of the accounting methods and techniques adopted by the company, which have a direct impact on the accuracy of the company's reporting. If an accounting policy is developed in accordance with current legal requirements in the field of accounting, the company's financial reporting is likely to be accurate, provided the accuracy is understood as compliance of reporting with regulations. However, we shall not forget that, first, the accounting policy correctness shall be validated at its development stage, and the process of accounting carried out by the company's employees or by an outsourcing service shall be checked for compliance with this policy. For this purpose, one uses the accounting policy audit and improvement of the internal audit system. We should also keep in mind that, in the absence of national regulations governing the accounting for certain business transactions, companies can develop similar methods on their own, guided by the current regulatory framework for accounting and the International Financial Reporting Standards. Particular attention should be paid to any changes in the company's accounting policy, as they carry the risks of misrepresentation of financial reporting in the absence or upon incorrect implementation of the necessary adjustments.

3 Research Results

In the course of our research, we analyzed the activities of a major oil company based on corporate governance principles. Based on the results of an external audit, the company's financial reporting was recognized to be accurate. This was confirmed by the analysis of financial reporting, the company's risk management, and internal audit systems. Also, we took into account the unconditionally positive opinion of the company's independent auditor on the accuracy of its reporting.

The analysis of the company's accounting policy showed that, although the assessment of reporting accuracy of the information provided is positive in general, it could be improved. To find a way to improve the accuracy of financial reporting by optimizing the accounting policy, we identified the main sources of risk of a decrease in the accuracy of reporting information. Those sources are presented in the Table 1 below and take into account the company's specifics.

Table 1. Risks of inaccuracy of the oil company's reporting information

Risk area	Possible area of reporting misrepresentation
Significant number of subsidiaries and affiliates	Income from property rental; accounts receivable and payable; financial investments; cost of sales (possible purchase of resources at inadequate prices)
Involvement in litigations as a defendant	Estimated liabilities
Specifics of the company's activities, which are prone to cause environmental damage	Estimated liabilities
Lack of intangible assets revaluation	Intangible assets
Accounting for financial derivatives at fair value according to the IFRS	Financial derivatives at fair value through profit or loss
Periodic one-off remuneration to managers	Estimated liabilities

Reducing the risk of inaccurate information in some of these areas can be achieved by making clarifying amendments to the company's accounting policy. It should be noted that the company provides a fairly detailed explanation of such an aspect of its accounting policy as the disclosure of information on related parties. As for the accounting of estimated liabilities, although this aspect is not covered sufficiently in the accounting policy, information on these accounting items is disclosed in detail in the comments to the company's financial reporting. Therefore, it is advisable to include this information on the types of possible estimated liabilities, contingent assets and liabilities into the accounting policy. On the other hand, the requirements of national standards for this aspect of accounting do not provide for many options. In this regard, the accounting policies coverage of this aspect of accounting is not mandatory. The company can maintain a high level of disclosure without changing its accounting policies. In this case, sufficient accuracy is ensured by compliance with the accounting standards.

Regarding the accounting of intangible assets, the relevant section of the company's accounting policy does not provide for any regular revaluation of intangible assets at current market value, although this option is provided for by national and international financial reporting standards. Valuation at historical cost may reduce the accuracy of information about intangible assets in the financial reporting without reflecting the current market situation. Incorporating the requirement for regular revaluation of intangible assets at current market value into the accounting policy can increase the accuracy of the company's reporting information.

Our analysis showed that the accounting policy could affect the accuracy of accounting information in some cases, as shown in the Fig. 1.

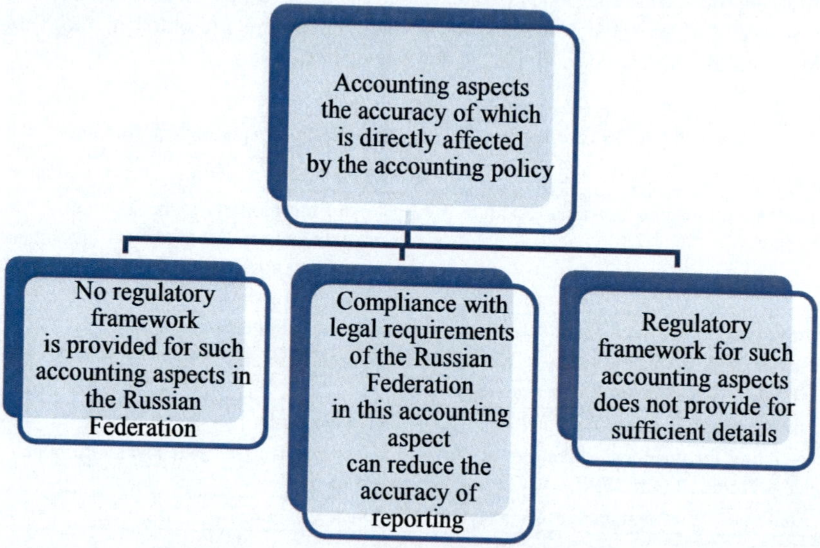

Fig. 1. Areas of the accounting policy influence on the accuracy of accounting information

Let us consider in detail the above three aspects and the mechanism of accounting policy influence on the accuracy of reporting in each case. In some situations where the accounting procedure for a particular activity is not provided for in accounting regulatory documents, the company may develop its own method of accounting in this field. In this case, accounting national regulations, as well as international financial reporting standards, are taken into account. In such a situation, the accuracy of reporting is determined by the chosen accounting methods and depends directly on the company's accounting policy. With appropriate coverage of such matters in the accounting policy, there should be no intentional or unintentional mistakes in reporting prepared in accordance with the accounting policy. Such reporting will be accurate in terms of both their form and contents. In terms of the form, this is due to the fact that national standards allow the company to introduce such rules independently, and in terms of contents, the fact that in such cases, the main goal of the accounting policy is an accurate reflection of certain events of the company's business life in its financial reporting.

A serious and unexplored problem is ensuring the accuracy of financial reporting if the company has to combine national and international standards when providing indicators in one reporting form. This is a difficult task that requires efforts of highly qualified specialists both in the development of accounting policy and in the compliance with its requirements as part the daily accounting process. With the company in question, an example of this aspect of accounting is financial derivative accounting. The company's balance sheet includes two separate items that are not provided for in the standard forms of this statement: short-term and long-term financial derivatives at fair value through profit and loss. Changes in the value of financial derivatives at fair value—arising from changes in the value of assets which such financial derivative is linked to—are presented in the company's profit and loss statement. In terms of the statement accuracy, the process of determining the fair value of such assets is crucial. The relevant section of the company's accounting policy does not specify what is meant by the fair value of various types of financial derivatives. On the other hand, the comments to the company's balance sheet and the profit and loss statement provide explanations on this matter. Thus, based on the above document, the company has effected currency swap transactions and foreign exchange forward transactions for the sale of U.S. dollars, which is part of the foreign-exchange-and-interest-rate risk management process. When calculating the fair value of such derivatives, the company is guided, as stated above, by the present value of future cash flows, using the information about the consensus forecast regarding exchange rates. The procedure for determining the fair value of financial derivatives linked to other assets (various securities, raw materials, etc.) is disclosed neither in the comments to the company's balance sheet and the profit and loss statement nor in the relevant section of the company's accounting policy. The company's accounting policy covers information only on the need and procedure for revaluating derivatives as well as the methods of presenting them and relevant processes in accounting. Given the company's scale and presence in the international market, it would be wrong to think that the only derivatives that will ever be acquired by the company are currency swaps and other instruments tied solely to the exchange rate. Accordingly, disclosure of the information on the procedure and mechanism for determining the fair value of financial derivatives linked to various types of assets would allow the company to increase the accuracy of its financial reporting. Moreover, the company's accounting policy also does not provide for determining the time when a financial derivative is recognized in its financial reporting. Under paragraph 3.1.1 of IFRS 9 "Financial Instruments", the time of recognizing such an instrument is considered the time when the company becomes a party to an agreement regarding this instrument and the rights thereto. Including this clarification in the accounting policy would allow establishing accurately the moment of derivative recognition in the company's accounting. We recommended adding the following paragraph in the section "Accounting for Financial Derivatives" of the company's accounting policy: the time of recognizing a financial derivative in accounting is considered the time when the company undertakes the contractual terms regarding such instrument.

The next type of situation where the company's accounting policy can directly affect the accounting information accuracy is a situation where the compliance with the regulatory legal requirements in this aspect of accounting can impair the statement accuracy. In this case, reporting will be considered accurate in terms of their form but not in terms of their contents. The company in question used this right in the matter of securities accounting. In its accounting policy, the company sets forth that directly following the rules of national standards and reflecting the security revaluation results at market value as part of financial performance may lead to the statement misrepresentation. Therefore, the accounting policy provides for recognizing such amounts as additional paid-in capital (if not planned to sell the securities in question). This provision of the accounting policy increases the accuracy of information about the company's financial situation. On the other hand, the company makes no direct reference to any specific international standard, arguing that "this approach complies with IFRS requirements as well." In this case, for a more detailed justification of the adopted accounting methods, the company shall provide a reference to a specific international standard, by which it was guided when making such a decision. That is, the company is invited to include the following item in its accounting policy: revaluation of financial investments in the form of securities is presented in accounting in compliance with the requirements of paragraph 5.2.2 of IFRS 9 "Financial Instruments."

Another aspect of an accounting policy that has a significant impact on the accuracy of financial reporting is the choice of accounting methods in matters that are not detailed sufficiently in national regulations. With the company under consideration, this problem arises in the recognition of research and development costs. Namely, national standards contain a list of criteria under which a company has the right to recognize R&D costs in accounting and confirm the amount of such costs. In matters of accounting for R&D costs with this company, one of the controversial aspects is the accounting of R&D objects created on its own. One of the criteria for their possible capitalization is "the ability to reliably estimate the costs associated with an R&D object during its development." The concept of a reliable estimate is not explained in any way by the company, and no criteria are provided for it. In this regard, it is unclear where the estimate is reliable and where it is not. Therefore, when accounting such objects, the accountant relies on his or her professional judgment; so, there is a risk of the human factor influencing the reporting information accuracy. In this case, the accuracy of reporting can be increased by adding the relevant clarification of the reliable estimate concept into the company's accounting policy. The company can draw up a list of documents that can serve as the basis for a reliable assessment of the costs of these activities. Such documents may be:

- an approved cost estimate;
- sales checks and other evidence of payment for the purchase of items required by the company for R&D;
- certificates of services rendered—if any part of the works was performed by a third party, but in general, the relevant R&D object is recognized as produced independently by the company;
- an approved opinion of an independent appraiser on the value of the R&D item;
- other official supporting documents.

4 Conclusion

The main goal of this paper is to assess the impact of the company's accounting policy on the accuracy of its financial reporting. When preparing the paper, we conducted an analysis of the accounting policy of a major oil company and assessed the impact of some elements of its accounting policy on the accuracy of financial reporting. Our analysis showed that increasing the accuracy of the company's financial reporting by amending its accounting policy is possible in cases where the accounting policy provides for insufficient detailing of some accounting aspects. In this case, the accounting policy has the greatest impact on the accuracy of financial reporting in those aspects of accounting that are either not governed by the national regulations or governed more effectively by international standards (in terms of the reporting information accuracy) or not governed by national standards and regulations in sufficient detail. Given this, our recommendations for improving the accounting information accuracy through changes in the accounting policy are of clarifications and additions, in their nature, without making any fundamental changes to this document. We are talking about fine tuning of the accounting policy to increase its effectiveness to provide more accurate information in financial reporting. The author hopes that the issues of the accounting policy effects on the accuracy of financial reporting will be the subject of active scientific discussions and will be reflected in the publications of researchers.

References

1. Dean, G., Clarke, F.: 'True and Fair' and 'Fair Value'—accounting and legal Will-o'-the-Wisps. Abacus **41**(2), i-viii (2005)
2. Australian Accounting Standards Board, Amendments to Australian Accounting Standards arising from the Trans-Tasman Convergence Project. https://www.aasb.gov.au/
3. Alexander, D., Eberhartinger, E.: The true and fair view in the European Union. Eur. Account. Rev. **18**(3), 571–594 (2009)

4. Dumitru, M., Florin, B.A.: Study regarding the quality of the true and fair view supplied by the financial statements from financial auditor's point of view. J. Fac. Econ.-Econ. **3**(1), 1298–1309 (2008)
5. Miller, P.B.W., Bahnson, P.R.: Faithful representation: it's much more than verification. Account. Today **21**(1), 15 (2007)
6. Smieliauskas, W., Craig, R., Amernic, J.: A proposal to replace 'true and fair view' with 'acceptable risk of material misstatement'. Abacus **44**(3), 225–250 (2008)
7. Pelger, C.: The return of stewardship, reliability and prudence–a commentary on the IASB's new conceptual framework. Account. Eur. **17**(1), 33–51 (2020)
8. Efimova, O., Rozhnova, O.: The corporate reporting development in the digital economy. In: The 2018 International Conference on Digital Science, pp. 71–80. Springer. Cham (2018)
9. Dombrovskaya, E.: Development of the conceptual framework for financial reporting in the context of digitalization. In: International Conference on Digital Science, pp. 334–344. Springer. Cham (2019)
10. Lugovsky, D., Kuter, M.: Accounting policies, accounting estimates and its role in the preparation of fair financial statements in digital economy. In: Antipova, T. (eds.) Integrated Science in Digital Age. ICIS 2019. Lecture Notes in Networks and Systems, vol 78, pp. 165–176. Springer, Cham (2020)
11. Kapić, J., Bašić, M.: Accounting policies and accounting estimates for the purpose of ensuring the satisfactory level of financial reporting quality. Bus. Consul./Poslovni Konsultant **5**(31) (2013)
12. Shlifer, E.V.: Effective integrated accounting policy as an element of the internal control system for implementing the goals of financial, tax and management accounting. In: Science Today: Problems and Solutions, pp. 151–153 (2016)
13. Tkhagapso, R., Kuter, M., Trukhina, A.: Liquidation financial reporting of russian companies in terms of digital economy. In: Antipova, T. (eds.) Integrated Science in Digital Age. ICIS 2019. Lecture Notes in Networks and Systems, vol 78, pp. 72–81. Springer, Cham (2020)
14. Tkhagapso, R., Trukhina, A., Khatkhokhu, S.: Development of methodology of valuation reserves information disclosure of russian companies in terms of digital economy. In: Antipova, T., Rocha, Á. (eds.) Digital Science 2019. DSIC 2019. Advances in Intelligent Systems and Computing, vol. 1114. Springer, Cham (2020)

Modern Aspects of Digital Technologies Development in Retail Networks

Elena Alexandrova(✉) and Anna Kochieva

Economy Department, Kuban State University, Stavropolskaya Street, 149, 350040 Krasnodar, Russia
al-helen@mail.ru

Abstract. The paper considers modern aspects of digitalization of the retail network companies. Retail networks are rapidly increasing the volume of digital services provided, introducing digital technologies into many business processes in order to increase their own competitiveness, speed up decision-making and develop new markets. Continuous development of digital technologies and consumer expectations regarding these technologies are the main prerequisites for digitalization of the retail. The article discusses the practice of using digital technologies, such as big data, e-commerce, and implementing omni-channel model for customer service. Big data is considered one of the key leading digital technologies in the retail sector, which includes geolocation analytics, video analytics, analysis and forecasting of demand for products. Not only the advantages of implementing digital technologies in the retail were emphasized in the article, but the problems associated with this process such as cyber attacks, personal data leaks, etc. were also paid special attention. Promising areas for research within the framework of this problem were formulated.

Keywords: Digital technologies · Digitalization · Retailing · Omni-channel retailing · Big data · E-commerce

1 Introduction

In modern companies of various sectors and industries, introduction of digital technologies is aimed at ensuring collection, processing and provision of the information required by the user [1]. Cloud computing, mobile technologies, sensors, Internet of Things (IoT), big data, Artificial Intelligence (AI), augmented reality technologies (AR), robotics, additive manufacturing (3D printing), drones and other technologies implemented in everyday environment allow you to automate many business processes and economic operations (including marketing; production, sales; finances; accounting [2]), they can extract information from physical devices (sensor readings containing information about the state of a physical device), quickly distribute it (using mobile technologies), store it in the cloud, and instantly analyze it (using big data and advanced analytics), thereby integrating products, services and processes [3]. Digital technologies allow delivering information to target consumers, on a global scale in almost real time mode. Specifics of development and implementation of digital technologies in companies depend on characteristics of the industry in which they operate.

T. Antipova (Ed.): ICCS 2020, LNNS 186, pp. 111–120, 2021.
https://doi.org/10.1007/978-3-030-66093-2_11

Retail networks have become one of the industries rapidly increasing the volume of the provided digital services and implementing digital technologies in many business processes in order to increase their own competitiveness and speed up decision-making.

Implementation and management of digital technologies in retail networks is facilitated by active development of digitalization, which is considered intellectual business and a process of creating value in electronic form using information and communication technologies [4]. Social media, big data and other emerging technologies (e.g., Artificial Intelligence (AI), virtual reality (VR), augmented reality (AR), blockchain, etc.) are transforming the retail business models. At present «increasing portions of the retail trade are shifted from store-based to internet-based formats, including pure plays, manufacturer online operations, and platforms» [5].

On the one hand, diffusion of digital technologies has rapidly increased in physical stores (stationary retailing), due to the continuous development of digital technologies and consumer expectations for these technologies [6]. Although stationary retailing has its roots in the physical retail, and the economic model of such retailers is primarily based on profitability of their offline operations, they nevertheless actively implement omni-channel strategies that synchronize data and information across all physical and digital channels of interaction with consumers. When applying such strategies, it is important to take into account the connection between digital and physical space in retail supply chains for adaptive synchronization of supply and demand, which ultimately leads to a reduction of the period of an order execution, transaction costs reduction, which are key performance indicators in today's competitive retail markets [7].

On the other hand, there are new players – digital retail platforms, such as Amazon, Alibaba, which "take over" a significant part of the trade and this is due to the fact that an increasing number of customers prefer the convenience of online purchases and door-to-door delivery, bypassing the space of a physical store [8]. Digital retail platforms allow a large number of sellers and buyers to interact effectively.

Digital technologies in the retail networks, implemented at various stages of value creation, are able to create powerful ecosystems of brands that interact with consumers by means of various applications, direct selling, engagement marketing and development of customer experience, as well as personalized communication which create completely new value offers [5]. Consumers want to use modern technologies that will help them interact with the stores at every stage of making a purchase from waiting for the products to be available in stock to being able to choose from several delivery options. Digital technology either enhances or disrupts the value co-creation process, where customers' needs and desires drive them to interact with service providers in the physical retail space [9].

Despite a number of obvious advantages associated with introduction of digital technologies, the retail sector, however, faces a number of problems and challenges, from adequate training of store employees to difficulties in measuring the return on digital investment. Many retailers do not put into practice the digital initiatives expected by consumers. It is important to understand and evaluate the reaction of retailers themselves to the actual level of digital technology spreading in various business processes and areas of activity. In addition, the introduction of basic digital technologies depends on the categories of retail trade, sizes and types of network retailers.

The need to follow omni-channel models and strategies in the retail leads to the necessity to make "right" decisions both by new market players (who initially build their business models taking into account the digitalization processes) and offline leaders (who have to transform their business models when implementing digital technologies) in order to fit into the new retail environment [10].

In the first part of the study the analysis of directions and prospects of implementation of digital technologies in the sphere of retail chains was performed, the features of implementation of the most important digital technologies for retailers (big data, e-commerce) were examined on the basis of certain individual retail chains, the main benefits and problems of omni-channel strategy were described. In the second part of the study, the main conclusions and results are presented, and areas for further work in this area are identified.

2 Main Directions of the Retail Digitalization: Experience, Prospects and Barriers

Global and national retail chain companies such as IKEA, Metro AG, Carrefour, Walmart, Costco, Home Depot, H&M, Inditex, ASOs, LaModa, X5 Retail Group, Magnit and many others, having done significant financial groundwork and having an objective need for digitalization and digital transformation, use a wide range of modern digital technologies in various business processes, such as e-commerce, Big Data, video analytics, face recognition, contactless payments, etc. Individual representatives of the retail sector, based on global digital platforms, as Google, Amazon, Facebook, Apple, Alibaba, Baidu, Tencent, have grown to the level of "superstars". Such companies, unlike the traditional ones, have a higher level of productivity, while the efficiency and operation on the global scale make them part of a small but more and more concentrating group of companies that generate value for the shareholders above the cost of the capital on the grounds of actively developing digital technologies in their business processes in the maximum possible range [11]. Figure 1 shows the key changes and expected effects from the introduction of digital technologies both in the business processes of chain retail companies and in their relationships with customers. Retailers – ecommerce and brick and mortar stores – are moving in a common direction to provide seamless consumer experience.

Digital technologies in retail networks allow creating a unique brand vision by ensuring interaction between the key components of a retail store forming the consumer experience and the value for customers:

- consumers (e.g., a mobile app that shows consumers where products are located in the store);
- employees (e.g., digital changing room);
- product (e.g., RFID "smart labels" on price tags);

Physical store space (e.g., robots inside stores to assist with inventory from shelves to carts).

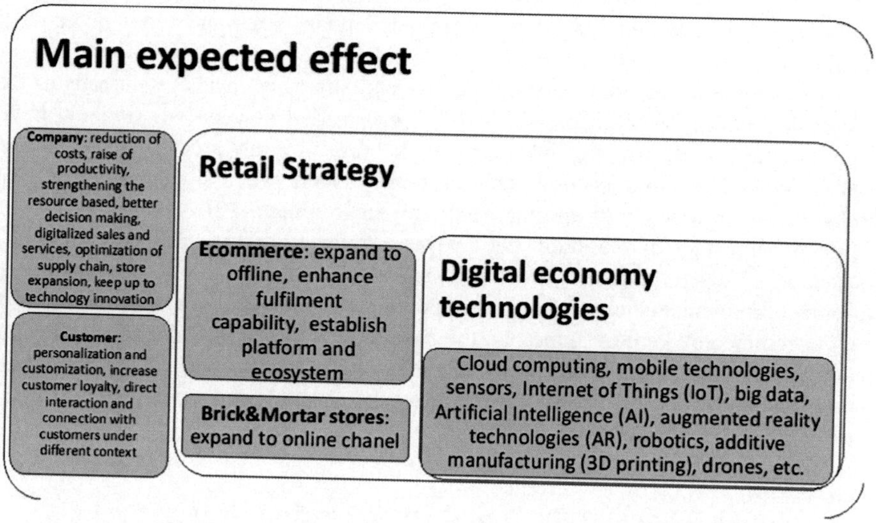

Fig. 1. Introduction of digital economy technologies into network retail companies

Growth of internet users in number sets the development vector for modern retail: more and more consumers shop online, thus the line between trading in physical stores and purchases using mobile apps, internet sites, marketplaces, social networks, etc. is gradually becoming less visible. The e-commerce segment, unlike traditional offline stores, has low costs connected with the maintenance of retail space and equipment, besides, the probability of illiquid inventory accumulation is reduced, which reduces the price of the product for the consumer. Despite the opinion of the experts that traditional physical stores will soon be displaced from the market by online stores, both categories have been coexisted and changing under the influence of the transformation of consumer preferences [12]. As of January 2020, the average number of internet users in Europe was 87%, in Asia – 61%, in North America – 88%, in Australia and Oceania – 70%, the minimum level of internet coverage is observed on the African continent and in South Asia. The global internet coverage rate is 59%. Mobile internet is used by 92% of all users. Up to 90% of internet users between the ages of 16 and 64, who were reported to perform any online activity in the past month (globally), visited online retail stores on the web, and 74% of them purchased products online using any device [13].

China's retail sector is a clear example of the digital technologies diffusion in retail sphere and transformation of traditional business models of physical stores. The retail sector is currently valued at $ 3.8 trillion, and is expected to reach $ 6.6 trillion by 2024. China's retail is characterized by blurring of the line separating the offline and the online retail (this model is called "new retail" in China). For instance, big e-commerce players, such as Alibaba, have established partnership with small local family stores, helping them to modernize by providing no- or low-cost retail management platforms, allowing partnering stores to optimize product procurement and sourcing [14].

Against the backdrop of the COVID-19 coronavirus pandemic, retail is also taking on new challenges and reshaping the structure of its functioning, adapting to changing shopping habits. During the global lockdown, online shopping was the only way to get a range of essential goods while minimizing contact with strangers. Thus, such innovations as, for example, curbside pick-up, robotic delivery, and "Just Walk Out" grocery shopping have been used [15]. Omni-channel communication has become widely used during the pandemic. Retailers who haven't adapted their activities to new realities are putting their market positions at risk and currently need completely new toolkit. Of course, the pandemic has made significant changes in the consumption process, and its consequences will be felt for the foreseeable future, therefore the introduced innovations will continue to be used.

The key element in the retail digitalization and the defining trend in its development are omni-channel strategies that combine all channels around a user. Consumers can easily switch between channels and still receive personalized service in the frame of such strategies. The current crisis compels both the retailers and consumers to resort to the omni-channel and adopt the online channels equally with the offline ones. One of the "effects of the pandemic" is the accelerated development and implementation of the omni-channel model where it was not used before, and the improvement and expansion of channels for brand interaction with consumers as well. The availability of a well-developed infrastructure for the prompt execution of online orders is one of the crucial growth factors for players on the market. Nowadays, customers use omni-channel interaction more and more actively, for example, an option of ordering products online, but picking them up in an offline store at convenient time (click & collect). Amazon, Sephora, and LVMH already use this model. E-commerce platforms (LaModa, Wildberries), networks X5 Retail Group, «Azbuka Vkusa» started using omni-channel model in Russia.

The customer's shopping requirements is becoming more and more diversified in retail sales. As part of the omni-channel strategy, brand retailers should follow and leverage the trend of customization and personalization to upgrade the manufacturing system to more digitalized and intelligent stage. Big Data is among the key leading digital technologies in the retail sector, the implementation of which in operational activities is currently a necessary element to ensure competitiveness of the retail. It should be noted that Big Data Analytics (BDA) is currently among the top 3 investment priorities across all IT and communications solutions [16].

Geolocation analytics, which includes both the process of choosing a location for a business and collecting data about the customer's location, is among promising areas of BDA for retail. There are various algorithms designed to recommend the best location for a business and identify areas lacking a certain type of business; the time spent on customer service at a particular point/territory is analyzed. That allows entrepreneurs to choose the location of a new business with the highest probability of reaching a potential client [17].

A common tool for collecting data (tracking) about consumers in the framework of geomarketing is GPS, which allows tracking the users through the global navigation satellite system and determining the coordinates of their location. Potential consumers can find stores or products on their mobile maps in real time mode. Retail may combine this information with other pieces of demographic information to keep in touch with users on a specific territory. Despite a number of obvious advantages, the retail does not use all the options of Location Intelligence. Thus, in 2018, companies mainly used geographical data analysis on a state, a region, a city, an index, or a country levels (76%, 72%, 69%, 68%, 64% of the number of the respondents respectively), while only 26% of the respondents used custom geography, 26% – virtual location, and 17% – block groups [18].

In retail networks, the Big Data technologies allow us to effectively analyze and forecast demand for various types of goods using a set of mathematical methods. Such technologies are being applied in H&M, Zara, etc. X5 Retail Group and Magnit retail chain use trade analytics technologies in Russia. Research into the market basket in retail is possible on the basis of associative algorithms, such as the Apriori algorithm. The role of the Apriori algorithm is to identify shopping habit patterns of consumers through shopping basket and transactions data analysis to select potentially needed assortment for the consumer [19]. For example, cash receipts are analyzed to determine which product category is purchased more often, which products are purchased together, and whether consumers supplement one product with another.

Modern video surveillance systems are software and hardware complexes with a set of intelligent tasks, such as detecting events, counting people, recognizing behavior, incidents, images, faces, and objects. In 2019, for instance, retail chains Magnit and X5 Retail Group introduced video analytics to control product availability on shelves and queues at checkout stands in a number of their stores, which reduced queues by 70% [20]. Many stores become aware of the need to implement self-checkout systems – Tesco, Carrefour and Magnit are among them, this reduces staff costs, selling space, and customer service time.

Internet of things technologies allow online retail companies to keep in touch with consumers throughout the entire life cycle of their products. For example, manufacturers of household appliances (Samsung, Miele) sell consumables using this technology, when, for example, washing machines automatically order detergents after a set number of washing cycles [5].

Immersive technologies allow a user to get a most realistic experience from interacting with virtual objects or even people. According to forecasts, about 70% of enterprises will experiment with immersive technologies, and about 25% will implement them in their activities by 2022 [21].

However, despite a number of obvious advantages and benefits from digitalization, modern digital technologies in the retail create new industry problems and risks, such as threats of the user's data leakage. Thus, scanning of face in the checkout queue should not go beyond its direct purpose and be used for other purposes, on other resources, in other programs without the consent of the consumer. It should be noted that there are problems such as threats of cyber attacks aimed at stealing personal data of companies and customers, the loss of consumer confidence due to misuse of their personal data or their unreliable protection; introduction of the big data technologies in business processes necessitates radical changes in the structure and activities of the company, which entails additional difficulties and problems; deficiency of highly qualified specialists in the field of digital technologies, lack of the development of new software, ensuring trouble-free operation and protection. In addition to these risks and problems, it should be noted that huge volumes of generated big data can lead to statistical problems, information noise, problems of scalability of information and problems of its interpretation in the retail chain companies [22, 23]. These barriers, in our opinion, can serve as a basis for further research and development in the field of digital technologies in the retail, which will help to neutralize threats and ensure progressive development of the industry.

3 Results

In a competitive environment among the offline retail companies actively implementing digital technologies and developing omni-channel strategies together with growing of the online gaming sphere (Amazon, Alibaba, etc.), those companies which were able to transform the consumer experience with the help of new technologies in to the best advantage, digitalize operations and key business processes, quickly adapt to rapidly changing consumer expectations, exert flexibility and vision in both online and offline environments, will be successful. Consumers are now becoming more demanding, therefore the entire process from product selection to its delivery should be as smooth as possible. Those consumers who are considered to be "digital natives" are used to convenience, high speed of products and services ordering and delivery (everything can be done through a mobile phone application). It is them who dictate terms to manufacturers, who have to constantly improve the availability and usability of their products and services, and make personal experience of consumers develop, taking into account their characteristic features and interests. Less than a third of consumers around the world prefer a store to an app, and for the many the two channels are equal. Consumers find mobile apps more useful than stores when looking for the information about products and promotions (55% vs 41%), comparing products (54% vs 46%) and providing feedback (53% vs 46%) [24].

The business model of modern retailers has been transformed towards engaging consumers, building digital communications with them and developing consumer experience to improve sales conversion rate under the influence of digital transformation and increasing digital awareness of consumers. Along with it shopping channel boundaries are getting dissolved, and traditional retailers need to transform traditional sales channel management into a seamless omni-channel service model [25]. Eventually, integration of internal processes, resources, and systems across all channels aimed at supporting omni-channel operations in retail networks is meant to improve consumers' shopping experience. The omni-channel retail has become common and shoppers can easily switch between different channels for a single purchase.

Implementation and use of big data technologies makes retail more competitive and allows it to adapt to market changes more effectively. However, with all the benefits that corporations receive from geomarketing, videoanalytics, data analysis, personalization, etc., there are challenges and threats, such as, for example, the threat of data theft and a shortage of specialists in this field. Big data is a rather useful tool for tactical planning and realization of operational planning when decisions are regularly made, for example, in the case of fine-tuning of the promotional calendar, shelf-space allocation, or the development of targeted offers to loyalty-card holders. However, strategic planning of a company's activity and development requires significantly more information, than it is provided by big data, thus, if «big data thinking» becomes the norm, there is a real danger that tactical thinking will take the fore over more strategic thinking in retail managers' time allocation [22].

4 Conclusion

Active development of digitalization in economic processes and various types of activities, increasing consumer awareness of digital technologies and their expectations, and the growing number of competitors in the offline and online environment call for the introduction of advanced digital technologies in retail chains. Along with it, diffusion of digital technologies goes rapidly in physical stores, and new players (digital retail platforms) have started to emerge. Those companies who actively use innovative technologies in their operating activities can claim leadership in the market, while belated mastering digital tools leads to a gradual loss of competitiveness and weakening of market standing.

So-called "digital mediation" is increasingly blurring the line between offline and online retailers, between manufacturers and sellers, and between suppliers and consumers. In their turn, consumers have evolved through digital transformation, as they expect more and more personalized purchases with minimal costs. Using the example of big data technology and its various applications, the article examines promising areas and individual problems of digital technology development in the retail sector. It is concluded that, despite all the changes directly or indirectly caused by the big data revolution and accompanying technologies, the basic conditions remain unchanged: a consumer buys the product which satisfies his needs to the fullest extent, using the retail channel that best meets his requirements.

Digital technologies implemented in retail chains form new sources of value creation, improve the accepted by consumers advantages of a particular retail network, and thus create a competitive advantage. However, many opportunities of the digital economy have not yet been realized in retail chains, which is caused by barriers to the new technologies, lag effects, and problems of commercialization.

The conducted research does not claim to fully characterize the promising areas and key problems of the implementation and use of digital technologies in modern retail chains. As areas for further research, we can highlight the following topics: ensuring the safety of data and personal information of consumers, the problem of excess workforce due to the total digital transformation of the retail.

References

1. Alexandrova, E.: Digital economy in competitiveness of modern companies. In: Antipova, T., Rocha, Á. (eds.) DSIC 2019. AISC, vol. 1114, pp. 114–125. Springer, Cham (2020). https://doi.org/10.1007/978-3-030-37737-3_11

2. Lugovsky, D., Kuter, M.: Accounting policies, accounting estimates and its role in the preparation of fair financial statements in digital economy. In: Antipova, T. (ed.) ICIS 2019. LNNS, vol. 78, pp. 165–176. Springer, Cham (2020). https://doi.org/10.1007/978-3-030-22493-6_15
3. Spremić, M.: Governing digital technology – how mature IT governance can help in digital transformation? Int. J. Econ. Manag. Syst. **2**, 214–223 (2017)
4. Schmidt, R., Möhring, M., Härting, R.C., Reichstein, C., Neumaier, P., Jozinović, P.: Industry 4.0 – potentials for creating smart products: empirical research results. In: International Conference on Business Information Systems, pp. 16–27. Springer, Cham (2015)
5. Reinartz, W., Wiegand, N., Imschloss, M.: The impact of digital transformation on the retailing value chain. Int. J. Res. Mark. **36**(3), 350–366 (2019)
6. Pantano, E., Vannucci, V.: Who is innovating? An exploratory research of digital technologies diffusion in retail industry. J. Retail. Consum. Serv. **49**, 297–304 (2019)
7. Pereira, M.M., Frazzon, E.M.: A data-driven approach to adaptive synchronization of demand and supply in omni-channel retail supply chains. Int. J. Inf. Manag. 102165 (2020)
8. Kapner, S.: Retail slump shows Amazon effect. Wall Street J. (2016). https://www.wsj.com/articles/kohls-shows-effects-of-retail-slump-as-some-ponder-the-amazon-factor-1463084218. Accessed 01 July 2020
9. Nöjd, S., Trischler, J.W., Otterbring, T., Andersson, P.K., Wästlund, E.: Bridging the valuescape with digital technology: a mixed methods study on customers' value creation process in the physical retail space. J. Retail. Consum. Serv. **56**, 102161 (2020)
10. Cai, Y., Lo, C.K.Y.: Omni-channel management in the new retailing era: a systematic review and future research agenda. Int. J. Prod. Econ. **229**, 107729 (2020)
11. Superstars; The dynamics of firms, sectors, and cities leading the global economy, McKinsey Global Institute, October 2018. https://www.mckinsey.com/~/media/McKinsey/Featured%20Insights/Innovation/Superstars%20The%20dynamics%20of%20firms%20sectors%20and%20cities%20leading%20the%20global%20economy/MGI-Superstars-Discussion-paper-Oct-2018.ashx. Accessed 23 June 2020
12. Gallino, S., Moreno, A.: Integration of online and offline channels in retail: the impact of sharing reliable inventory availability information. Manag. Sci. **60**, 1434–1451 (2014)
13. Digital 2020: Global Digital Overview. We Are Social, Hootsuite. https://datareportal.com/reports/digital-2020-global-digital-overview. Accessed 11 July 2020
14. The 2019 Global Retail Development Index. Kearney. https://www.kearney.com/global-retail-development-index/2019. Accessed 15 July 2020
15. Colleen, P.K., Rifkin, L.S.: I'll trade you diamonds for toilet paper: consumer reacting, coping and adapting behaviors in the COVID-19 pandemic. J. Bus. Res. **117**, 124–131 (2020)
16. Post-pandemic Growth Opportunity Analysis of the Big Data Analytics Market. Frost & Sullivan. https://ww2.frost.com/news/press-releases/big-data-analytics-among-top-three-deployment-priorities-for-enterprises-says-frost-sullivan/. Accessed 11 July 2020
17. Kumar, A., Shankar, S., Reddy, P., Gadekallu, T., Al-Ahmari, A., Abidi, M.: Location based business recommendation using spatial demand. Sustainability **12**(10), 4124 (2020). https://doi.org/10.3390/su12104124
18. State of Location Intelligence. Carto. https://carto.com/state-of-location-intelligence-2018/#download. Accessed 14 July 2020
19. Kaini, G.H.: Determining profitable products in the retail market with consideration of cash limitation and exhibition periods. J. Retail. Consum. Serv. **55**, 102079 (2020). https://doi.org/10.1016/j.jretconser.2020.102079

20. “Magnit” showed the first store of the “superstore” format. Retailer. https://retailer.ru/magnit-pokazal-pervyj-magazin-formata-superstor/. Accessed 14 July 2020
21. Top 10 Strategic Technology Trends for 2020. Gartner. https://www.gartner.com/en/publications/top-tech-trends-2020. Accessed 08 July 2020
22. Dekimpe, M.G.: Retailing and retailing research in the age of big data analytics. Int. J. Res. Mark. **37**(1), 3–14 (2020)
23. Alexandrova, E., Poddubnaya, M.: Digital technologies development in industry sectors and areas of activity. In: Antipova, T. (ed.) ICIS 2020. LNNS, vol. 136, pp. 112–124. Springer, Cham (2021). https://doi.org/10.1007/978-3-030-49264-9_10
24. Making the Digital Connection: Why Physical Retail Stores Need a Reboot. Digital Transformation Institute. https://www.capgemini.com/wp-content/uploads/2017/01/report-making-the-digital-connection.pdf. Accessed 07 July 2020
25. Transformation of traditional retailer in the era of New Retail. Deloitte Analysis. Deloitte China, October 2017. https://www2.deloitte.com/content/dam/Deloitte/cn/Documents/cip/deloitte-cn-cip-omni-channel-retail-white-paper-en-171107.pdf. Accessed 07 July 2020

Check for updates

Technology of Express Analysis of Bank Profitability in the Digital Age

Liudmyla Lakhtionova[1(✉)], Svitlana Kalabukhova[2], Olga Kuzminska[2], and Oksana Isai[2]

[1] National Aviation University, 1, Liubomyr Husar Avenue, Kyiv 03067, Ukraine
Ludmilala@i.ua
[2] Kyiv National Economic University Named After Vadym Hetman, 54/1, Peremohy Avenue, Kyiv 03057, Ukraine

Abstract. The article examines the profitability of the bank from the standpoint of assessing the prospects for development and risks of banking activities by interested external and internal users of financial information for decision making. A comprehensive technological procedure for «rapid» analysis of the bank's profitability has been developed, which includes four stages, starting with trend analysis of net profit and ending with forecasting the level of interest rate risk in the conditions of digital computer technologies. Forms of analytical tables are proposed, the use of which will contribute to the unification of the management document "Management Report" of banking institutions; provide stenciling of analytical conclusions; strengthen the possibility of instant comparability of analytical information on changes in the profitability of entities in the banking sector; increase the information culture of documenting changes in profitability in the process of managing banking activities; increase the clarity and informativeness of analytical conclusions about the profitability of banks.

Keywords: Digitalization of the banking sector · Profitability · Rapid analysis · Bank

1 Introduction

The fourth industrial revolution brings rapid changes in all spheres of human life and has no analogues in its previous experience. Its results will fundamentally change the way people live, work and communicate (communicate) with each other. There are tremendous technological breakthroughs: artificial intelligence, robotics, unmanned vehicles, three-dimensional printing, nanotechnology, biotechnology and more.

In 2016, the 46th World Economic Forum on the topic of The Fourth Industrial Revolution took place in Davos. Klaus Schwab, the founder of the forum, a Swiss economist and one of the main theorists of the Industry 4.0 phenomenon, spoke about its preconditions and challenges.

According to a survey of 800 technology company leaders conducted specifically for the Davos forum, key drivers of change will be cloud technology, the development of Big Data collection and analysis, crowdsourcing, the sharing economy and

T. Antipova (Ed.): ICCS 2020, LNNS 186, pp. 121–131, 2021.
https://doi.org/10.1007/978-3-030-66093-2_12

biotechnology [1]. 45% of respondents believe that in 2025 in the boards of directors of large companies may be present artificial intelligence [1].

In the forum in his speech Klaus Schwab noted: «The capabilities of billions of people connected to each other by mobile devices with gigantic power and memory, providing access to all the knowledge of mankind, are truly limitless. And these opportunities will be multiplied many times over by new breakthroughs in the fields of artificial intelligence, robotics, the internet of things, autonomous transport, nanotechnology, materials science and quantum computers. Artificial intelligence is already here in the form of autonomous machines, drones, virtual assistants, translators» [1]. On the same Klaus Schwab emphasized in his book «The Fourth Industrial Revolution» [2, p. 7; 3, p. 7]. In 2018, a book by Klaus Schwab and Nicholas Davis was published (Shaping the Fourth Industrial Revolution) [4]. The main topic of the meeting in Davos in 2019 was «Globalization 4.0: the formation of global architecture in the era of the fourth industrial revolution» [5].

In Ukraine, in 2019, a book translated into Ukrainian by Klaus Schwab will be published «The fourth industrial revolution. Shaping the Fourth Industrial Revolution» [6]. Comments on it note that Klaus Schwab has created a real «alphabet of the fourth industrial revolution» , which provides practical advice for entrepreneurs and anyone who wants to understand future changes, understand trends and join in creating a new one [7].

It should be understood that the break in the technological paradigm brings not only new perspectives, but also new challenges associated with the transformation of banking.

Much more publications of Ukrainian and foreign scientists are devoted to the solution of theoretical and practical issues of analysis of profitability of economic entities in the field of material production than to the study of methodological and organizational provisions of the analysis of profitability of banks. Through a comprehensive study of the financial results of economic activity, and especially profitability, considers the assessment of the financial condition of the enterprise French scientist Jacques Richard [8]. American researchers G. Carlin, A. McMinn offer approaches to the analysis of the statement of financial performance based on the principles of GAAP [9], some aspects of the analysis of the financial statement of profits and losses of enterprises are studied in the work of American scientists K.R. Subramanyam and John J. Wild [10].

Problems of financial management of banking business, questions of definition of net accounting profit of bank through size of a net interest margin, accounting and economic models of the analysis of bank efficiency are covered in work of the American prof. J. Sinki [11]. Problems of financial analysis of banks in the developing environment are revealed in the work of Australian researchers H.W. Collier, C.B. McGowan, J. Muhammad [12]. South African scientists M. Cumbrirai and R. Webb proposed a detailed system of analytical indicators of profitability to conduct a coefficient of financial analysis of the bank [13].

In the works of these scientists in-depth theoretical, methodological and organizational provisions for the analysis of financial performance of banking institutions.

The complexity and unification of procedures for analyzing the profitability of banks in the scientific literature are insufficiently covered.

Critical evaluation of information from professional literature sources confirmed the need for in-depth study of the development of a comprehensive unified procedure for «rapid» analysis of bank profitability in digital computer technology, due to the importance of this indicator in bank management and limited time users of financial information for making management decisions.

The aim of the article is to develop in terms of digital computer technology a comprehensive procedure for «rapid» analysis of bank profitability and construction of forms of analytical tables, the use of which in the formation of the Bank Management Report will provide analytical conclusions and allow instant comparability of analytical information on changes in profitability activities in the banking sector.

2 Research Method

In the course of the research the methods of scientific generalization, comparison (when selecting key indicators of profitability of banking institutions for the needs of express analysis, development of analytical information carriers on various aspects of bank profitability) were applied; methods of general scientific abstraction, induction and deduction, analysis and synthesis (in determining the main stages of the process of rapid analysis of bank profitability); methods of trend and factor analysis (in the development of technology for rapid analysis of bank profitability in absolute and relative terms).

3 Research Results

A feature of the fourth industrial revolution is the extremely rapid introduction of new technologies and their widespread use. The innovatively active market is becoming a place of intense struggle for many companies around the world, regardless of their size. They all compete for the opportunity to be the first among those who will be able to bring a new product, a new service to market and win the favor of customers. This is also true for the banking sector, where the traditional way of doing business is threatened by innovative financial and technological companies.

James Haycock (founder of the digital agency Adaptive Lab) and journalist Shane Richmond (formerly of the Daily Telegraph) in his book «Goodbye Banks?: How Retail Banks Are Displaced, Shrinking and Misinformed by Technical Startups and What They Can Do to Survive in Kindle Edition» (2016) describes the attack to which classic financial institutions are exposed by a growing number of fintechstartups and digital companies [14]. According to the authors, to keep market share classicoretical, methodological and organizational provisions of the analysis are deepened in the works of these Retail banks will have to give up much of their business processes, technologies, and sometimes staff that go back in time. Rapid evolution and digital transformation, balanced work with multichannel and the search for new ways to become a part of customer life - these are the keys to the survival of classic banks [14].

There has been lots of discussion of digital and open banking, banking-as-a-service, banking platforms, FinTech and TechFin and more, over the past decade. This all indicates that we are in a decade of rapid cycle change that presents huge challenges and huge opportunities. Billion-dollar unicorns appear rapidly, whilst internet giants achieve global domination. How are banks dealing with these changes and are any banks showing leadership? Well yes, a few are. With all the gloom merchants saying that traditional banking is doomed, a few banks have made radical moves to adapt and survive.

Chris Skinner, a world-leading commentator on banking and technology, has selected five of those banks - JPMorgan Chase (USA), BBVA and ING (Europe), and DBS and CMB (Asia) - to share their experiences. In detailed interviews, and with wide-ranging commentary, he has discovered the secrets of how not just adapt and survive, but how to thrive in this sea change of finance and technology [15].

The future of banking services is revealed in Brett King's book "Bank 4.0: Banking Everywhere, Never at a Bank" [16]. The role and features of modern work of analysts are reflected in the work of Mark Bradshaw, Yonca Ertimur, Patricia O'Brien [17].

A practical guide for analysts and investors on the analysis of financial statements of banks is contained in the book of the eminent German scientist Thomas Radberg [18].

The book covers: the specific accounting rules that apply to banks; how to analyse bank segment reporting; the ratios to use when analysing bank financial statements; how to analyse bank profit and loss accounts; equity analysis and stock analysis of banks [18].

Recent years have shown the risks which can evolve from banks, but normal instruments of financial statement analysis are not sufficient to analyse banks and locate these risks: different methods are needed.

Therefore, in modern conditions of introduction and development of digital computer technologies it is necessary to develop a comprehensive procedure for «quick» analysis of bank profitability and construction of forms of analytical tables, the use of which in the formation of the Bank Management Report will provide analytical conclusions and allow instant comparability of analytical information profitability of entities in the banking sector.

The financial result of the bank (profit, loss) for the reporting period is determined by comparing income with expenses. It is a generalized cost indicator that characterizes the efficiency of banking. Qualitative economic information collected by the bank's accounting system on the result of activities should provide the opportunity to make sound management decisions. Financial and performance information about banking activities is collected for the reporting period and its changes and deviations from the planned ones are determined. Management decisions on the profitability and profitability of banks should be aimed at optimizing the functioning and increasing the profitability of the banking business.

The limited time of users of financial information for management decisions necessitates the development of a unified technology for "rapid" analysis of bank profitability

Let's consider the technology of express analysis of the bank's profitability according to the data of financial statements. Financial statements are often called accounting, as they are compiled from accounting data.

The first stages of development of accounting took place in Italy. It was from these times that many accounting terms were fixed, which are of Italian (Latin) origin and are translated as: debit - "it must"; credit - "loan, debt", etc. The names of Italians are associated with the development of accounting: accountant Nicolo d'Anastasio, theoretician and lawyer Francesco Villa, Fabio Besta and others.

A significant contribution to the development of the theory of accounting was made by the American school of bookkeeping, one of the representatives of various directions of which were I. Fisher and D. Scott. Irving Fisher (1867–1947) - economist, creator of the modern theory of monetary circulation, statistician, who formulated the main provisions of the theory of indexes, had a great influence on the theory of accounting.

At the beginning of the twentieth century, both in Europe and in America, there was a belief that the balance sheet is a central category that explains all the details of accounting and is considered the basis of all financial statements. Fisher does not place the central place in the financial statements, not the balance sheet, but the income statement.

Today's archival research by M.I. Kuter and M.M. Gurskaya (2017–2020) allows you to visually see the content of early reporting) [19], perform an analysis of the structure of medieval balances [20].

Let's consider the technology of express analysis of the bank's profitability according to the data of financial statements.

In the banking sector, «bank profitability» is defined as a system of absolute and relative indicators that characterize the economic efficiency of the bank for the reporting period.

Express analysis of the bank's profitability is proposed to begin with a trend analysis of the bank's net profit using the following analytical form (Table 1).

Table 1. Trend analysis of the bank's net profit

Levels of a number of dynamics, n	The value of net profif of the bank, X	Absolute change, thousand UAN		Progress, %	
		basic	chain	basic	chain
(0) - 4 years ago	349 406	-	-	-	-
(1) - 3 years ago	(3 416 362)	- 3 765 768	- 3 765 768	x	x
(2) - 2 years ago	(2 847 715)	- 3 197 121	+ 568 647	x	120,0
(3) – previous period	954 124	+ 604 718	+ 3 801 839	273,1	x
(4) – reporting period	1 021 365	+ 67 1959	+ 67 241	292,3	107,0
Analytical operations:					
Calculation of growth trend of bank's net profit: $\overline{T_{ЧП}} = \sqrt[4]{X_4 / X_0}$		$\overline{T_{ЧП}} = \sqrt[4]{1021365/349406} = \sqrt[4]{2{,}9231} = 1{,}308$ or 130,8%; average annual growth for 5 years – 30,8%			

(Source: annual report of PJSC "OTP Bank" for 2016–2017 from the official website at https://ru.otpbank.com.ua/about/informations/annual_reports/ [21–23]).

According to the rules of the comparative financial analysis, if there is a negative amount in the base period and a positive in the next (reporting) period (or vice versa), you cannot calculate a significant percentage change. Therefore, for some time series there will be no growth rate and growth of the bank's net profit, while when calculating the average growth trend, these missing values of the chain growth rate are not affected. In the reporting period the bank's net profit to increase (by +67241 thousand UAN), its growth rate was 7%, which indicates a positive trend. Over the last 5 years, the average overall growth trend of the bank`s net profits is also positive and averages 30,8% growth annually.

At the second stage of the express analysis of the bank's profitability, it is proposed to conduct a factor analysis of the formation of the bank's return on capital (ROE) and the formation of pre-tax profit (PBT) of the bank in the reporting period using the analytical form shown in Table 2.

Table 2. Factor analysis of the formation of bank profitability

Key indicators	Reporting period, (1)	Previous period, (0)	Absolute change, Δ	Relative change, Δ%
1. Net profit (NP)	1 021 365	954 124	+ 67 241	+ 7,05
2. Profit before tax (PBT)	1 264 883	868 722	+ 396 161	+ 45,6
3. Income (E)	4 529 499	4 227 012	+ 302 487	+ 7,16
4. Average assets (A)	27 332 388	22 993 422	+ 4 338 966	+ 18,87
5. Average equity (EK)	3 095 441	2 103 574	+ 991 867	+ 47,15
Calculations of auxiliary relative indicators:				
6. Operating profit margin (OPM), % (line 2 / line 3)	27,93	20,55	+ 7,38	+ 35,9
7. Total Asset Turnover (TAT) (line 3 / line 4)	0,166	0,184	- 0,018	- 9,8
8. Equity Multiplier (EM) (line 4 / line 5)	8,83	10,93	- 2,1	- 19,2
9. Taxable Profits Structure (TPS) (line 1 / line 2)	0,807	1,098	- 0,291	- 26,5
10. Return on Equity (ROE), % (line 1 / line 5)	33,0	45,36	- 12,36	- 27,3
Analytical operations:				
1. Construction of 4-factor multiplicative model of bank return on capital	ROE = OPM × TAT × EM × TPS $ROE_{(0)} = 45,36 \approx 20,55 \times 0,184 \times 10,93 \times 1,098$ $ROE_{(1)} = 33,0 \approx 27,93 \times 0,166 \times 8,83 \times 0,807$ ΔROE = 33,0 – 45,36 = - 12,36%			
1.1) calculation of the impact of factor 1 - changes in operating profitability of the bank: $\Delta ROE_{(F1)} = \Delta\ OPM \times TAT_{(0)} \times EM_{(0)} \times TPS_{(0)}$	(+7,38) × 0,184× 10,93 × ×1,098 = + 16,29%			
1.2) calculation of the impact of factor 2 - changes in resource efficiency: $\Delta ROE_{(F2)} = OPM_{(1)} \times \Delta\ TAT \times EM_{(0)} \times TPS_{(0)}$	27,93 × (-0,018) × 10,93 × ×1,098 = - 6,03%			
1.3) calculation of the impact of factor 3 – changes in the structure of sources of financing the bank's activities	27,93 × 0,166 × (-2,1) × ×1,098 = - 10,69%			

$\Delta ROE_{(F3)} = OPM_{(1)} \times TAT_{(1)} \times \Delta\ EM \times TPS_{(0)}$	
1.4) calculation of the impact of factor 4 - changes in tax differences: $\Delta ROE_{(F4)} = OPM_{(1)} \times TAT_{(1)} \times EM_{(1)} \times \Delta TPS$	27,93 × 0,166 × 8,83 × × (-0,291) = - 11,91%
1.5) verification of the reliability of calculations of the influence of factors: ΔROE	ΔROE = - 12,36% ≈ (+16,29) + (-6,03) + (-10,69) + (-11,91)
2. Construction of a 4-factor multiplicative model of pre-tax profit of the bank: PBT = EK × OPM × TAT × EM	PBT = EK × OPM × TAT × EM $PBT_{(0)}$ = 868 722 ≈ 2 103 574 × 20,55/100% × 0,184 × 10,93 $PBT_{(1)}$ = 1 264 883 ≈ 3 095 441 × 27,93/100% × 0,166 × 8,83; Δ PBT =1 264 883 - 868 722 = + 396 161
2.1) calculation of the impact of factor 1 - changes in the amount of capital of the bank: $\Delta\ PBT_{(\Phi 1)} = \Delta EK \times OPM_{(0)} \times TAT_{(0)} \times EM_{(0)}$	(+991 867) × 20,55/100%× × 0,184 × 10,93 = = + 409 924 thousand UAN
2.2) calculation of the impact of factor 2 - changes in operating profitability of the bank: $\Delta\ PBT_{(\Phi 2)} = EK_{(1)} \times \Delta OPM \times TAT_{(0)} \times EM_{(0)}$	3 095 441 × (+7,38)/100% × 0,184 × 10,93 = = + 459 427 thousand UAN
2.3) calculation of the impact of factor 3 - changes in resource efficiency: $\Delta\ PBT_{(\Phi 3)} = EK_{(1)} \times OPM_{(1)} \times \Delta\ TAT \times EM_{(0)}$	3 095 441 × 27,93/100% × × (-0,018) × 10,93 = = - 170 093 thousand UAN
2.4) calculation of the impact of factor 4 - changes in the structure of sources of financing of the bank: $\Delta\ PBT_{(\Phi 4)} = K_{(1)} \times OPM_{(1)} \times TAT_{(1)} \times \Delta\ EM$	3 095 441 × 27,93/100% × ×0,166 × (-2,1) = = -301 384 thousand UAN
2.5) verification of the reliability of calculations of the influence of factors: Δ PBT	Δ PBT = + 396 161 ≈ (+409 924) + 459 427 + (- 170 093) + + (- 301 384)

(Source: annual report of PJSC "OTP Bank" for 2016–2017 from the official website at https://ru.otpbank.com.ua/about/informations/annual_reports/ [21–23]).

In the reporting period, the positive impact on the formation of the bank's profitability was caused by: +16,24% or 459,427 thousand UAH - increasing the operating profitability of banking activities; at +409 924 thousand UAH - increase in the bank's capital, which indicates effective management of costs and capital of the bank. The negative impact on the formation of the bank's profitability was caused: by −6,03% or −170,093 thousand UAH - reduction of the bank's resource efficiency; by −10,69% or −301 384 thousand UAH - change in the structure of sources of financing of the bank, by −11,91% - change in tax differences, which indicates inefficient management of assets, liabilities and taxes of the bank. Management staff first of all needs to look for reserves through the new tax planning, as well as to optimize the structure of liabilities taking into account the WACC indicator.

In the third stage of the express analysis of the bank's profitability, it is proposed to assess the multiplier effect of capital and value added of the bank in the reporting period using the following analytical form (Table 3).

Table 3. Assessment of the multiplier effect of capital and value added of the bank

Key indicators	Reporting period, (1)	Previous period, (0}	Absolute deviation, Δ
1. Operating profit (OP)	2 728 231	2 269 281	+458 950
2. Average assets (A)	27 332 388	22 993 422	+4 338 966
3. Average capital (K)	3 095 441	2 103 574	+991 867
4. Average rate of interest expenses (SPV)	4,58	5,86	-1,28
5. Capital multiplier (EM)	8,83	10,93	-2,1
6. Return on Equity (ROE),%	33,0	45,36	-12,36
Auxiliary calculation indicators:			
7 Economic return on bank assets (Basic Earning Power, BEP),%, (line 1 / line 2 × 100%)	9,98	9,87	+ 0,11
Analytical operations:			
Determination of the multiplier effect of the bank's capital (MEK): MEK = (BEP - SPV) × EM	(9,98 - 4,58) × 8,83 = = + 47,7%		
Determination of the added value of the bank (Economic Value Added, EVA): EVA = (ROE - SPV) / 100% × K	(33,0 - 4,58) / 100% × ×3 095 441 = 879 724 thousand UAN		

(Source: annual report of PJSC "OTP Bank" for 2016–2017 from the official website at https://ru.otpbank.com.ua/about/informations/annual_reports/ [21–23]).

In the reporting period, the bank effectively attracts and uses "paid" resources – liabilities on which interest is paid. This does not violate financial stability. In particular, the increase in the average "interest" liabilities of the bank provided an increase in the average balance of the bank's capital by +47,7%, which indicates the effective management of the spread. In the reporting period, the bank's management generated the added economic value of the bank, which amounts to UAH 879,724 thousand. This indicates: preservation of the bank's financial capital; The "value" of past banking transactions by management; increasing the value of the bank's business; growth of the market value of its own capital by this amount; improving the quality of management activities of the bank's management.

At the fourth stage of the express analysis of the bank's profitability, it is proposed to identify the level of interest rate risk of the bank using the analytical form shown in Table 4.

Conclusions: since interest income and interest expenses are the basis of bank profits, the net interest margin and its change indicate trends in profitability. The Bank has a moderate interest rate risk of loss of profitability (16,25%). Therefore, the bank's management needs to focus on interest rate policy and search for opportunities to increase the resource base.

Table 4. Identification of the average level of interest rate risk of the bank

Year of activity	Interest income, thousand UAH (II)	Interest expenses, thousand UAH (IE)	Net interest income, thousand UAH (NII)	Average assets, thousand UAH (A)	Net interest margin,% (NIM)
1	2	3	4 = (col. .2 - col. 3)	5	6 = col. 4 / col. 5 × × 100%
4 years ago	2 786 490	874 664	1 911 826	20 046 180	9,54
3 years ago	3 369 008	1 092 372	2 276 636	20 152 506	11,30
2 years ago	3 122 679	1 421 060	1 701 619	20 999 337	8,10
Previous period (PP)	3 088 820	1 199 459	1 889 361	22 993 422	8,22
Reporting period (RP)	3 035 594	1 087 530	1 948 064	27 332 388	7,13

Analytical operations:

Calculation of the average net interest margin for 5 years	$M(X) = \frac{\sum_{i=1}^{n} x_i}{n}$	$(9{,}54 + 11{,}30 + 8{,}10 + 8{,}22 + 7{,}13) / 5 = 8{,}86\%$
Calculation of the standard (rms) deviation of the net interest margin annually:	$\sigma(X) = \sqrt{\frac{\sum_{1}^{i}(x_i - M(X))^2}{n}}$	$\sqrt{\frac{(9{,}54-8{,}86)^2+(11{,}3-8{,}86)^2+(8{,}10-8{,}86)^2+(8{,}22-8{,}86)^2+(7{,}13-8{,}86)^2}{5}} = 1{,}44\%$
Calculation of the coefficient of variation of the net interest margin:	$V(X) = \frac{\sigma(X)}{M(X)}$	$1{,}44 / 8{,}86 = 0{,}1625$ or $16{,}25\%$
Identification of the level of interest rate risk of the bank:	V(X) < 10% - weak risk; V(X) > 30%- significant risk	10% < 16,25% < 30% (moderate risk)

(Source: annual report of PJSC "OTP Bank" for 2016–2017 from the official website at https://ru.otpbank.com.ua/about/informations/annual_reports/ [21–23]).

4 Conclusion

The innovation of the obtained results is the development of a comprehensive technological procedure for «rapid» analysis of bank profitability, the main purpose of which is achieved through the formation of understanding of development prospects and risks of banking by external and internal users of financial information.

The scientific novelty of the obtained results is the development of a comprehensive technological procedure for "rapid" analysis of bank profitability, the main purpose of which is achieved through the formation of understanding of development prospects and risks of banking by external and internal users of financial information.

This complex technological procedure (express-analysis) will be easily carried out in the conditions of digital computer technologies and further introduction of artificial intelligence in banking.

Theoretical and Practical Significance of the Study. A comprehensive technological procedure for «rapid» analysis of bank profitability, which includes four stages, starting with trend analysis of net profit and ending with forecasting the level of interest rate risk, allows monitoring the effectiveness of bank profitability management strategies, assessing prospects for banking, forecasting financial instability in the banking sector. The proposed forms of analytical tables will contribute to the unification of the management document «Management Report» of banking institutions and provide instant comparability of analytical information on changes in profitability of banking entities in decision-making by interested external and internal users of financial information.

Socio-economic Effect that Arises as a Result of the Introduction of Scientific Results. The obtained scientific results allow to increase the information culture of documenting changes in profitability in the management of banking, to increase the clarity and informativeness of analytical conclusions about the profitability of banks, significantly reduce the complexity of analytical work of bank management, especially in digital computer technology.

Prospects for Further Research. Promising areas of research in the field of analysis of bank profitability should be procedures for gap-analysis of the sensitivity of bank operations to changes in interest rates.

References

1. Chetvertaya promyshlennaya revolyutsiya: internet veshchey, tsirkulyarnaya ekonomika i blokcheyn (Fourth Industrial Revolution: Internet of Things, Circular Economics and Blockchain). http://www.furfur.me/furfur/changes/changes/216447-4-aya-promyshlennaya-revolyutsiya. Accessed 29 June 2020
2. Klaus, S.: The Fourth Industrial Revolution. World Economic Forum, Cologny/Geneva (2016)
3. Klaus, S.: The Fourth Industrial Revolution. World Economic Forum, Cologny/Geneva (2017)
4. Klaus, S., Nicholas, D.: Shaping the Fourth Industrial Revolution. World Economic Forum, Cologny/Geneva (2018)
5. Ekonomichnyy forum v Davosi 2019: holovne (Davos Economic Forum 2019: the main thing). https://www.rbc.ua/rus/news/ekonomicheskiy-forum-davose-2019-glavnoe-1548090888.html. Accessed 29 June 2020
6. Klaus, S.: Chetverta promyslova revolyutsiya. Formuyuchy chetvertu promyslovu revolyutsiyu (The Fourth Industrial Revolution. Shaping the Fourth Industrial Revolution). Klub Simeynoho Dozvillya, Kyyiv (2019)
7. Pro shcho knyha «Chetverta promyslova revolyutsiya. Formuyuchy chetvertu promyslovu revolyutsiyu» avtora Klaus Shvab
8. (What is the book «The Fourth Industrial Revolution. Forming the Fourth Industrial Revolution» by Klaus Schwab), https://nashformat.ua/products/chetverta-promyslova-revolyutsiya.-formuyuchy-chetvertu-promyslovu-revolyutsiyu-916180. Accessed 29 June 2020
9. Zhak, R.: Audit i analiz khozyaystvennoy deyatel'nosti predpriyatiy (Audit and analysis of economic activities of enterprises). Audit, UNITY, Moscow (1997)

10. Karlin, T.R., Makmin, A.R.: Analiz finansovykh otchyotov (na osnove GAAP) (Financial Report Analysis (Based on GAAP)). INFRA-M, Moscow (2001)
11. Subramanyam, K.R., Wild, J.J.: Financial Statement Analysis. McGraw-Hill/Irwin, New York (2009)
12. Sinki, J.: Finansovyy menedzhment v kommercheskom banke i v industrii finansovykh uslug (Financial Management in a Commercial Bank and in the Financial Services Industry). Al'pina Biznes Buks, Moscow (2007)
13. Collier, H.W., McGowan, C.B., Muhammad, J.: Financial analysis of financial institutions in an evolving environment. In: Proceedings of the Meeting of the Decision Sciences Institute, March 2006. Southwest Region, Oklahama City (2006)
14. Kumbirai, M., Webb, R.: A financial ratio analysis of commercial bank performance in South Africa. African Rev. Econ. Finance **2**(1), 30–53 (2010)
15. James, H., Shane, R.: Bye Bye Banks?: How Retail Banks are Being Displaced, Diminished and Disintermediated by Tech Startups and What They Can Do to Survive Kindle Edition. Adaptive Lab, UK (2016)
16. Chris, S.: Doing Digital: Lessons from Leaders. Singapore, Marshall Cavendish International (Asia) Pte Ltd (2020)
17. Brett, K.: Bank 4.0: Banking Everywhere. Never at a Bank. Wiley, New York (2018)
18. Mark, B., Yonca, E., Patricia, O.: Financial Analysts and Their Contribution to Well-Functioning Capital Markets (Foundations and Trends(r) in Accounting). Now Publishers Inc, Boston-Delft (2018)
19. Padberg, T.: How to Analyse Bank Financial Statements, A Concise Practical Guide for Analysts and Investors. Harriman House, Petersfield (2017)
20. Kuter, M., Gurskaya, M., Andreenkova, A., Bagdasaryan, R.: The early practices of financial statements formation in Medieval Italy. Acc. Hist. J. **44**(2), 17–25 (2018)
21. Kuter, M., Gurskaya, M., Aleinikov, D.: The early practice of analytical balances formation in F. Datini's Companies in Avignon. In: Antipova, T. (eds.) Integrated Science in Digital Age. ICIS 2019. Lecture Notes in Networks and Systems, vol 78, pp. 91–102 (2020) https://doi.org/10.1007/978-3-030-22493-6_10. Accessed 15 Jul 2020
22. Banka, O.: Godovye otchety 2019 – 2013 (Bank reports. Annual reports 2019 – 2013). https://ru.otpbank.com.ua/about/informations/annual-reports/. Accessed 15 July 2020
23. Finansova zvitnist' ta zvit nezalezhnoho audytora za rik, yakyy zakinchyvsya 31 hrudnya 2017 roku (Financial statements and independent auditor's report for the year that has ended 31 December 2017. https://ru.otpbank.com.ua/upload/medialibrary/e82/2017_ua.pdf. Accessed 15 Jul 2020
24. Finansova zvitnist' ta zvit nezalezhnoho audytora za rik, yakyy zakinchyvsya 31 hrudnya 2016 roku (Financial statements and independent auditor's report for the year that has ended 31 December 2016. https://ru.otpbank.com.ua/upload/medialibrary/d57/2016_ua.pdf. Accessed 15 Jul 2020

Admiralty Regulation: Mercantilist Impact on the Evolution of Accounting in the Eighteenth-Century Russia

Dina Lvova(✉)

Saint Petersburg State University, St. Petersburg 199034, Russia
d.lvova@spbu.ru

Abstract. This paper concerns and articulates the impact of Peter the Great (reigned 1682–1725) reform on formation of Russian accounting by studying one of the most valuable law act of the epoch. The study highlights the role of economic and legal doctrines of the time in the changes that affected the Petrine public administration. It has two main objectives. Firstly to reveal interrelation between historical phenomena of mercantilism, regular army and systematic accounting in the context of the eighteenth-century Russia. Then to discover the sources of Russian accounting and to trace its linkages with European patterns. The study proves that the first Russian accounting rules came from the French sources and provides a review of accounting procedures established by the Admiralty Regulation. It also recognizes military influences, which played an important role in the evolution of governmental and even private sector accounting in Russia in the eighteenth century. Accounting discourse supplements the conception of Peter the Great's epoch, its reformative character.

Keywords: Peter the Great · Admiralty Regulation · History of accounting

1 Introduction

Reforms instituted by Peter the Great transformed public administration. He introduced new configuration of government bodies'—the colleges (*kollegii*), which had a clearly defined scope of responsibilities; at his direction, the first state budget was developed. Peter, who recognized accounting as an important function of government and control, personally contributed to the development of accounting rules prescribed to one of the state authorities, the Admiralty. In 1722 he created and issued the Regulation on Admiralty and Shipyards Management (the Admiralty Regulation), a document that is considered to be the first Russian legal act containing detailed accounting instructions.

It was only in the mid-twentieth century that Russian accounting historians showed interest in the Admiralty Regulation. The first to discover it as an accounting artefact was a Leningrad scholar Shiroky [1], who authored the pioneering study of accounting in the eighteenth century Russia.[1]

[1] Regrettably, Shiroky's PhD dissertation [1] entitled "The Balance-sheet in Law, Literature and Accounting Practice of the Eighteenth-Century Russia" has not been published yet.

T. Antipova (Ed.): ICCS 2020, LNNS 186, pp. 132–143, 2021.
https://doi.org/10.1007/978-3-030-66093-2_13

In 1972, another Soviet accounting historian, Mazdorov [2], highlighted the reformative nature of the Admiralty Regulation, which laid the foundation for 'a transition to Western European accounting forms' (p. 7). Sokolov also gave a brief summary of the Regulation in his famous accounting history monographs [3, 4]. He considered the publication of the Admiralty Regulation to be 'the greatest event in the history of Russian accounting'. However, neither Shiroky nor Mazdorov nor Sokolov conducted any special research into the Admiralty Regulation. The reason for that was justified: in the twentieth century, when accounting history was just germinating in Russia (and in the Soviet Union) and no one studied it apart from the above-mentioned authors, it was above all necessary to give a general overview of accounting evolution. Though the number of research on Russian accounting history has significantly increased over recent years, there are still too many blank pages.

Peter the Great directed main reform efforts at reorganization of the army on a regular basis and establishment of the Navy. Those were the times of war, and the first Russian accounting rules were designed for military purposes. Mercantilism, which Peter regarded by as the economic basis for his reforms, also assumed in this age a certain military quality. These considerations offer an explicitly military perspective on the study of Russian eighteenth-century accounting. Indeed, such studies of accounting in connection with the war and the army gained, from the beginning of the century, noticeable presence in accounting historiography, especially in English-speaking countries, but also in Italy and France [5]. Accounting historians such as Black [6], Chwastiak [7, 8] and Funnell [9, 10], investigating the twentieth-century accounting concepts for the fighting services, have focused on the British and the U.S. army cost accounting, and Zambon and Zan [11] pursued the same goal, focusing on Venice's sixteenth- seventeenth-century arsenal accounting. Funnell [12, 13], Cobbin, Burrows [14] research British army, and navy accounting practices aimed at enhancing military efficiency in connection with social environment and political struggles.

This paper presents the first study of Russian military accounting, more precisely 'with the navy as the first line of defense' [6, p. 147] and as the driver of the Russian accounting history. This investigation sought to analyze the Russian eighteenth-century fleet accounts as a seminal development of mercantilist and absolutist-styled accounting model. In this regard, the current article also adds to the research of Littleton [15], Chatfield [16] and Legay [17, 18], by considering the impact of mercantilism on accounting, and to the research of Lemarchand [19], Legay [17, 18] as well as Gomes, Carnegie and Rodrigues [20], by exploring the use of merchant accounting in public institutions of European countries in the eighteenth century.

The hypothesis of this research is that Emperor Peter the Great, who started the reforms, motivated by the purpose of constructing the modern Navy, designed the accounting legacy based on mercantilist ideology and absolutist power. Peter believed that governmental endeavors serve as models to be copied in people's private lives and extended the totalitarian, army-style methods of government to all spheres of economic activity.

The paper proceeds as following. The first section explores the historical facts in relation to the Admiralty Regulation, its creation and origin, without which it appears impossible to determine the distinctive features of the first Russian government accounting system. The second section shows the direct connection between French mercantilism, merchant accounting and the Admiralty Regulation. It also discusses the implementation of the new fleet accounting procedure as influenced by the reforms in merchant accounting practices.

2 Origins of the Admiralty Regulation

The origins of the Admiralty Regulation lie in the military and administrative reforms undertaken by Peter the Great and in the setting up of the Russian Fleet. The first warships appeared in Russia in 1696. In the same year on October 20 (O.S. November 4), the Boyar Duma adopted a famous resolution: ‘Sea-going vessels shall be’ [21, p. 52] and approved the first shipbuilding program for construction of 170 ships [22, p. 13]. From this time on, Peter’s fleet kept expanding and by the year when the Regulation was published (1722) it had become impressive in size: more than 700 ships, including 46 large battleships, were built during that time.[2]

In 1712, Peter the Great transferred the administration of the Navy from Moscow to St. Petersburg, the new capital of the Russian Empire. In his ‘Northern Capital’, Tsar established the Chancellery of the Navy, which took over the functions of Moskovsky Prikaz (a central administrative department in Muskovy, the prototype of modern ministries). In 1717, he reorganized the entire imperial bureaucracy into a college system and transformed all prikazy into colleges. The Admiralty College received the authority to administrate the Navy and shipyards.

Together with the establishment of the Colleges, Tsar ordered to issue their charters known as Reglamenty (Regulations). It was a type of legislative act introduced into legal practice during the reign of Peter the Great, which gradually disappeared in the early nineteenth century. Regulations determined the structure, the scope of work, the staff and responsibilities of a respective public body, including its paperwork management [26, p. 711; 27, p. 27].

Regulations were comprehensive, detailed documents combining ‘important legislative provisions with administrative orders and instructions’. In terms of modern law, the legal status of Regulations was rather low – no more than that of an ‘act of internal statute’ [28, p. 449] or the charter of a respective college, in this case, the Admiralty College. However, in the early eighteenth century, the hierarchy of legal acts was not yet strictly established and the real historical role of the Admiralty Regulation had proven to be more significant than its formal status.

The Admiralty Regulation had one special characteristic that distinguished it from regulations issued by other colleges. It contained instructions not only for economic but also for military administration of the Navy [29, p. 39].

Another important point is that the legislator extended provisions of the Regulation, immediately after its issuance, to other, quite peaceful sectors of state administration. Peterson [30, p. 408] notes after Miliukov that there is no doubt that the Tsar

[2] According to the data published by Soloviev [23, pp. 524–525] and Tsvetaev [24, p. 108]. Altogether, no less than 1,104 ships and other vessels were built during the Petrine era [25, p. 118].

considered the Admiralty Regulation an exemplary product when, in May 1722, he ordered:

> Regulations in accordance with that of the Admiralty College are to be drawn up in all the colleges, although terms and names are to be changed where necessary, but the organization is to be entirely the same in all ways, except that the colleges shall ignore those things that they do not have.

The Instructions for the Revision Office (future Revision College) suggested a corresponding provision on accounting:

> Letters and books… Shall be kept in due order and clean so as to see the receipts and expenses for each day: in order to keep record of all the receipts and expenses of money and other items, books shall be kept in the forms prescribed by the Admiralty Regulation… Should any Colleges or Offices submit accounting statements without veritable examination and not signed by the whole College, and not in the forms prescribed by the Admiralty Regulation, such incorrect accounting statements shall not be accepted into the Revision Office.[3]

However, as Anisimov [25, p. 150] noted:

> For Peter – the reformer of the state – there was a characteristic urge to transfer military principles into the sphere of civilian life and state administration. This was manifested both in the direct extension to civilian administration of military legislation and in the attribution to laws defining the work of civilian institutions of the significance and force of military codes.

Consequently, the history of the Admiralty Regulation starting from the moment it was created and up to the time it was abrogated (in 1765) places it among the most important documents of its age.[4] It is a well-known fact that the Naval Fleet was the favorite project of Peter the Great. Therefore, Peter, who was always very attentive to anything connected with the Navy, personally participated in the drafting, first, of the Navy Regulations (Ustav Morskoi 1720) and then, of the Admiralty Regulation, which was not the case with the other Regulations.[5] The Admiralty Regulation always attracted interest of scholars investigating the history of the army and the Navy (see also [22, 32–34]. The first academic and special study of the Regulation dates back to 1855; it was carried out by the academician and public official Count Tolstoy.[6]

Starting from the late nineteenth century, the Admiralty Regulation, alongside with other Petrine statutes and codes, became an object of legal studies. These documents interested legal historians as the first attempt in Russia to copy the European legislative model. The early explanation of the document's origin maintained that regulations of Peter's colleges were borrowed from the Swedish legislative practice. This explanation has its history. In the eighteenth century, Peter the Great's administrative reforms received a mixed reaction in Russia and Europe. Even Frederick the Great criticized the

---

[3] Polnoe Sobranie Zakonov Rossiiskoi Imperii (hereafter – PSZ), Vol. 6, № 4127.

[4] Encyclopaedia Brokgauz i Efron, s.v. "Reglament".

[5] In the work schedule of the Emperor for the year 1721, four days per week — from Monday through Thursday — were set aside for the drafting of the Admiralty Regulation [32, p. 85]. Peter mentioned his contribution to the work on the Regulation '…done all by hard work of His Majesty, not just by orders [to someone else] but by the sovereign's own labor created, not only in the morning but in the evening, twice daily it was done in different time' [32, p. 87].

[6] Count Dmitry Tolstoy, cited here, was a remote relative of the writer Count Leo Tolstoy.

Russian reforms. In one of his writings, the King of Prussia upbraided Peter for his 'mindless, senseless imitation of Sweden' [35, p. 5].

One can conclude, following Peterson [30, p. 406], that 'the English legislation was to serve as the point of departure and each point in the English text was to be matched by the comparable articles in the other foreign naval regulations.' Later, however, Peter showed more interest in the French codes, making them the main source of his maritime law. This fact has yet to find a plausible explanation, but one can assume that the French laws, which bore a clear impress of the 'enlightened absolutism', were more in tune with his philosophy.

It was the academician Voskresensky, the author of the most significant study into the legislative process of the first quarter of the eighteenth century, who finally put an end to the discussion about the origin of the Admiralty Regulation. Through the comparative analysis, Voskresensky [36, p. 108] proved that the original sources of the Admiralty Regulation lay in two French naval ordinances: the 1681 *Grande Ordonnance de la Marine d'Août* (The Sea Laws) and, to a much greater extent, the 1689 *Ordonnance pour les Armees Navales et Arsenaux de Marine* (The Naval Ordinance). Unfortunately, the results of Voskresensky's research on the origins of the Admiralty Regulation are still unpublished and kept as manuscripts in the archives of the Russian National Library. That is why Peterson, unfamiliar with this source, had to conclude [30, p. 406]:

> Whether any foreign legislative acts, and if so which ones, were used as a basis for drawing up the instructions for the admiralteiskaia kollegiia [Admiralty College] is a question which can only be answered after a study of the preliminary working papers of the commission charged with that task.

Pilgrim [37, p. 236] called The Naval Ordinance a 'detailed synopsis of the assumptions and policies of the two Colberts'. It is an obvious characteristic of the document whose author was Jean-Baptiste Antoine Colbert, Marquis de Seignelay (1651–1690), the Navy Minister of Louis XIV and the son of the other Jean-Baptiste Colbert (1619–1683), Louis XIV's Minister of Finance. De Seignelay, who, just like his famous father, was an advocate of the mercantile system, created an act where all measures of government regulation of the economy typical for the doctrine of mercantilism were applied to the naval forces. By borrowing the provisions of this document, Peter the Great reorganized the administration of the Russian naval fleet in line with the familiar and understandable French mercantilism and absolutism.

3 Admiralty Regulation, French Mercantilism and Merchant Accounting

The French Naval Ordinance—the original source of the Admiralty Regulation—clearly manifested mercantilist character of its regulatory requirements, prescribing the procedures for the purchase of victuals, munitions and other kinds of stocks. As Pilgrim [37, p. 237] notes, 'it was constantly emphasized in these years that France's strategic and economic interest required maximization of national self-sufficiency wherever possible'. For example, Livre XVIII of Ordinance phrased such a requirement as follows (translated from the Russian version of the text made in 1715):

> This has to be observed when buying merchandise, arms and ammunition, so that all of these should be made in the home country, and if it is the case that it is necessary to get these from foreign lands, and then contracts should be made with the merchants of the French people, who will place them in the arsenals [37, p. 386].

Peter literally copied this provision and included it in the Admiralty Regulation in the following wording:

> It is required to check in awarded contracts and in purchases of stocks, arms and ammunition that all of these are from our country except for those that cannot be found in our country, or even if findable, then of bad quality; and if possible, order these from foreign countries through our merchants.[7]

Mercantilism provided a new direction for the development of accounting. It encouraged widespread implementation of requirements that obliged merchants to maintain trade books and to account for the goods and assets that they had at their disposal. The 1673 *Ordonnance sur le Commerce* (also known as the Code Savary) created under the guidance of Colbert Sr. Included a requirement that obliged business people to collect the information in the trade books and submit periodically a summarized report of the property at their disposal. Article eight of the third title, for instance, provided for the following requirement:

> During the same six month period [after the date of the publication of the Ordinance], all merchants are obliged to draw up an inventory of all their movable and immovable property, all their receivables and liabilities, which are to be inspected and updated every two years [38, p. 37].

The impact of this document spread far beyond France; it laid the basis for commercial and accounting law in the countries of continental Europe. The same accounting requirements mercantilist by their origin appeared shortly in the state-owned economy. The Naval Ordinance defined the requirement for officials of the French fleet and port towns to maintain accounting books and draw up report sheets. Similar rules were later instituted by the Navy Regulation and then by the Admiralty Regulation of Peter the Great.

The Prussian diplomat Johann Vockerodt (1693–c. 1756), in his sketches on Russia written in 1737 and intended for Voltaire, says that the new system of Admiralty accounting was organized along the 'mercantile' lines[8] [1, p. 40]. It is necessary to add that its mercantile character had European features. Another diplomat, Chaadaev [39, p. 77], wrote about Peter the Great that 'he has introduced western expressions into our language … using western books we learned how to spell out the names of things'. Peter's Admiralty Regulation introduced the concepts of 'debit', 'credit', and 'accountant' into the naval fleet accounting, the terms borrowed from West-European

[7] PSZ, Vol. 6, № 3937, p. 528.

[8] In an earlier law act called the Military Code (1716), a short section on accounting entitled 'On Merchants' Books and Reports'. One can hardly find a better indication that, in the spirit of the time, Peter did not see the difference between merchant accounting and governmental accounting.

trade practices via the French Naval Ordinance. Registers (books) of systematic entries established by the Admiralty Regulation also had 'commercial' origin, and the type of records, with their use of the linear method for entering debit and credit, was in essence representational of the European written accounting practice.

The presence of 'mercantile' accounting in the Admiralty Regulation was so evident that it made certain researchers assume that double entry bookkeeping and balance sheets became usual in governmental accounting of Petrine Russia. Such hasty conclusions rely on the existence of multi-column books of accounts.

Figure 1 shows one of such books that, according to the Regulation, the admiralty paymaster was obliged to maintain (sample books were included in the Regulation). The book in question summarized the information on current operations with monetary resources and for drawing up monthly and annual reports. The latter prepared for the Admiralty College and were a kind of periodic reporting.

<table>
<tr><td colspan="12">The book to be maintained from the book of receipts and charges for the purpose of submitting monthly and yearly reports to the College as follows</td></tr>
<tr><td colspan="2">Page 1</td><td colspan="4">Debit for 1722</td><td colspan="6">Credit</td></tr>
<tr><td>Months, dates</td><td>№ of order</td><td></td><td>Pages</td><td>Roubles</td><td>Kopecks</td><td>Months, dates</td><td>№ of order</td><td></td><td>Pages</td><td>Roubles</td><td>Kopecks</td></tr>
<tr><td colspan="12"></td></tr>
</table>

Fig. 1. Model example of a systematic register with linear notes for money accounting (Source: PSZ, Vol.6, № 3937).

Mazdorov, the researcher who we can criticize for his hasty conclusions, linked the beginning of European-style regulation of accounting with the use of double entry bookkeeping and the balance sheet. He wrote [2, p. 7] that 'the new order of accounting made it possible to regularly determine, with sufficient accuracy, the available funds in the commercial unit at a given moment; the balance sheet, previously unknown in Russia, was introduced in the accounting practices'. Further on, '... double-entry method of accounting and drawing up of balance sheets was introduced, first in the navy department and then in the mining industry, in particular, at the mining plants of the Urals'.

The only reason for the conclusion about the use of double entry bookkeeping can be found in the terms 'debit' and 'credit' appeared in the table headers, but this cannot be considered a sufficient criterion. The Regulation did not use accounts per se and did not define the rules according to which one should maintain books. The money to be recorded in the book shown in Fig. 1 was seen as part of the assets (as an item of property) and not as a kind of accounting measurement; and the rules for money accounting were similar to the rules of, for instance, stock accounting. Let us compare it with the book of stocks accounting shown in Fig. 2, which the store officer had to maintain. The layout of the book is almost the same; the difference lies in the units of measure – in the latter case, these are Russian puds (sixteen kilograms) and pounds.

The book of all stocks with different names in alphabetical order, which are kept in the store, in accordance with the Regulation to record the quality and the quantity of those, and what is received, to be maintained as follows:											
Stocks to be in the store. Year 1722						Received into the store					
Months, dates	№ of order		Pages	Puds	Pounds	Months, dates	№ of order		Pages	Puds	Pounds

Fig. 2. Model example of a systematic register with linear notes for stocks accounting (Source: PSZ, Vol.6, №3937).

The Regulation stated clearly that the officer had to record stocks only in quantitative terms. It said nothing at all about the monetary value of stocks. There was also no mention of the balance sheet as a method for summarizing all records and certainly nothing about a combined table. The Admiralty College received reports on the cash on hand (daily), statements of receipts, expenses and the balance of money (weekly, monthly and annually), reports on stocks consumption (weekly), reports on items produced (weekly), and other similar reports. They did not represent a financial statement from the modern point of view.

4 Conclusion

Anisimov [25, p. 5] perfectly formulated the main feature of the Petrine historiography: 'Disputes about the significance in Russian history of Peter's personality and his reforms have not subsided for three centuries, and it is doubtful that they will ever end. The historiography of the Petrine transformations is exceptionally broad and many-sided'. This paper has examined only one but previously unknown side of the interests and activity of Peter as a first-hand creator of Russian accounting. The archival documents remain to testify to Peter's diligence in developing the Admiralty Regulation, including an extensive section of accounting rules, which no one addressed before him. The Tsar permanently fought stealing and continued deficit and very likely focused on accounting as a possible effective measure. Several aspects make this attempt noteworthy. The Regulation improved Russian accounting system, bringing it in line with the European one. The French source considerably enriched it by introducing accounting methods previously unknown in Russia. The new objects of accounting, terms 'debit' and 'credit', the linear notes method of the registers, and other innovations, prescribed by the Regulation, turned out to be of great significance.

Peter treated regulations as military codes and officials as military men, bringing military principles into public administration and civil life. He did not see anything absurd about giving a detailed description of accounting rules in a military statute first and then ordering to extend them to other spheres of administration, even civil ones. However, scholars did not see anything unusual about this approach even much later. Galagan, for instance, who authored the first book on accounting history in Russia, pointed out [41, p. 129–130] that Russia's economic development became quite significant in Peter's times, giving rise to the need for the institutes which would facilitate proper administration of the strengthened industry and trade, including in terms of accounting. He then highlights that Peter did respond to that demand by introducing an accounting institute, but in which sphere did he do it? He did not do it in the industry or trade (showing no interest in Colbert's famous Ordinance de Commerce for some reason), but in the army.[9]

Out of other countries adhering to the mercantilist model of economy and administration, France was the closest one to Peter. Having strong absolute monarchy with an extensive bureaucratic apparatus, it implemented the policy of mercantilism, which reached its climax under Colbert. This policy characterized by especially active involvement of government entities in the economy (étatisme) and direct state engagement in setting up new industrial enterprises as well as pedantic control over all economic activities etc. (e.g. the *Conseil de Commerce Reglements* were observed in all their detail through the manufacturing process). An expression 'colbertism' coined to stand for such type of active mercantilist policy [43, pp. 371–372; 44, pp. 526–527]. The Russian version of mercantilism, as exemplified by the Admiralty Regulation, was in perfect harmony with the autocratic will to transform private law into public law, and vice versa, and to use detailed acts of public law to regulate all fields of activity, including accounting. The history of the Admiralty Regulation gives us a paradoxical example of commercial accounting rules introduced in public administration and then, in turn, brought from governmental accounting into commercial one for the simple reason that both spheres were based on the same principles. Finally, it illustrates to perfection that mercantilism, accounting and the naval fleet can have something in common.

Acknowledgments. The reported study was funded by RFBR, project number 20–21-00001.

[9] Galagan was not familiar with the Admiralty Regulation, although he noticed the short section called 'On Merchants' Books and and Reports' in Peter's War Code of 1716 (For more details on this outstanding Russian theorist and accounting historian, see [42]).

References

1. Shyroky, V.F.: Balans v zakonodatel'stve, literature i schetnoi praktike Rossii XVIII v. (Баланс в законодательстве, литературе и счетной практике России XVIII в.) PhD thesis, Leningradsky Institute Sovetskoy Torgovli Imeni F. Engelsa, Leningrad (1939)
2. Mazdorov, C.A.: Istoriya razvitiya buhgalterskogo ucheta v SSSR (1917–1972) (История развития бухгалтерского учета в СССР. (1917–1972)). Finansy, Moscow (1972)
3. YaV, S.: Istoriya razvitia bukhgalterskogo utcheta (История развития бухгалтерского учета). Finansy i Statistika, Moscow (1985)
4. Sokolov Ya.V.: Bukhgalterskyi utchet: Ot istokov do nashih dney (Бухгалтерский учет: от истоков до наших дней). Audit, UNITI, Moscow (1996)
5. Funnell, W.N., Walker, S.P.: Accounting for victory. Account. Hist. Rev. **24**(2–3), 57–60 (2014). https://doi.org/10.1080/21552851.2014.970799
6. Black, J.: Full circle: the cost accounting experiment in the British army 1917–25 and the corps of military accountants. J. Soc. Army Hist. Res. **79**(318), 145–163 (2001)
7. Chwastiak, M.: Taming the untamable: planning, programming and budgeting and the normalization of war. Acc. Organ. Soc. **26**(6), 501–519 (2001). https://doi.org/10.1016/S0361-3682(01)00010-1
8. Chwastiak, M.: Rationality, performance measures and representations of reality: planning, programming and budgeting and the Vietnam War. Crit. Perspect. Account. **17**(1), 29–55 (2006). https://doi.org/10.1016/j.cpa.2004.05.006
9. Funnell, W.N.: Pathological responses to accounting controls: the British Commissariat in the Crimea 1854–6. Crit. Perspect. Account. **1**(4), 319–335 (1990)
10. Funnell, W.N.: Military influences on public sector accounting and auditing 1830–1880. Account. Hist. **2**(2), 9–31 (1997)
11. Zambon, S., Zan, L.: Controlling expenditure, or the slow emergence of costing at the Venice Arsenal, 1586–1633. Account. Bus. Finan. Hist. **17**(1), 105–128 (2007). https://doi.org/10.1080/09585200601127731
12. Funnell, W.N.: National efficiency, military accounting and the business of war. Crit. Perspect. Account. **17**(6), 719–751 (2006). https://doi.org/10.1016/j.cpa.2004.11.008
13. Funnell, W.N.: Social reform, military accounting and the pursuit of economy during the liberal apotheosis, 1906–1912. Account. Bus. Finan. Hist. **21**(1), 69–93 (2011). https://doi.org/10.1080/21552851.2011.548549
14. Cobbin, P.E., Burrows, G.H.: The British navy's 1888 budgetary reforms. Account. Hist. **15**, 153–172 (2010). https://doi.org/10.1177/1032373209359321
15. Littleton, A.C.: Accounting Evolution to 1900. Russell and Russell, New York (1966)
16. Chatfield, M.: Mercantilism. In: Chatfield, M., Vangermeersch, R. (eds.) The History of Accounting: An International Encyclopedia, p. 415. Garland Publishing, New York (1996)
17. Legay, M.-L.: The beginnings of public management: administrative science and political choices in the eighteenth century in France, Austria, and the Austrian Netherlands. J. Mod. Hist. **81**(2), 253–293 (2009). https://doi.org/10.1086/598924
18. Legay, M.-L.: La science des comptes dans les monarchies française et autrichienne au XVIIIe siècle. Histoire Mesure **XXV-1**, 231–260 (2010)
19. Lemarchand, Y.: Introducing double-entry bookkeeping in public finance: a French experiment at the beginning of the 18th century. Account. Bus. Finan. Hist. **9**(2), 225–254 (1999)
20. Gomes, D., Carnegie, G., Rodriges, L.L.: Accounting change in central government: the adoption of double entry bookkeeping at the Portuguese Royal Treasury (1761). Account. Audit. Account. J. **21**(8), 1144–1184 (2008). https://doi.org/10.1108/09513570810918797

21. Elagin, S.: Istoriya russkogo flota. Period azovsky, V Tipografii Komissionera Imperatorskoy Akademii Hudozestv, Gogenfeldena I Co, St. Petersburg (1864)
22. Veselago, F.F.: Ot nachala morehodstva (IX vek) do konchyny Pavla I, vol. 1. Kratkaya istoriya russkogo flota of F.F. Veselago. Tipografiya V, Demakova, St. Petersburg (1893)
23. Soloviev, S.M.: Sochineniya. Istoriya Rossii s drevneishih vremen (Современная история России с древнейших времен), vol. 18. Golos, Kolokol-Press, Moskow (1995–97)
24. Tsvetaev, D.V.: Osnovanie russkogo flota. Istoricheskii ocherk (Основание русского флота. Исторический очерк). Russkaya Starina **87**(7) (1896)
25. Anisimov, E.V.: The reforms of Peter the Great: progress through violence in Russia. Translated with the introduction by J.T. Alexander. Routledge, London (2015)
26. Belyaev, I.D.: Lekcii po istorii russkogo zakonodatel'stva (Лекции по истории русского законодательства). Tipo- Litografiya S.A Petrovskogo and N.P Panina, Moscow (1879)
27. Latkin, V.N.: Uchebnik istorii russkogo prava perioda imperii (XVIII i XIX st.) (Учебник истории русского права периода Империи (XVIII и XIX ст.)), 2nd edn. Tipografia Montevida, St. Petersburg (1919)
28. Giovannoni, E., Riccaboni, A., Platonova, N.V.: Peter the Great's government reforms and accounting practice in Russia. Account. Hist. **14**(4), 437–464 (2009). https://doi.org/10.1177/1032373209342476
29. Chubinski, V.G.: Istoricheskoe obozrenie ustroistva upravleniya Morskim vedomstvom v Rossii (Историческое обозрение устройства управления Морским ведомством в России). Tipografiya Morskogo Ministerstva, St. Peterburg (1869)
30. Peterson, C.: Peter the Great's Administrative and Judicial Reforms: Swedish Antecedents and the Process of Reception. Nordiska Bokhandeln, Stockholm (1979)
31. Pavlenko, N.I.: Petr I (k izucheniu social'no-politicheskih vzgliadov). Rossiya v period reform Petra I (Петр I (к изучению социально-политических взглядов). Россия в период реформ Петра I. Izd-vo "Nauka", Moskva (1973)
32. Puzyrevsky, A.K.: Razvitie postoyannyh reguliarnyh armii i sostojanie voennogo iskusstva v veke Ludovika XIV i Petra Velikogo (Развитие постоянных регулярных армий и состояние военного искусства в веке Людовика XIV и Петра Великого). Typografia C. S. Balasheva, St. Petersburg (1889)
33. Zhitkov KG (1912) Istorija russkogo flota: period Petrovskij. 1672–1725 (История русского флота: период Петровский. 1672–1725). Komis. po zavedyvaniju kapitalom im. gr. S. A. Stroganova, St. Petersburg
34. Beskrovny, L.G.: Russkaya armiya i flot v XVIII v. (ocherki) (Русская армия и флот в XVIII в. (очерки)). Voenizdat, Moscow (1958)
35. Berendts, E.A.: Neskol'ko slov o "kollegiah" Petra Velikogo. Rech', proiznesennaia na godichnom akte Demidovskogo juridicheskogo liceya 30 Avgusta 1895g. ordinarnym professorom Je. Berendtsom (Несколько слов о «Коллегиях» Петра Великого. Речь, произнесенная на годичном акте Демидовского юридического лицея 30 августа 1895 г.). Tipo- Litografia E.G. Falk, Yaroslavl
36. Voskresensky, N.A.: Petr Velikii kak zakonodatel'. Issledovanie zakonodatel'nogo processa v Rossii v epohu reform pervoichetverti XVIII veka. Okonchatel'naia redakciya (Петр Великий как законодатель. Исследование законодательного процесса в России в эпоху реформ первой четверти XVIII века). National Library of Russia, the Manuscripts Department, file 1003, manuscript 14.
37. Pilgrim, D.: The colbert-seignelay naval reforms and the beginnings of the war of the League of Augsburg. French Hist. Stud. **9**(2), 235–262 (1975)

38. Ustav o voiskah morskih i o ih arsenalah. Ludovika chetvertogo na desiat' korolia francuzskogo i navarskogo. Napechatalsa Poveleniem Carskogo Velichestva (Устав о войсках морских и о их арсеналах. Людовика четвертого на десять короля Французского и Наварского. Напечатался повелением Царского Величества), St. Peterburg (1715)
39. Kommercheskii kodeks Francii (Коммерческий кодекс Франции). Predisl., per. s fr., dopolnenie, slovar'-spravochnik i kommentarii VN Zahvataeva. Walters Kluwer, Moscow (2008)
40. Chaadaev, P.J.: Apologija sumasshedshego (Апология сумасшедшего). Voprosy filosofii i psihologii **4**(84) (1906)
41. Galagan, A.M.: Sthetovodstvo v ego istoritcheskom razvitii. Gosizdat, Moscow and Leningrad (1927)
42. Kuter, M.I., Gurskaya, M.M., Kuznetsov, A.V.: Aalexander Galagan: Russian titan of the enlightenment in the history of accounting. Account. Hist. **24**(2), 293–316 (2019). https://doi.org/10.1177/1032373218761129
43. Korhov, Y.A., Troitsky, S.M.: Mercantilism. In: Sovetskaya Istoricheskaya Enciklopediya, Malta – Nahimov, vol. 9, pp 369–372. Sovetskaya Enciklopediya, Moscow (1966)
44. Wilson, C.H.: Trade, society and the state. In: The Cambridge Economic History of Europe. Vol. IV. The Economy of Expanding Europe in the Sixteenth and Seventeenth Centuries. Cambridge University Press, Cambridge

Review of Due Diligence Best Practices Widespread Outside Russia

Nadezhda Antonova(✉)

Kuban State University, Stavropolskaya St., 149, Krasnodar 350040, Russia
antonova_n@list.ru

Abstract. Due diligence is a modern professional service provided by audit and consulting companies. In Russia, the practice of conducting due diligence service is a new direction and is gaining popularity. It should be noted that more and more publications are presented every year. For example, the national bibliographic database "RSCI' contains about three hundred works. But this is almost hundred times less than the number of foreign articles. The study of foreign experience is advisable, as it allows you to use knowledge and skills, as well as to implement new ideas for the theory and practice of auditing. The author examines certain aspects of the foreign experience of due diligence. Foreign scientists express conclusions regarding the definition, identifies the features of the organization and conduct of due diligence procedure. The author of the article notes Russian scientists who are engaged in the development of audit and consulting activities. The article also identifies areas for further research. It is advisable to develop a methodology for planning and performing of due diligence to provide a unified approach to conduct procedure and become a guide for auditors and consultants.

Keywords: Due diligence · Investment · Best practices · Audit · Consulting

1 Introduction

Due diligence as a professional service is only becoming widespread in audit and consulting practices in Russia. At the same time, due diligence has long been known, recognized, and applied in many foreign countries for many years. In addition, a lot of studies about due diligence are presented in international multidisciplinary bibliographic and abstract databases. While the national bibliographic database of scientific citation "RSCI" contains about three hundred works of Russian scientists and practitioners, which is almost **a hundred times less** than the number of publications of foreign colleagues. Therefore, the study of foreign experience seems to be expedient from the point of view of the fact that it will expand knowledge about the procedure under study, provide a better understanding of existing problems in the field of due diligence, create conditions for the **implementation of new ideas** to improve the theory and practice of audit activity. In addition, the study of foreign experience and best practices is necessary to identify the features of specific approaches and methods, **create new tools and procedures, develop standards and recommendations** that can be adapted and applied in practice in today's constantly changing economic conditions.

T. Antipova (Ed.): ICCS 2020, LNNS 186, pp. 144–152, 2021.
https://doi.org/10.1007/978-3-030-66093-2_14

In this article author examines certain aspects of foreign experience in the field of due diligence. Foreign definitions of due diligence are analyzed. The points of view of foreign scientists and practitioners in relation to the peculiarities of due diligence, to the subjects conducting the procedures, to the approaches used in the process of due diligence are revealed. Attention is paid to the issues of identifying and assessing the risks of an investment transaction and formalizing the results of due diligence. Conclusions regarding the definition and characteristics of due diligence implementation are formulated.

2 Previous Studies and Techniques

There is enough foreign scientific works, publications, books, other information, and resources that disclose characteristics and interpretations of due diligence definition. In this article, it seems important to consider and compare the points of view put forward by foreign scientists and practitioners in determining due diligence.

R. Shain singles out a separate chapter in his book devoted to the questions: what due diligence is and what the types of due diligence are. The author claims that the idea of what is due diligence is revealed through the variety of existing definitions [1].

R. Shain discusses due diligence as a process of systematic assessment of information to identify the risks and problems associated with the proposed investment transaction.

In the book of M. Ramsinghani, a specialist in the field of venture investments, there is chapter #16, which is called "The Art of Due Diligence". In this chapter, the author notes that due diligence is an "art" that determines investment opportunities: potential and risks [2]. According to M. Ramsinghani, the importance of "art" lies in understanding what information needs to be requested to perform due diligence procedure, but more importantly, how and in what form information should be requested [2].

W. Gole and P. Hilger define due diligence as a procedure that allows to reveal an unacceptable level of unforeseen risk [3].

The opinion on what due diligence is of A. Sherman is interesting. The author believes that in the coming years due diligence will be associated mainly with international transactions [4]. A. Sherman, like M. Ramsinghani, comes to the conclusion that it is necessary to accept the idea that due diligence is both an "art" and a science. A. Sherman, like R. Shain, believes that the investigated procedure is a process, not an event.

Due diligence, according to D. Denison and I. Ko., is an all-encompassing process of studying and evaluating business opportunities in mergers and acquisitions [5].

The authors also refer to other scientists and cite the following statements: due diligence almost always includes a review and analysis of statistical information about the company's products, financial position, business model, technologies, mainly, issues are considered legal and financial nature. If due diligence is carried out correctly, then the procedure allows you to identify the key problems that may interfere with the proposed capital transaction [5]. Based on the foregoing, due diligence is a complex procedure.

The point of view of A. Sacek coincides with the position of D. Denison and I. Ko. A. Sacek also talks about due diligence as a procedure performed in the process of mergers and acquisitions of companies. The author notes that due diligence focuses on the current critical factors. According to A. Sacek, the acquisition of a target company in one industry reduces the risk of canceling the deal, since the acquirer knows all the critical factors and can prevent them at the stage of due diligence. In addition, the possibility of studying the target company, taking into account the previous acquisition experience, allows to develop knowledge about the processes of mergers and acquisitions and the skills of due diligence, contribute to the successful implementation of synergistic effects after the acquisition [6].

D. Cumming and S. Zambelli in their research talk about the types of due diligence, who should carry out the specified procedure, what is its value. The authors define due diligence as the process of investigating the proposed investment in a target company. Due diligence includes a thorough assessment of several factors such as: management skills, industry characteristics of the target company, competitors, opportunities and prospects, financial forecasts, strategy, operational and financial risks of the company. D. Cumming and S. Zambelli argue that the nature of due diligence does not necessarily depend on the type of the proposed capital transaction, however, it can vary depending on the emerging problems and factors disclosed during due diligence [7].

A review of the official websites of the Big Four companies made it possible to establish that due diligence is described as a professional service. Below are some characteristics of due diligence.

According to EY specialists [8], due diligence procedure allows analyzing internal documents and various reports of the target company to identify trends in the financial condition. Within the framework of due diligence, the structure and dynamics of income are investigated in detail through the analysis of the main risks and financial indicators for previous periods. All this is necessary to determine the feasibility of continuing the activity and understanding how the constantly changing conditions of the internal and external environment can affect the activities of the target company.

Pwc experts argue [9] that due diligence can be carried out in order to identify threats to a deal breakdown. In addition, the procedure can be carried out for a more detailed analysis of the financial condition of the target company, in order to establish a framework for negotiating in the process of an investment transaction, to assess the potential for synergies and possible risks.

The goal of due diligence, according to KPMG professionals [10], is to assess the significant problems that the target company faces, as well as the investor's ability to get a correct understanding of the business carried out by the company.

Deloitte specialists [11] present due diligence as a service that allows you to get up-to-date information about the current value and possible risks of the target company and increase the likelihood of a successful transaction. Therefore, due diligence plays an important role in the process of assessing the risks of a proposed investment transaction, in compliance with legal requirements, in the implementation of financial and tax planning, in determining the future value of cash flows, in identifying additional costs in the investment process.

Based on the considered foreign interpretations of the definition and characteristics of due diligence, important features can be distinguished (see Fig. 1).

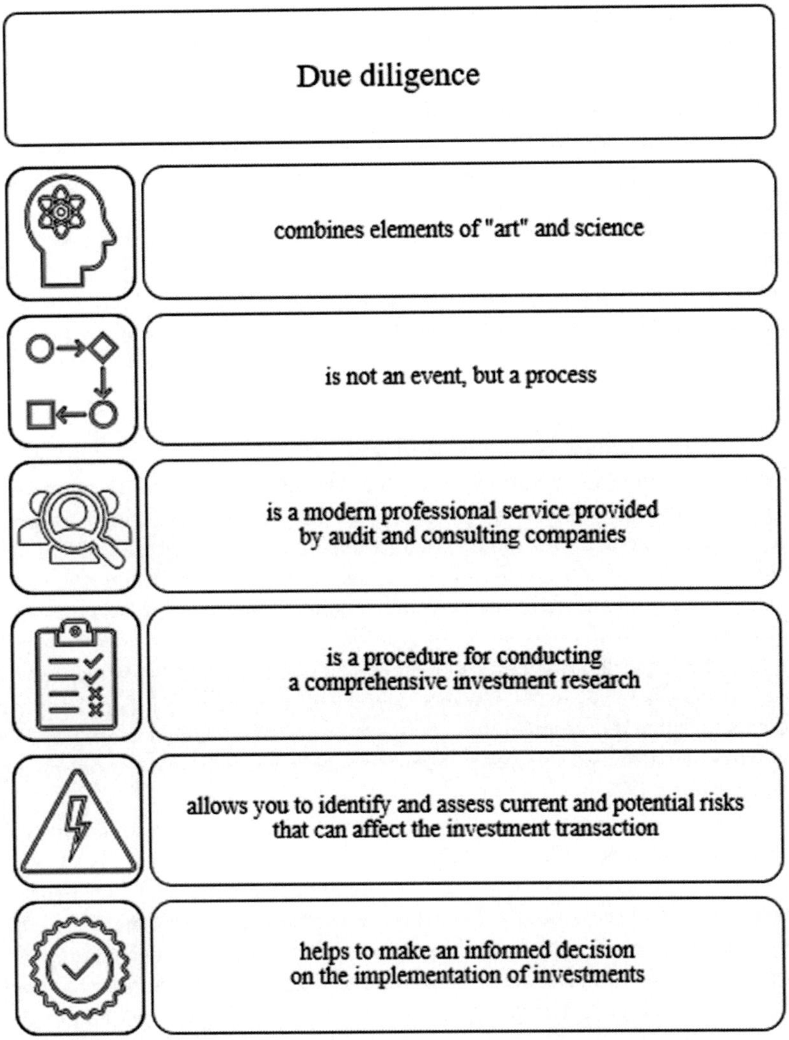

Fig. 1. Important features of due diligence.

Thus, these features make it possible to obtain a generalized idea of the essence of the procedure under study.

3 Results and Discussion

Due diligence as a professional service has a number of features in the context of planning and conducting.

Due diligence involves certain costs on the part of the investor. In addition, a team of specialists requires a certain amount of time, during which it is planned to carry out the necessary procedures, prepare reports and form the necessary recommendations for the investor. If the time budget is reduced, then the investment research may not be carried out in detail and thoroughly, in such an outcome, important factors may be overlooked. So R. Shain has concluded that due diligence is important to pay attention to details, and not look for ways to save time and money [1].

M. Ramsinghani makes an interesting remark: at present, due diligence is rarely carried out according to a specific algorithm, the procedure is usually carried out in a free form [2]. Indeed, there are no international generally established requirements that would strictly regulate the planning and conducting of due diligence. At present, the main service providers are audit and consulting companies, whose activities are regulated by international auditing standards and other regulatory documents. At the same time, there are no international standards that would reveal the algorithm for planning and conducting due diligence as a modern service.

The question of who should carry out the due diligence procedure becomes relevant. There is no consensus on this issue. For example, W. Gole and P. Hilger, discussing the professionals who should conduct due diligence, are of the opinion that the due diligence team should consist of specialists, each of whom is entrusted with the responsibility of studying the business from the point of view of their field of activity or functionality [3].

D. Cumming and S. Zambelli also believe that due diligence can be carried out by different specialists (consultants, lawyers, accountants, auditors) [7]. At the same time, the authors note that if investment fund managers were involved in due diligence, better results would be achieved. However, due diligence is currently carried out mainly by outside specialists [7].

W. Gole and P. Hilger also raise the issue of approaches to investment research in the form of due diligence. The authors note that all due diligence specialists look at the target company through the prism of their individual professional experience. The results obtained while conducting investment research by areas are critical for effective due diligence. However, these results do not provide a sufficient understanding of what is happening. Therefore to conduct effective due diligence it is necessary to perform cross-functional analysis. It means to carry out all the verification procedures in conjunction to ensure an integrated approach [3].

It is interesting to note the opinion of A. Sherman. The author believes that due diligence requires a more creative and strategic approach, rather than the mechanical implementation of a specific methodology. When conducting due diligence, a deeper study of the company in respect of which it is planned to make a transaction is necessary, study of the mission, values, culture and other aspects of the target company, and not a simple performance of a formal review of key agreements and other corporate documents [4].

D. Denison and I. Ko also note that due diligence should not include traditional approaches [5].

A. Sacek draws attention to the fact that in the modern literature, assumptions about what factors should be investigated in the due diligence process cover a number of critical aspects of the target company, since the complexity of mergers and acquisitions is amplified mainly due to trends globalization [6]. The problem is that all critical factors are investigated on an individual basis. But in modern economic conditions, the process of determining the cost of acquiring a target company requires a more complex and comprehensive approach.

J. McGee and J. Byington investigated the problems of due diligence in China [12]. The authors have concluded that the procedure should at least include a discussion of such issues as: strategic positioning, current financial condition, finance, taxes, legal assessment. Also, J. McGee and J. Byington note that for Chinese companies the due diligence procedure is a sufficient new economic phenomenon [12].

G. Bing has written a book devoted to the analysis and techniques of due diligence, as well as the issues of decision making based on the results of due diligence. According to the author, effective comprehensive due diligence allows you to obtain information and form a position to avoid costly and sometimes disastrous consequences for investment transactions [13].

R. Bell argues that when conducting due diligence, it is necessary to use the following five sources of information: personal knowledge, answers to requests in the form of documents, information and information obtained during observation, information from official sources to which there is free access, information obtained through interviews [14]. The author also believes that even though the disclosure process is the main technique of any audit, valuation or transaction, no conclusion is presented during due diligence. Within the framework of due diligence, the process of disclosing information is necessary to understand the existing problems as some characteristics [14].

Taking into account the presented points of view of foreign scientists regarding due diligence, it can be concluded that today there is no consensus on how to plan and conduct this procedure. Considering the difference of opinions of foreign researchers, the author of this article has formulated some general conclusions (see Fig. 2).

Consider the points of view of foreign scientists and practitioners in relation to identifying risks and formalizing the results of due diligence.

R. Shain discusses due diligence and compares the procedure with the analysis of existing debts of borrowers carried out by banks in order to identify and assess risks, with an independent analysis of the financial situation and the future of the company on the eve of major capital investments or entering stock exchanges, etc. [1].

According to M. Ramsinghani, a well-organized due diligence helps the specialists carrying out this procedure to identify the main risks and determine the growth potential of all opportunities. The author believes that, being in conditions of uncertainty, due diligence can be used to obtain the necessary information, but not all. And the specialists performing the procedure must accept the fact that there are conditions of uncertainty [2].

The point of view of W. Gole and P. Hilger should be noted. The authors believe that financial and accounting due diligence (as separate areas of the procedure) comprehensively allows the study of the target company. The leader of the expert due diligence team should look at the deal as a whole. The application of a goal-oriented approach provides the necessary basis for a team of specialists to confirm their understanding of the business, determine the potential and drivers of growth, and assess the overall risks [3]. W. Gole and P. Hilger also note that the key aspect is the

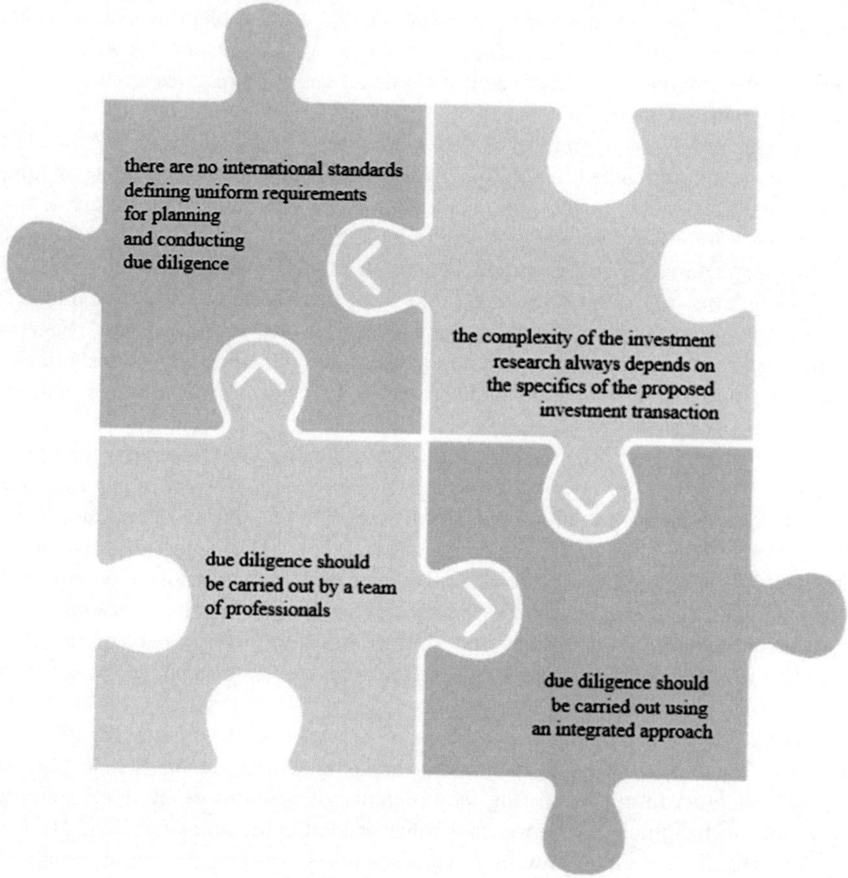

Fig. 2. Author conclusions on due diligence.

assessment of the risks associated with obtaining a synergistic effect. These risks can take many forms.

According to A. Sacek, due diligence is becoming a theoretically solid basis for a successful capital transaction through the analysis of identified risks and the study of critical factors. From a practical point of view, the results of the ongoing comprehensive investment research provide the necessary information and provide the investor with a proven model for the preliminary assessment of candidates for acquisitions [6].

W. Gol and P. Hilger present in their book a separate chapter, which is devoted to the issues of reporting on the results of due diligence and making decisions based on the results obtained. The authors argue that, based on the results of the procedures performed, the team of due diligence specialists can conclude that the level of risk is unacceptably high and give recommendations not to make a deal with capital [3]. The due diligence report should summarize all measures proposed by the team of specialists to assess potential investments in the company and identify any unforeseen risks. The report should also provide information and recommendations that determine the nature and degree of influence of factors that can seriously affect or disrupt the transaction [3].

Thus, according to foreign researchers, conceptually, in the process of performing due diligence, the identification and assessment of risks is carried out, which may have an impact on the implementation of an investment project. After completing all the procedures, the expert group should prepare information in the form of a comprehensive report on the identified risks in order for the investor to get a general idea of the investment object and make an informed decision on capital investment.

4 Conclusion

In this article the positions of foreign authors in relation to aspects of the due diligence procedure are analyzed. Generalizations and valuable conclusions are presented, considering the points of view of foreign scientists, allowing to establish the existing problems that need to be solved through science. It is important to note that in Russia great attention is paid to the study of theory and practice of due diligence in the studies of Y.N. Guzov [15, 16].

As a prospect for further research, it is important to continue to study international experience on the subject of due diligence, to search, develop and implement new procedures, methods and tools for due diligence, thereby ensuring the development of theoretical and methodological foundations of audit activity.

It should also be noted that thoughts on further directions of development of the theory and methodology of auditing, improving the quality of auditing and consulting services were expressed in the articles of M.V. Melnik [17], M.I. Kuter [18] and V.Y. Sokolov [17, 18].

Taking into account the above, it is advisable to develop methodological recommendations describing the requirements for risk-based planning due diligence, which will provide a unified understanding of the specifics of the procedure and will become a convenient guide for professionals providing this service in Russia.

References

1. Shain, R.: Hedge Fund Due Diligence: Professional Tools to Investigate Hedge Fund Managers. Wiley, Hoboken (2008)
2. Ramsinghani, M.: The Business of Venture Capital: Insights from Leading Practitioners on the Art of Raising a Fund, Deal Structuring, Value Creation, and Exit Strategies, 2nd edn. Wiley, Hoboken (2014)
3. Gole, W.J., Hilger, P.J.: Due Diligence: An M&A Value Creation Approach. Wiley, Hoboken (2009)
4. Sherman, A.J., Welch, J.E.: Due diligence in a new era of accountability. J. Corp. Account. Financ. **20**(65), 70 (2009)
5. Denison, D., Ko, I.: Cultural due diligence in mergers and acquisitions. Adv. Mergers Acquis. **15**, 53–72 (2016)
6. Sacek, A.: Critical factors of pre-acquisition due diligence in cross-border acquisitions. Contemp. Stud. Econ. Financ. Anal. **98**, 111–134 (2016)

7. Cumming, D., Zambelli, S.: Due diligence and investee performance. Eur. Financ. Manag. **9999**(9999), 1–43 (2016)
8. Ernst and Young. https://www.ey.com/Publication/vwLUAssets/EY-Due-diligence-in-growth-strategy/$FILE/EY-Due-diligence-in-growth-strategy.pdf
9. PricewaterhouseCoopers. https://www.pwc.com/vn/en/deals/transaction.html
10. KPMG. https://home.kpmg.com/au/en/home/services/advisory/deal-advisory/services/transaction-services.html
11. Deloitte. https://www2.deloitte.com/us/en/pages/mergers-and-acquisitions/topics/transaction-due-diligence.html
12. McGee, J.A., Byington, J.R.: Due diligence issues in China. J. Corp. Account. Financ. **20** (25), 30 (2009)
13. Bing, G.: Due Diligence Techniques and Analysis: Critical Questions for Business Decisions. Quorum Books, Westport (1996)
14. Bell, R.: Basic due-diligence. Environ. Claims J. **13**(4), 123–140 (2001)
15. Guzov, Y.N., Savenkova, N.D.: Due diligence: theory and practice. Audit J. **7**, 56–69 (2015)
16. Guzov, Y.N., Savenkova, N.D.: Due diligence: identification and assessment of investment object risks. Audit Financ. Anal. **2**, 201–206 (2013)
17. Melnik, M.V., Sokolov, V.Y.: The future of audit: meeting with specialists from the international auditing and assurance standards board (IAASB). Audit J. **11**, 62–75 (2015)
18. Kuter, M.I., Sokolov, V.Y.: Russia. In: A Global History of Accounting, Financial Reporting and Public Policy: Eurasia, the Middle EST and Africa, pp. 75–106. Emerald Group Publishing Limited, Sydney (2012)

COVID-19 Impact

Digital View on COVID-19 Impact

Tatiana Antipova(✉)

Institute of Certified Specialists, Perm, Russia
antipovatatianav@gmail.com

Abstract. COVID-19 pandemic has produced such issues as drastic changes in people's way of life and work, housing instability, economic shock, privacy issues and health care crisis. This paper aims to study COVID-19 Impact on worldwide Life by digital approach. Searched some digital libraries and other electronic sources (e.g. ABI/Inform Global, DBLP, Elsevier, IEEE, Nature, Plos, Springer, Taylor & Francis, Wiley) on 'COVID-19 Impact' topic we have found data that were analyzed and systemized. Total confirmed cases, total deaths caused by COVID-19 pandemic were analyzed based on Reports of the World Health Organization. Based on hypothesis that COVID-19 spreading is subject to the law of normal distribution we tried to predict the length of this pandemic. Using Excel function, we predicted some values of COVID-19 spreading for the nearest future.

Keywords: Coronavirus · COVID-19 · 2019-nCov · Coronavirus impact · Lockdown · Impact · Prediction of COVID-19 spreading · Crisis · Pandemic

1 Introduction

The name "coronavirus" was given by researchers because this virus under microscope visually looks like it is surrounded by a crown. This virus has been officially named COVID-19. Also, the virus affecting people now is what's called a novel coronavirus (nCoV) because it's the first time this particular strain is being seen in humans. So, we have two officially abbreviations for Coronavirus: COVID-19 and 2019-nCoV (for laboratory tests) [1].

Since 11.03.2020 COVID-19 was declared as a pandemic by World Health Organization (WHO) [2, report 51, 71] we have had an unthinkable time in human history. The COVID-19 pandemic has created unprecedented changes to all aspects of our life. So, COVID-19 is now recognized as the one of the most tempting challenges and largest tragedy of the century after the Second World War [3] and can be qualified as 'Black Swan' event [1]. Humans are social animals, and we rely on each other for our health and well-being [4]. As such, social disconnect can have serious consequences for our physical and mental health [5]. This is especially true during times of uncertainty and distress when social contact can act as a buffer against adversity and suffering. The COVID-19 pandemic is having an unprecedented impact on people around the world and can be viewed as a global stressor induced by a threat to health, economic consequences, and a disruption of daily routines. Social connections provide us with support when dealing with negative emotions, such as feelings of distress and

T. Antipova (Ed.): ICCS 2020, LNNS 186, pp. 155–164, 2021.
https://doi.org/10.1007/978-3-030-66093-2_15

worry, especially in such times of adversity and uncertainty [6]. However, to limit the spread of the virus, most countries have instituted varying degrees of social distancing measures (and in particular physical distancing measures), some that require large swaths of the population to stay home and restrict physical proximity to others [7]. But COVID-19 does not affect everyone equally. It magnifies pre-existing class differences and reveals a social gap among individuals living in the same country, even in the same city, but experiencing completely different living conditions. In some countries the main issue is the high concentration of people in precarious living conditions, with high demographic density per room and inadequate basic sanitation services [8]. There are signs that governments' authorities are still trying to conceal the true scale of the problem with Coronavirus, but at this point the virus appears to be more contagious than the pathogens behind diseases such as Ebola or SARS—though some experts say SARS and coronavirus are about equally contagious. Events like the coronavirus epidemic, and its predecessors—such as SARS, Ebola and MERS—test Health Care systems and force us to think about the unthinkable. [9] Crisis caused by COVID-19 pandemic is not only an economic crisis, but also – and perhaps more fundamental – a health crisis [10].

2 Method and Methodology

Worldwide spreading of COVID-19 we analyzed based on data of World Health Organization (WHO) mostly. In addition, searched some digital libraries and other electronic sources (e.g. ABI/Inform Global, DBLP, Elsevier, IEEE, Nature, Plos, Springer, Taylor & Francis, Wiley) on 'COVID-19 Impact' topic we have found data and information that were analyzed and systemized.

The previous version of this paper was presented at International Webinar on "Quantum Mathematical Modelling On Covid-19 (QMMC)" organized by the Department of Physics in collaboration with IQAC, Derozio Memorial College, Kolkata, India that held online on 30th September 2020. The author analyzed the questions and comments expressed at the end of this presentation and then corrected the research approach and some insights on COVID-19 distribution. During this work was used systematic approach by digital, visual, and intuitive views on considering data and the information similar to some scientists in [11–16].

Author believe these insights are not only valuable for the ongoing mitigation of the current pandemic, but may also serve to inform governments' and public health organizations' information dissemination and infection control strategies in the future. One of the manifestations COVID-19 Impact on human life have so-called lockdown.

3 COVID-19 Lockdown

As coronavirus cases continue to be reported around the world, WHO says countries still have a chance of containing the outbreak. Officials have also sought to differentiate Covid-19 from other viruses, as part of efforts to quell public panic. And the most of countries apply lockdown in their countries.

What is Coronavirus Lockdown?
While "lockdown" isn't a technical term used by public-health officials, it can refer to anything from mandatory geographic quarantines to non-mandatory recommendations to stay at home, closures of certain types of businesses, or bans on events and gatherings. Within a some period of time, the population was required to largely withdraw from their normal lives and practice physical distancing, while the social-, financial-, and health-related consequences of COVID-19 were becoming rapidly apparent.

This period is characterized by unprecedented physical distancing (often communicated as social distancing) measures resulting in significant changes to people's usual social lives [7].

Lockdown is stay-at-home orders for elderly people aged 65+ and often includes:

- borders closures to travelers in most of countries;
- restricting all movement to only work, and family emergencies;
- shopping for essentials;
- ban on all public gatherings (including cinemas, gyms, weddings, funerals, sports events);
- moving schools & universities education from physical to virtual mode (so-called online education);
- mask wearing and compliance with social distancing between people in public places (usually 1,5 m);
- all large events, museums, cultural centers, swimming pools, and spas have been closed at least in Springer 2020.

How Many People Are on Lockdown?
As of March 2020, the world population is 7.8 billion people. As of May 26, 2020 more, then one third of the world population lives in one form or another under the blockade caused by coronavirus. More than three billion people in almost 70 countries and territories asked to stay at home. Thus, about one third of the world population was blocked in spring 2020.

But as testing is not readily available in many places, this number is likely to be much higher[1], especially after 'second wave' that started in the fall of 2020.

[1] https://www.express.co.uk/news/world/1260709/lockdown-which-countries-are-in-lockdown-how-many-people-coronavirus-cases.

4 Coronavirus Spreading

WHO daily reports incidence of confirmed 2019-nCoV cases on https://www.who.int/emergencies/diseases/novel-coronavirus-2019/situation-reports/. The first Situation Report was dated as of 20.01.2020. It is necessary to note that new WHO Situation Reports remark that "Data as received by WHO from national authorities". It means that there are reported just data provided by authorities. Bearing in mind that "Everybody lies" it means that we can see the picture by wishes of authorities and we cannot check how truly it is because data reflection sometimes depends on political situation/game in each separate country.

In addition, there are many questions about how to calculate the number of confirmed 2019-nCoV cases and deaths from coronavirus. These methods differ from country to country. For example, in Italy, the authorities considered only the number of a severe form of disease (15% of infected people, according to WHO data [2]), but in China - all who received a positive 2019-nCoV test result. In the method of calculation also matters the degree of the population's coverage by testing for coronavirus and antibodies to it. As of the current date, there is no uniform methodology for calculating the number of deaths from coronavirus that all countries would apply equally. So, we can consider just approximately data that represented by national authorities then counted by WHO.

In addition, from a general point of view, the outputs returned by the simulation fit quite well the data reported by the WHO. In particular, they estimated reasonably the date and magnitude of the peaks corresponding to the number of new cases, new deaths and amount of hospitalized people. This indicates that the proposed methodology can be used as a useful decision tool for policy makers. However, considering an estimation performed at an early stage of the epidemic could produce poor results. Focusing on the study of the peaks of some relevant curves of the epidemic (e.g. new exposed people, new cases, new deaths and hospitalized people), results show that the control measures in China were efficient to quickly reduce the magnitude of the epidemic. However, due to the transition rates between the disease compartments, the visible impact for the society has a delay of two weeks after the beginning of those control measures [17].

The author has analyzed bimonthly WHO situation/operational reports to define trend of Coronavirus spreading. As a result of those reports' study, total numbers of registered cases of Coronavirus infection and total deaths are shown on Fig. 1.

To date (Fig. 1), the total number of confirmed Covid-19 cases globally topped over 33 025 642 people, causing about one million (996 342) deaths as of 28.09.2020 and covering more than 230 countries.

Intuitively, we can assume that COVID-19 spreads according to the laws of normal distribution (Gaussian curve), since the behavior of COVID-19 distribution curve is similar to the Gaussian curve (Fig. 2).

Moreover, this assumption has historical roots. The Great Plague of London, lasting from 1665 to 1666, was the last major epidemic of the bubonic plague to occur in England. It happened within the centuries-long Second Pandemic, a period of intermittent bubonic plague epidemics which originated from Central Asia in 1331, the first

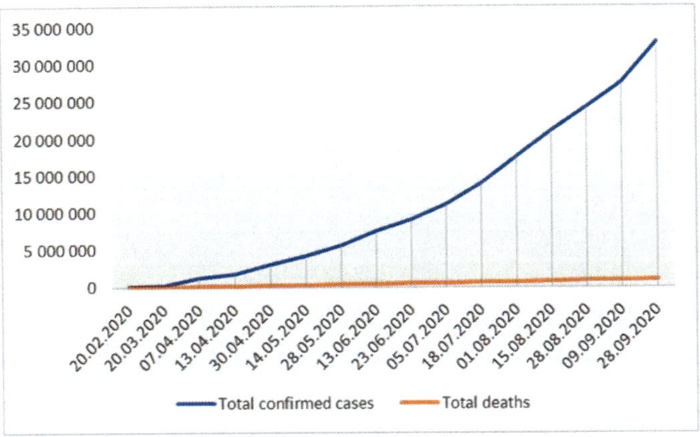

Fig. 1. Total confirmed cases of coronavirus infection causing total deaths. *Source: Author's analysis based on WHO Situation/Operational Reports* [5]: *1–145.*

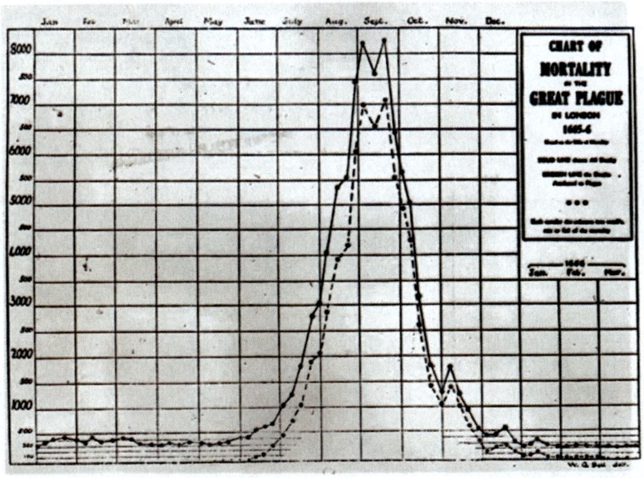

Fig. 2. The plague spreads during the Great Plague of London from 1665–1666. The solid line shows the total deaths and the broken line deaths attributed to plague. Source: https://www.history.com/news/plague-pandemic-great-fire

year of the Black Death, an outbreak which included other forms such as pneumonic plague, and lasted until 1750 [18].

The Great Plague killed an estimated 100,000 people—almost a quarter of London's population—in 18 months, based on the data of plague pandemic (Great Plague in 1665–1666 in London) [19, 20].

As we can see on Fig. 2 this graph looks like Gaussian curve in general (see Fig. 3).

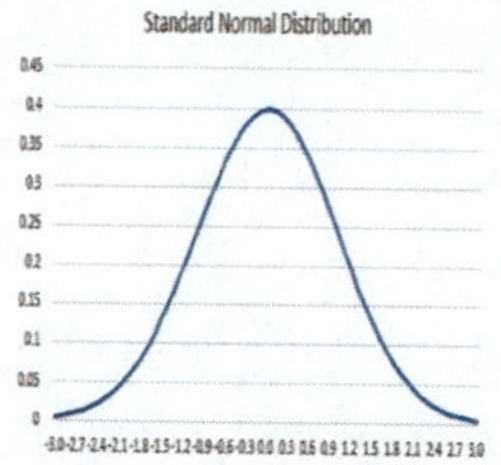

Fig. 3. Gaussian curve.

Also, we can see the similarity between Gaussian and COVID-19 distributions curves if we place them side by side (Fig. 4).

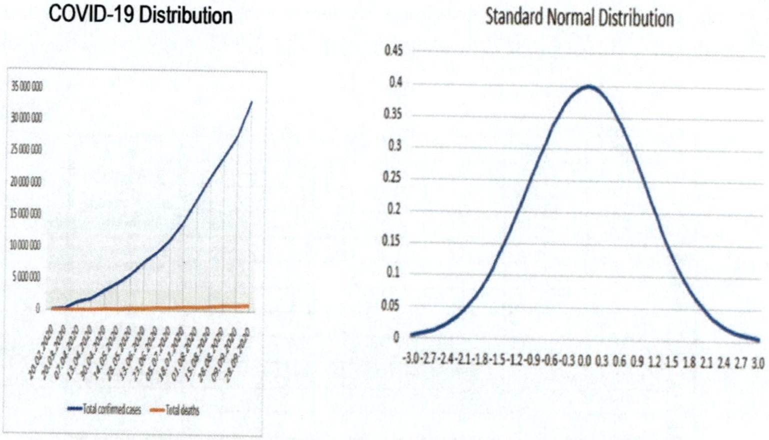

Fig. 4. Comparison COVID-19 distribution with Gauss's curve.

Figure 4 shows visual similarity of COVID-19 distribution and Gaussians curves. Anyway, we can see that curve of total COVID-19 confirmed cases has no downward trend yet. Bearing in mind that in many European countries came "second wave" of COVID-19 we cannot expect close decreasing. Moreover, given that the most favorite holiday of the year is Christmas, we suppose that COVID-19 distribution would increase until the Christmas week at least, when it is customary to meet with the whole family in most countries due to the historical and cultural traditions of most countries.

According to visual representation and similarity of two curves behavior (Fig. 4) we can suppose that the length of COVID-19 spreading takes about 1/3 of whole path and taking into account the well-known symmetry of Gaussian curve about maximum, the extent in time would also take one third. At this moment, the COVID-19 is currently spreading for nine months (January–October) officially. Consequently, we can expect the duration/length of the COVID-19 curve for approximately 27 months. This hypothesis is true if the coronavirus spreads complain with Gaussian law and other conditions. Also, it is important to notice that our analysis and forecasts assumed that the data are accurate. To define when will occur the peak of COVID-19 spread we need more accurate data and take into account many other factors.

To forecast COVID-19 distribution in the nearest future we can try to use Excel function FORECAST.ETS. Having the curve's behavior for COVID-19 distribution based on WHO situation reports data from 20.01.2020 to 28.09.2020 we can continue this type of curve until 20.12.2020 by implementing simple Excel function FORECAST.ETS for prediction. Using Seasonable index 1 we have received following data and curve (Table 1 and Fig. 5). Table 1 shows data from WHO situation reports until 28.09.2020 and then predicted values by Excel function FORECAST.ETS (in red) from 12.10.2020 to 28.12.2020 with the keeping of the trend line after break for forecasting.

Table 1. COVID-19 distribution data (with prediction in red)

Date	Total confirmed cases	Total deaths
20-01-2020	282	3
9-02-2020	37 558	813
23-02-2020	78 811	2 471
3-03-2020	89 435	3 071
13-03-2020	132 758	4 955
20-03-2020	234 073	9 840
31-03-2020	750 890	36 405
13-04-2020	1 773 084	111 652
30-04-2020	3 090 445	217 769
14-05-2020	4 248 389	292 046
28-05-2020	5 593 631	353 334
13-06-2020	7 553 182	423 349
23-06-2020	8 993 659	469 587
5-07-2020	11 125 245	528 204
18-07-2020	13 876 441	593 087
1-08-2020	17 396 943	675 060
15-08-2020	21 026 758	755 786
28-08-2020	24 257 989	827 246
9-09-2020	27 486 960	894 983
28-09-2020	33 025 642	996 342
12-10-2020	35 692 452	1 066 341
27-10-2020	39 422 184	1 143 956
12-11-2020	43 400 563	1 226 745
26-11-2020	46 881 646	1 299 186
12-12-2020	50 860 026	1 381 976
28-12-2020	54 838 406	1 464 766

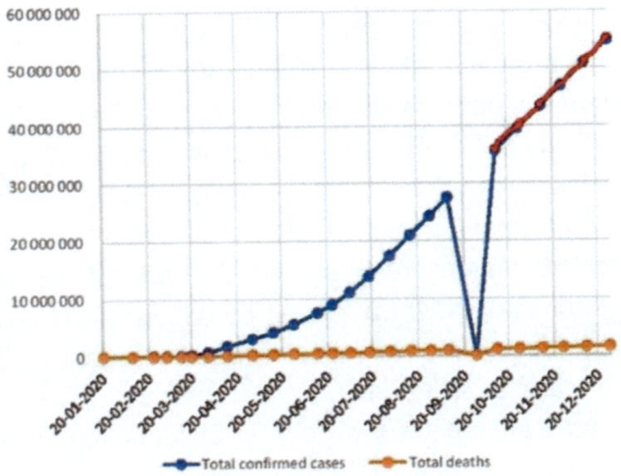

Fig. 5. COVID-19 distribution with prediction (in red)

Month later, after receiving fact values as of 27.10.2020 author compared these with the prediction (Table 2).

Table 2. Comparison of predicted and actual values

	Prediction		Fact	
Date	Total confirmed cases	Total deaths	Total confirmed cases	Total deaths
09.10.2020			36 361 054	1 056 186
12.10.2020	35 692 452	1 066 341		
27.10.2020	39 422 184	1 143 956	43 341 451	1 157 509

The difference in the dates of Table 2 is due to the cancellation of WHO situation reports. Instead of daily situation reports, WHO posts operational reports on its website (who.int) weekly and we compare the closest dates. As you can see from Table 2, the number of deaths is closer to the predicted values than the number of confirmed cases. The accuracy of prediction for total deaths value is 10% as of 27.10.2020.

Surely, to make any conclusion about the accuracy/imprecision of our prediction we need long-term period of observation and more accurate data. Doubts in fair representation of COVID-19 data were evidenced by some questions from above mentioned webinar participants: Participant 1. Many governments are trying to hide and manipulate COVID-19 data. How much damage this tendency is doing for proper analysis? And does this data manipulation harm initiatives that are based on those data?

Participant 2. The COVID-19 spread is increasing day by day but the data of infected persons may not be the actual/truly because the total number of infected persons that are getting must be dependent on the coronavirus testing but how can we define those who are infected but not get tested?

Bearing in mind these questions we can suppose that there are some issues/challenges with COVID-19 data representation. An international organization with broad powers and due diligence can verify the facts of testing on COVID-19 and complying with recognized by the most governments methodology for defining the number of confirmed cases of COVID-19. But this is the topic for future researches.

5 Conclusions

The COVID-19 pandemic has a dramatic Impact that presents an unprecedented challenge to public health, food systems and the world of work. The economic and social disruption caused by the pandemic is devastating: tens of millions of people are at risk of falling into extreme poverty, while the number of undernourished people, currently estimated at nearly 690 million, could increase by up to 132 million by the end of the year. Now is the time for global solidarity and support, especially with the most vulnerable in our societies, particularly in the emerging and developing world. Only together can we overcome the intertwined health and social and economic impacts of the pandemic and prevent its escalation into a protracted humanitarian and food security catastrophe, with the potential loss of already achieved development gains [21].

In this paper, the author considered the signs of a lockdown as one of the main forms of COVID-19 Impact, analyzed the data of WHO about the COVID-19 spread in order to approximately predict some values in the near future. As a result of this work, proposed simplest model by using Excel function FORECAST.ETS can thus be used dynamically to approximately predict the spread of COVID-19 in the nearest future. Author believe these insights are not only valuable for the ongoing mitigation of the current pandemic, but may also serve to inform governments' and public health organizations' information dissemination and infection control strategies in the future.

References

1. Antipova, T.: Coronavirus pandemic as black swan event. In: Antipova, T. (ed.) ICIS 2020. LNNS, vol. 136, pp. 356–366. Springer, Cham (2021). https://doi.org/10.1007/978-3-030-49264-9_32
2. Coronavirus Disease 2019 (COVID-19): Situation report, 1–77. https://www.who.int/docs/default-source/coronaviruse/situation-reports/
3. Gautam, S.: The influence of COVID-19 on air quality in india: a boon or inutile. Bull. Environ. Contam. Toxicol. **104**, 724–726 (2020). https://doi.org/10.1007/s00128-020-02877-y
4. Snyder-Mackler, N., et al.: Social determinants of health and survival in humans and other animals. Science **368**, eaax9553 (2020). https://doi.org/10.1126/science.aax9553
5. Bzdok, D., Dunbar, R.I.M.: The neurobiology of social distance. Trends Cogn. Sci. **24**, 717–733 (2020). https://doi.org/10.1016/j.tics.2020.05.016
6. Holt-Lunstad, J.: Why social relationships are important for physical health: a systems approach to understanding and modifying risk and protection. Ann. Rev. Psychol. **69**, 437–458 (2018). https://doi.org/10.1146/annurev-psych-122216-011902
7. Nitschke, J.P., et al.: Resilience during uncertainty? Greater social connectedness during COVID-19 lockdown is associated with reduced distress and fatigue. Br. J. Health Psychol. (2020). https://doi.org/10.1111/bjhp.12485
8. Cardoso, E.H.S., et al.: Characterizing the impact of social inequality on COVID-19 propagation in developing countries. IEEE Access **8**, 172563–172580 (2020). https://doi.org/10.1109/ACCESS.2020.3024910
9. Briançon, P.: Are markets putting too much hope in ECB's capacity to help europe deal with coronavirus crisis? Barron's (Online) (2020)
10. Budding, T.: The impact of the COVID-19 crisis on public sector accounting – what can be learned from previous crises? CIGAR Newsl. **11**(4), 1–2 (2020)
11. Tsvirko, S.: Informational technologies for the efficiency of public debt management in Russia. In: MosITS 2017, AISC 724, pp. 104–113 (2018). https://doi.org/10.1007/978-3-319-74980-8_10
12. Rozhkova, D., Rozhkova, N., Blinova, U.: Digital universities in Russia: prospects and problems. In: DSIC 2019, AISC 1114, pp. 252–262 (2020). https://doi.org/10.1007/978-3-030-37737-3_23
13. Popov, E., Mingaleva, Z.: The digital technologies for improving the operational efficiency: case of Russian industry of ferrous and non-ferrous metals scrap. In: ICIS 2019, LNNS 78, pp. 351–363 (2020). https://doi.org/10.1007/978-3-030-22493-6_31
14. Voskanyan, Y., Shikina, I., Kidalov, F., Andreeva, O., Makhnovskaya, T.: Impact of macro factors on effectiveness of implementation of medical care safety management system (2021). https://doi.org/10.1007/978-3-030-49264-9_31
15. Antipova, T.: Nobel prize roots in Russia. J. Digit. Art Humanit. **1**(1), 33–41 (2020). https://doi.org/10.33847/2712-8148.1.1_4

16. Konyavsky, V., Ross, G.: New method for digital economy user's protection. In: ICIS 2019, LNNS 78, pp. 221–230 (2020). https://doi.org/10.1007/978-3-030-22493-6_20
17. Ivorra, B., et al.: Mathematical modeling of the spread of the coronavirus disease 2019 (COVID-19) taking into account the undetected infections. The case of China. Commun. Nonlinear Sci. Numer. Simul. **88**, 105303 (2020)
18. Haensch, S., et al.: Distinct clones of yersinia pestis caused the black death. PLOS Pathog. **6** (10), e1001134 (2010). https://doi.org/10.1371/journal.ppat.1001134. PMC 2951374, PMID 20949072
19. The Great Plague of London: Contagion, historical views of diseases and epidemics. Harvard University (1665)
20. DNA in London Grave May Help Solve Mysteries of the Great Plague, 08 September 2016
21. https://www.who.int/news/item/13-10-2020-impact-of-covid-19-on-people's-livelihoods-their-health-and-our-food-systems. Accessed 13 Oct 2020)

University Communication in Times of Covid-19: The Ecuadorian Case

Arturo Clery[1(✉)], Lilian Molina[1], Soraya Linzán[1], Ricardo Zambrano-Maridueña[2], and Ana Córdova[3]

[1] Universidad Estatal Península de Santa Elena, La Libertad, Ecuador
clery@upse.edu.ec

[2] Unidad Educativa Liceo Naval Salinas Capitán Rafael Morán Valverde, Salinas, Ecuador

[3] Doctoranda Universidad de Málaga, Málaga, Spain

Abstract. The COVID-19 outbreak has become a worldwide concern for everyone, and specifically for Higher Education Institutions, their students, teachers and those responsible for university public policies. Social distancing is now the primary consideration for any student when deciding where and how to study, reducing their willingness to educate themselves in physical settings with others who may carry the virus. In this context, this article analyzes how the Covid-19 pandemic affects Ecuadorian universities in the form of a communication and marketing trial, and its influence on university actors. The Covid-19 crisis has put the processes of social structures to the test. Communication and marketing are examples of these processes that face an unprecedented and unforeseen situation. In this sense, it is intended to reflect on how the pandemic affects and will condition the practice of university communication, both internal communication and its link with society. Finally, emphasis is placed on the importance of the universities assuming that reputation is a risk, and that this risk must be managed with the techniques and tools of communication, since the pandemic demonstrated higher education institutions the importance of managing relationships of trust with their teachers and students.

Keywords: Communication · Education · Academy

1 New Realities of University

Four months ago, no one would think of a global standstill on such a scale. The Covid-19 pandemic presumably begins in China, spreading through networks of global human mobility (Sirkeci and Yucesahin 2020), sending shock waves around the world, halting life in almost every countries of the world.

Not only the anxiety and fear of the spread of a deadly virus, but also the measures taken against it, changed the lives of consumers, academics and researchers. Here a new approach for universities emerges, a new standard in development, which includes academic marketing (López and Antropólogo 2020). Airlines, tourism, out-of-home entertainment and the hotel industries, among others, have collapsed under quarantine

T. Antipova (Ed.): ICCS 2020, LNNS 186, pp. 165–174, 2021.
https://doi.org/10.1007/978-3-030-66093-2_16

and isolation measures (Gössling et al. 2020). It is important to recognize the probability that long periods of confinement will lead to behavioral changes (Brouder 2020).

Marketing theory offers some insights into the potential impact on universities of the Covid-19 disruption, such as institutional loyalty, behavior change, sustainability concerns of teachers and students (Gössling et al. 2020). At the same time, universities have had to suddenly switch to digital, becoming a process that tested their resistance. Some of these universities may disappear, while others, more agile and ingenious, succeed in the process.

In this scenario, university actors will develop different interests and habits, although at present the evidence is not yet conclusive (Tucker and Yu 2020). Their return to the use of traditional services is not guaranteed, with social distancing, keeping many people at home, there are important changes in behavioral trends (Balis 2020). So the effect of the pandemic will only be felt in the coming months, being important the creation of new university models, which must emerge from this crisis, regardless of our reactions as teachers or students (Xifra 2020).

The Ecuadorian university must adapt to this new reality, along with the need to take advantage of the pulse of the consumer, their reactions and ideas, to prepare for the future, since their sense of trust and loyalty will be affected by the environment of rapidly changing marketing (López and Anthropologist 2020).

In this sense, the Ecuadorian university communication channels have changed, from an education called face-to-face, to being completely online, through the Internet, with the use of applications such as Zoom, Teams, in addition to social networks and other computer programs (Villón 2019).

This is how we find ourselves, with an academic community that has been working remotely from their homes, since the beginning of March 2020. This is part of the new Ecuadorian university reality. Which leads us to many and varied questions about the ways in which universities should be talking to their students: Where to invest in academic marketing? How to organize work between teams? What are the keys to ensure the permanence of the academic institution? What should be the university position on the sustainability of the planet, teachers, students, companies, communities and families? One thing is certain, that much research will be needed to assess the new reality and develop appropriate strategies to meet the objectives of the universities.

2 Some Ecuadorian University Initiatives

The outbreak of the COVID-19 coronavirus pandemic has become a global concern for universities, faculty, students, and public policy makers worldwide. As of June 15, 2020, more than 7.9 million cases of coronaviruses had been reported globally, corresponding to 46,751 cases reported to the Republic of Ecuador, reaching to this date a number of deaths of 433,472 worldwide, of which 3,896 they were Ecuadorian, according to Johns Hopkins University. To delay the spread of the virus and protect people's health, the World Health Organization and the Centers for Disease Control and Prevention have advised people to practice social distancing by staying home, avoiding social gatherings in crowded places and keeping at least 2 m apart.

The Centers for Disease Control and Prevention suggests limiting face-to-face contact with others is the best way to reduce the spread of coronavirus disease. Social distancing has become a major factor guiding people's decisions about where and how to educate or eat. In response to growing health concerns, many of Ecuador's universities have issued a series of measures that reduce the risk of studying. For example, some have posted signs on their entrances and on their floors to remind their users to keep a safe distance from each other. They have also added online studies and other options to serve students who are concerned about going to physical classrooms (Li and Li 2020).

Ecuadorian universities such as Espírito Santo, in collaboration with the Faculty of Medical Sciences and the School of Psychology, made a medical helpline available to the community, as an alternative to decongest health centers (Hakovirta and Denuwara 2020). The Ecuadorian government enabled telephone line 171, but this service was insufficient given the number of calls and constant complaints from users who were not receiving attention.

In this scenario, the Espiritu Santo Specialties University requested authorization to carry out tests in its laboratories. If a person suspects or experiences any symptoms of COVID-19, they can contact their call center, where more than eighty doctors take calls from citizens from Friday, March 27, 2020. The service is available from 09:00 until 18:00.

The dean of its Faculty of Medical Sciences, commented that the professionals approved a course of the Pan American Health Organization. That is, they can guide patients who feel discomfort from COVID-19 such as dry cough and fever. They attend them by telephone, tell them how to act in the face of the emergency and reinforce the sanitary measures that they must follow (Hakovirta and Denuwara 2020). The institution considers that the quarantine decreed to stop the spread of the coronavirus can cause feelings of stress, frustration or anger. For this reason, this university also provides psychological assistance services for people who require it. The objective is to provide emotional support to those who feel that they are going through episodes of anxiety, depression, burnout syndrome, product of the current situation, etc.

This service has the participation of teachers from the School of Psychology of this university and students from the last year of the degree. It can be accessed by dialing the same number of the institution. The service is from Monday to Sunday between 08:00 and 23:00.

The Salesian Polytechnic University is another of the universities that aims to help Ecuadorian citizens who live in fear and stress during quarantine. The professors of the Psychology degree from the Salesian Polytechnic University on the Guayaquil campus have developed the accompaniment project called "Share what you feel confidential", to provide the university community and the general public with psychological advice during the health crisis that crosses the country, due to COVID-19 (Xifra 2020).

The project seeks to mitigate in the population symptoms such as stress, anguish, fear and uncertainty caused by feeling at risk of permanent and latent death as a consequence of the current scenario that the city and the country are experiencing.

The support received by those interested is completely free and is carried out through the Zoom platform, via WhatsApp, whose link allows you to find the name of the professional, their telephone number and hours of operation.

Most of the professionals who lead the project are clinical psychologists with experience in the field, who can make obvious pathological diagnoses, said the director of the Psychology degree at this study center.

In addition to this field, the Psychology degree also has services such as psychoeducation that involves the rules of mental hygiene, facilitating strategies for the individual to cope with the current situation, he stressed (Cortez-Clavijo 2018).

Another service is coaching, which empowers you to set goals to achieve what you want. In addition, there is spiritual advice from the University Pastoral of the headquarters. During the emergency, this university has also collaborated in the development of masks.

The Computer Science Engineering, Multimedia specialization of the Escuela Superior Politécnica del Litoral in Guayaquil, has created the web application #QuédateEnCasa, in the context of the current health emergency experienced by Ecuador by COVID-19. It is an application that helps Ecuadorian families find medicines, food, health centers and all kinds of services that are useful at the moment (Hakovirta and Denuwara 2020). With this application, people can find services near their homes. In addition, most of the elements that appear on this platform make home deliveries. This application is very useful in these times when many want to stock up on medicine without leaving home.

Noting that many people shared useful information (mainly service directories or the like) in WhatsApp groups, it seemed to the University a good idea to collect all this information for later management, geolocation and validation. In addition to making it available to the entire community (Linzan 2020).

#QuédateEnCasa also has benefits for business owners, since it allows registered users to add, at no cost, service points on the map. It also has an option to search for initiatives and projects that help combat the problems caused by COVID-19.

The working model in which this application is framed is called 'crowdsourcing', and it consists of a massive and open collaboration.

For now there are only localities in Guayaquil, but the application allows you to add services in any city; being able to feed it from any part of Ecuador or the world for their own country. This tool has also been presented to different municipalities and prefectures of the country. The application can be found at: stayteencasa.app and on Twitter: @makubexs.

In addition, a group of volunteers from Espol University are also working on the manufacture of face protection masks, which will be delivered to health professionals who are in charge of patients with coronavirus. These instruments are manufactured through 3D printing and laser cutting. This work is part of the citizen initiative #HackingCOVID19ec that seeks to combat the virus and reduce the risk of contagion in medical personnel. Another group of polytechnic graduates, together with professionals from Latin America and Europe, are creating a low-cost solution that allows patients with severe cases of COVID-19 to access an artificial respirator, whose value is between \$350 and \$400. In the market this can cost from 3,500 to \$100,000.

3 For the Ecuadorian Government Higher Education is not a Priority

Higher education was hit with a new budget cut in May 2020. The Constitutional Court of Ecuador asks to stop it but the government insists on the decrease for public universities and polytechnic schools.

The budget cut to 26 public universities, three postgraduate universities and eight co-financed individuals, rather than a financial exercise, is a declaration of principles: the government is not taking into account the fundamental role that higher education plays in facing the pandemic. Without adequate financing, education cannot have the quality it needs. Without quality, the country is risking its future to, supposedly, overcome the situation (Clery 2020).

In Colombia, Mexico and Chile, public universities are doing research to help fight the spread of the coronavirus. But in Ecuador, the contribution of the academy was taken into account very late in the coronavirus crisis, which was used, rather, as a new pretext to underfund them.

And although the budget cut to public universities was provisionally stopped by order of the Ecuadorian Constitutional Court and a court of the city of Ambato, if it were to be reinstated, it would hit the higher education institutions and the country hard.

There are effects on the well-being of society, the main one is to deepen the ignorance of the role that the academy plays in progress. The cut deepens the deterioration in the quality of higher education, increasing the factory of graduates in the small Ecuadorian labor space (Santa-María 2019), widening the gap between private and public universities, the relevance of provincial universities will almost disappear again (what that destroys the relationship between the university and the environment), with overload of class hours for teachers and almost not without conducting scientific research (Molina 2016).

The 98 million raised by reducing the government to the budget of public universities had already been approved and allocated at the end of 2019. That money, according to the Ecuadorian Higher Education Agency, comes from the collection of the value added tax and the Tax on the rent. With the collection falling due to the economic paralysis caused by the health emergency due to Covid-19, the agency argues, there are no funds and the economic cut was necessary.

The funds withdrawn correspond to the item of expenses destined to the salary payment for teachers, administrative and service personnel. Many institutions said the decline put even more pressure on their operation. When it became known about the suspension of the cut, the Central University of Ecuador reported that it will reopen the registration process on May 18, 2020, which had been canceled on May 5, 2020, as a result of the reduction in funds.

But several experts disagree commenting that the cut is unconstitutional, because according to the Constitution, during a state of emergency, it is prohibited to use the money destined for education and health.

It is not only a matter of constitutional mandate. There are also various technical considerations. A second agency, the Higher Education Quality Assurance agency, had to issue a prior technical report. But neither in the official gazette nor on social networks has the document been published. So the Ministry of Economy and Finance made the decision in a disoriented way (Rey-Martin 2019).

The budget cut will affect the development of scientific knowledge in universities. Doing research is expensive and time consuming. But with the budget cut, the investigation would stagnate.

To the director of the first Agency it does not seem serious. In a television interview, he said that with the new normality due to the covid-19 pandemic, there are investigations that cannot be carried out, because people cannot travel or cannot go out into the field, they are not affected, certain priorities are prioritized. investigations (López and Anthropologist 2020). What we do not share from this research, since without research projects, universities cannot deliver knowledge or well-prepared professionals to society.

Reducing the financing of public universities will have medium and long-term effects and, in addition, it puts us at a disadvantage for the new normality. Ecuador should bet on the academy as one of its post-Covid-19 axes of progress, through innovation and technology. For this, the country needs research centers, the production of scientific articles to achieve intellectual freedom and evidence-based policies (Li and Li 2020).

To avoid budget cuts to universities, student associations submitted proposals to the government, asking to declare a moratorium on foreign debt, or collecting back taxes from 500 companies that appear as debtors to the treasury. The problem is that many of these obligations date back to 2005 and 2006, and are from companies that no longer operate in Ecuador or have no assets or patrimony, to respond to those charges that would represent lengthy and tortuous processes, before the Treasury Service and the judicial system.

The organization also proposes to implement a 2% estate tax on the 265 wealthiest families in the country and another 5% tribute to companies that earned more than a million dollars in 2019. It is difficult to know whether after the economic effects of the pandemic, these companies will have the liquidity for such a tax, if it would not reduce their productive capacity or, in the worst case, would not cause them to close permanently. Some experts have suggested that cuts, if necessary, be redirected and applied to military spending. What is clear is that an effort for public education is indispensable.

4 Public Higher Education in Ecuador at Risk

After the decision of the Constitutional Court to suspend the pre-budgetary cut destined to the payment of payroll of the public universities, the panorama in the Academy remains somewhat uncertain. The worst evil was avoided, however, some of the 33 public universities in the country already faced another cut at the beginning of the year, to which is added a debt of $131 million that the Ministry of Economy and Finance maintains with the Universities and Polytechnic Schools.

Part of the budget that the Government allocates to universities is obtained based on a projection of collection made by the Ministry of Finance and the Tax Agency, made up of 10% of the net collection of the Purchase Tax and 11% of the net collection of the Income Tax. At the end of the fiscal year, if what is actually collected is higher than projected, that balance must be reliquidated to the universities, according to the Law of the Permanent Fund for University and Polytechnic Development.

A technical study prepared by the director of the Latin American Faculty of Social Sciences; and the rector of the Technical University of the North, ensures that there is a loss to the Universities of $131 million for the pending re-liquidations corresponding to the years 2017, 2018 and 2019.

This damage would amount to more, if we consider the cut of about $138 million suffered by certain universities in January 2020. In the case of the Escuela Superior Politécnica del Litoral, this adjustment meant $4,770,065.23 of its budget, and for the National Polytechnic School was over $5 million.

This means that there was already accumulated damage, prior to COVID, and that rather the issue of the pandemic is aggravating the matter, a kind of attack on the Ecuadorian University with continued lack of budget reductions.

Value assigned to the Fund of Ecuadorian Universities versus the corresponding value (in thousands of dollars)

Year	Real collection - income tax	Real collection tax on purchases	Fund allocated	Fund according to actual collection	Difference
2015	4.732	6.274	993.075	1.043	50.428
2016	3.547	5.739	965.033	868.547	(96.486)
2017	3.625	6.319	871.234	925.411	54.171
2018	4.5	6.304	968.652	1.02	51.78
2019	4.31	6.241	968.652	994.273	25.621

Source: Ministry of Finance

In 2016, when there was a drop in tax collection due to the earthquake, the universities were going to be reduced in the budget, but in order not to affect the Academy, the values owed for 2012, 2013 and 2015 were re-settled. The same could It should have been done this year, but it was not (Vall 2017).

Thanks to a constitutional action led by various groups and university students, it was possible to suspend a cut called "criminal" of $98 million, which seriously endangered faculty and administrative staff. But the reality of public universities remains precarious. The percentage of spending on education according to the Gross Domestic Product of Ecuador, is lower than that of other 14 countries in Latin America, below Costa Rica, Argentina, Colombia, Mexico, Brazil, Chile, Peru, among others. In addition, since 2015 the resources allocated to universities have not increased.

We are the only country that cuts education in a crisis in which what is most needed is support so that universities and schools continue to function.

The cuts and the lack of investment in the Academy also threaten the growth of scientific research taking place in the country. For example, if we add up the annual publications of all Ecuadorian public universities, we cannot even exceed the scientific production of the University of Chile.

Dozens of investigations are jeopardized by government cuts. Currently, for example, the Escuela Superior Politécnica del Litoral is investigating how to counter the banana fungus called 'Fusarium Raza 4', a pest that appeared in Colombia and that could soon reach Ecuador, where it would destroy the plantations. They are also developing mechanisms to improve the standards of fine aroma cocoa, the export of which is at risk by failing to meet European standards. In this university, there are countless other research projects in areas of human, animal and plant health; climate change; renewable energy; technological development; big data; artificial intelligence (Gössling et al. 2020). From the beginning of the pandemic, the Biomedicine laboratory was ready to carry out diagnostic tests for COVID-19 and now they are offering them to the public.

From the National Polytechnic School, it is feared that because of the cuts it will end up affecting human talent. They also receive criticism about the number of administrative personnel that the universities have (in the case of Ecuadorian public universities the average is one teacher for each administrative). From the universities it is ensured that although they are called "administrative", many of these workers are laboratory technicians, doctors and specialists, who perform substantial functions in the institution.

In the Ecuadorian public sphere, they are classified as administrative, because there are no other classifications, for example, laboratory technicians, who work in analysis, who work, for example, in the Geophysical Institute doing field research, technicians from Nuclear Sciences departments, from Continuing Education Centers, which provide services and help to work in self-management, staff of student welfare, etc.

Art is not left out of this equation. The financial crisis that the country is experiencing due to the coronavirus pandemic has also generated uncertainty in the authorities of the University of the Arts and its students. The announcement of the cut in April was followed by the departure of the members of the Management Commission, ordered by the Government's Education authorities, among other teachers and administrators.

Their sudden departures, without any notification, mean that within hours of starting the new semester there is an absence of teachers. It is feared that the current crisis will trigger more cuts that will force the closure of this institution created in 2013.

5 University Autonomy in Danger?

The adverse reality of the coronavirus has overwhelmed health systems and devastated the economies of dozens of countries, including Ecuador, which also faces a historic decline in the oil market. The situation is critical for all sectors of the country, however, there are various proposals to avoid affecting the pillars of society: health and education.

Among several alternative measures, the rector of the University of Guayaquil insists that the reduction of the size of the public sector is essential, as well as the adjustment and ordering of remuneration. But it also talks about targeting the subsidy on fuels and guaranteeing foreign and national investment, creating incentives and benefits for new investors who decide to bring fresh resources to invest in the country, with a period of tax grace and establishing new forms of labor contracting.

One of the great challenges of the Ecuadorian Government has been the optimization of public spending, which in the previous administration reached up to 44% of the Gross Domestic Product. Currently the average size of the State measured through total public spending is around 36%, something that for some experts is still oversized. There has been no significant effort in the size of public sector spending, the State is still absolutely oversized and obese, there have been some reductions, but unfortunately they are still very small, spending on what is payroll is still 9,000 million dollars.

One of the assembly members of the Education Commission of the National Assembly, stated that there has been a mismanagement of finances during the health emergency, allocating funds to unnecessary or non-essential items in the current scenario. The legislator revealed a series of contracts in public institutions that could well have been done after the emergency, or whose funds could have been prioritized for health and education.

On March 30, the Ministry of Finance awarded $188,237 in contracting travel services for the issuance of air tickets on national and international routes, that is, in the month that quarantine was decreed in Ecuador.

With this scenario, on April 30, 2020, the Ecuadorian Government authorities unilaterally resolved to cut the amount of 51 public universities, which directly affects the teaching and administrative staff. Just a month earlier, the Secretary of Higher Education had agreed to work the cuts together with each university, to optimize and rationalize spending. But this never happened. The news caught the rectors by surprise the next day.

This decision, taken inconsistently, violates the autonomy of the universities, that essential principle that does not defend a political question, but the academic freedom to teach and research (Li and Li 2020).

The dialogue between the Government and the Academy for decision-making that affects the universities is essential. University autonomy has been reduced because it has been made impossible for the university to decide which items can work with the Minister of Finance herself or with the Higher Education Agency, depriving universities of decision-making capacity (Cevallos 2017).

It was a violation of university autonomy that an entity such as the Ministry of Finance tells a university where it has to cut its economic income, that is, deciding university policy.

This autonomy, which embodies the democratic values of free countries, gives public universities independence in the performance of their functions, guarantees them the right to organize their academic structure and the free administration, disposition and control of their resources.

References

Balis, J.: Brand marketing through the coronavirus crisis. Harvard Business Review (2020). https://hbr.org/2020/04/brandmarketing-through-the-coronavirus-crisis. Accessed 6 June 2020

Brouder, P.: Reset redux: possible evolutionary pathways towards the transformation of tourism in a COVID-19 world. Tour. Geogr. **22**, 1–7 (2020)

Cevallos, D., Molina, L.: La evaluacion de la calidad de las bibliotecas universitarias. Mar Azul: Ecuador (2017). https://issuu.com/marabiertouleam/docs/la_evaluacion_de_la_calidad_de_las_

Clery, A., Rey, C.: Communication of the quality of higher education institutions of Ecuador. In: Developments and Advances in Defense and Security, pp. 487–495. Springer, Singapore (2020)

Cortez-Clavijo, P., Cochea-Panchana, G.: Aprendizaje servicio en el conocimiento de los estudiantes universitarios mediado por TIC. Un enfoque teórico. Revista Científica y Tecnológica UPSE **5**(2), 101–110 (2018)

Gössling, S., Scott, D., Hall, C.M.: Pandemics, tourism and global change: a rapid assessment of COVID-19. J. Sustain. Tour. **29**, 1–20 (2020)

Hakovirta, M., Denuwara, N.: How COVID-19 redefines the concept of sustainability. Sustainability **12**, 3727 (2020). https://doi.org/10.3390/su12093727

Li, K.J., Li, X.: COVID-19 pandemic: social distancing, public policy, and market response. Public Policy, and Market Response (2020)

Linzán, S., Parrales, G.: Information society: access to public information. In: Marketing and Smart Technologies, pp. 350–362. Springer, Singapore (2020)

López, A.E.G., Antropólogo, U.C.V.: La Pandemia en la cotidianidad: El COVID-19 y las nuevas dinámicas globales. In: Covid-19 Caos 2.0: Ensayos desconfinados. Ideas de debate para la post pandemia, pp. 35–50. AnthropiQa (2020)

Molina, L., Vall, A.: La evaluación de las instituciones de Educación Superior. Revista Científica y Tecnológica UPSE **3**(3), 80–89 (2016)

Rey-Martin, C., Molina-Benavides, L.: The quality of communication in higher education institutions of Ecuador. INVESTIGACION BIBLIOTECOLOGICA **33**(81), 13–39 (2019)

Santa-María, G., Clery, A.: La calidad de la comunicación de las Instituciones de Educación Superior del Ecuador. Investigación bibliotecológica **33**(81), 13–39 (2019)

Sirkeci, I.: Marketing and consumers in an era of disruption caused by COVID-19 pandemic. Transnatl. Mark. J. **8**(1), 1–6 (2020). https://doi.org/10.33182/tmj.v8i1.954. ISSN 2041-4684. e-ISSN: 2041-4692

Sirkeci, I., Yucesahin, M.M.: Coronavirus and migration: analysis of human mobility and the spread of Covid-19. Migr. Lett. **17**(2), 379–398 (2020). https://doi.org/10.33182/ml.v17i2.935

Tucker, C.E., Yu, S.: The early effects of coronavirus-related social distancing restrictions on brands. SSRN (2020). http://dx.doi.org/10.2139/ssrn.3566612. Accessed 5 May 2020

Vall, A., Rey, C.: La evaluación de las instituciones de educación superior. Revista Electrónica En Educación Y Pedagogía **1**(1), 43–58 (2017). https://doi.org/10.15658/rev.electron.educ.pedagog17.09010103

Villón, H., Sojos, M., Mendoza, C., Guarda, T.: Pharming y Phishing: Delitos Informáticos Penalizados por la Legislación Ecuatoriana. Revista Ibérica de Sistemas e Tecnologias de Informação, (E17), 671–677 (2019)

Xifra, J.: Comunicación corporativa, relaciones públicas y gestión del riesgo reputacional en tiempos del Covid-19. El profesional de la información **29**(2), e290223 (2020). https://doi.org/10.3145/epi.2020.mar.20

Comparative Analysis of the International Development Banks' Activities During COVID-19 and Beyond

Svetlana Tsvirko(✉)

Financial University Under the Government of the Russian Federation,
Leningradskiy prospekt 49, 125993 Moscow, Russian Federation
s_ts@mail.ru

Abstract. The aim of the study is to carry out comparative analysis of the international development banks' activities in the current situation and to develop forecasts regarding the directions of their further performance, as well as some recommendations. The paper considers and summarizes the approaches of a number of international development banks to participation in anti-crisis measures, connected with COVID-19. The stages in the provision of financial resources for the fight against the coronavirus crisis are highlighted. So, first of all, these financial institutions carried out measures such as rapid response and support programs in the fight against COVID-19 pandemic. Going forward, based on the initial anti-crisis measures, new assistance programs are needed to ensure economic growth and sustainable development. To expand their activities, international development banks have increased the issue of their bonds. Activities of the Eurasian Development Bank (EDB) in the fight against coronavirus, as well as difficulties and problems, including the need for additional capitalization are studied. The practical significance of the study is expressed in the possibility of using the experience of some international development banks such as the Inter-American Development Bank, the Asian Development Bank and others in the context of a pandemic to improve the activities of the EDB. The analysis also provides some information about debt instruments on the international capital markets under the conditions of COVID-19 pandemic.

Keywords: International development banks · World bank · Pandemic · Crisis · Debt · Bonds · Social responsible investments · Sustainable development

1 Introduction

International development banks play an important role in the system of institutional structures that affect the functioning of the world economy [1, p. 12]. The mission of these international financial organizations is that they are instruments of joint action in solving global issues. International development banks work to reduce poverty, achieve shared prosperity and foster sustainable economic growth.

International regional development banks were created in addition to such international financial institutions as the IMF and the organizations that make up the World

T. Antipova (Ed.): ICCS 2020, LNNS 186, pp. 175–185, 2021.
https://doi.org/10.1007/978-3-030-66093-2_17

Bank group - the International Bank for Reconstruction and Development (IBRD), the International Finance Corporation (IFC), the International Development Association (IDA), Multilateral Investment Guarantee Agency (MIGA), International Centre for Settlement of Investment Disputes (ICSID). Many international regional development banks have been created and function like the IBRD.

Much attention is paid to the study of the problems of functioning of international development banks. In this area, the works of the following researchers were published: O. A. Grishina, E. A. Zvonova [1], V. V. Antropov [2], L. A. Badalov [3], G. A. Bunich, Y. A. Rovensky [4], R. Faure, A. Prizzon, A. Rogerson [5], R. M. Nelson [6], etc. At the same time, a number of issues related to the latest trends in the activities of international development banks remain poorly studied and are of scientific interest. So far, an aspect of the transformation of the activities of international development banks under the influence of the pandemic crisis in the world economy has not been sufficiently reflected in specialized publications. The purpose of this study is to analyze the activities of international development banks in the current situation, to develop forecasts regarding the directions of their performance and to formulate some recommendations. The findings can be useful for the participants of the financial markets (issuers, investors, financial authorities).

2 Methodology of Research

The methodological basis of the research are system approach, fundamental provisions of economic theory and the theory of international economic relations, methods of comparative analysis, expert assessments. Statistical data of international economic and financial organizations, as well as publications of foreign and domestic authors were used as sources of information.

3 Overview of the Situation in the Global Economy in 2020

As it was revealed by T. Antipova [7] coronavirus pandemic can be considered as Black Swan event for the world. Global economy is currently experiencing a combination of a severe health crisis, a deep recession in many countries and regions, and unprecedented social consequences for all countries. According to the World Bank's calculations in April 2020, this year a pandemic and a partial shutdown of the economies of various countries will plunge 60 million people into extreme poverty, and as a result, the world's population living on less than $1.90 a day will reach 700 million people, and the global poverty indicator will grow for the first time since 1998. The World Bank representatives emphasize that later these poverty indicators will turn out to be even higher [8]. The COVID-19 pandemic has negatively affected a significant number of sectors of the economy and threatened the existence of a large number of companies around the world.

The situation in many economies around the world is characterized by a significant decline in business activity. The most significant losses in the COVID-19 situation were incurred by service sector [9]. Deterioration is noted in the wholesale and retail trade, the tourism industry and the restaurant business, sectors providing other services. Additional negative factors were the changes in labor markets caused by the pandemic, associated with an increase in underemployment and a decrease in income of the population, and a decrease in the inflow of remittances in some developing countries. As a result, domestic consumer demand declined; there is a "cooling" of credit activity and a general decline in the indicators of enterprises. A significant decrease in production is observed in industrial sectors focused on investment demand (mechanical engineering, production of electronic and electrical devices, etc.). In general, the following factors are negatively affecting the economy: weakening domestic and external demand, measures of social isolation and limited communication between states, uncertainty about the prospects for economic development, deteriorating business sentiment, as well as disruption of value chains.

Current situation with coronavirus infection and quarantine measures carried out by most countries of the world and the widespread significant reduction in economic activity is unprecedented. Accordingly, there is no experience in creating conditions for recovery from a shock of this magnitude. Both developed and developing countries urgently need significant countercyclical funding to sustain economic activity, especially jobs.

International development banks are one of the tools that can be useful in achieving the anti-crisis effect [10]. These institutions can provide significant financial resources, minimize economic downturn, support recovery and facilitate structural transformation. In general, development banks are called upon to overcome market "failures", to finance structural transformations that lead to a more equitable and sustainable economy. They are most likely to operate in areas where the market is partially failing or lacking, such as financing small businesses, promoting innovation, building infrastructure. International development banks generally provide long-term financing for specific projects [2, 4]. Thus, international development banks face challenges in responding to the COVID-19 pandemic.

4 International Development Banks' Activities in the Context of the Coronavirus Crisis

4.1 World Bank Group

The largest and the oldest multilateral development bank is the World Bank with its subdivisions [6]. First of all, the World Bank has launched the set of emergency support operations. Such projects focus on:

- preventing and limiting local transmission, through laboratory equipment, improved surveillance systems, and training of front-line responders;
- goods and services such as gloves, masks, and portable ventilators;
- building or expanding clinical care facilities, such as refurbishing intensive care units or inpatient facilities in hospitals and preparing quarantine facilities;
- building systems for real-time community-based disease surveillance and through proactive, evidence-based citizen engagement;

- strengthening collaboration for research and response to facilitate the development of vaccines, therapeutics and other measures [11].

The amount under a fact track facility is $14.5 billion: IFC is providing $8 billion financing, with close to 300 companies requesting support; MIGA is launching $6.5 billion facility to support private sector.

The funding from IFC is important to micro, small and medium-sized enterprises. The IFC response will: 1) go to client banking institutions, so they can continue to offer trade financing, working-capital support and medium-term financing to private companies impacted by the pandemic; 2) help existing clients in economic sectors directly affected by the pandemic-such as tourism and manufacturing—to continue to pay their bills and employees.

MIGA's $6.5 billion facility redirects MIGA's capacity toward the purchase of urgent medical equipment, providing working capital for small and medium enterprises and supporting governments' short-term funding needs.

In February, 2020 the World Bank has announced that it is planning to provide up to $160 billion over the next 15 months to support measures that will help countries respond to immediate health consequences of the pandemic and bolster economic recovery, including over $50 billion of the IDA resources on grant and highly concessional terms [11].

According to the information from the World Bank, they provide four thematic pillars of their crisis response:

1) support to health interventions for saving lives threatened by the virus;
2) social response for protecting poor and vulnerable people;
3) economic response for saving livelihoods, preserving jobs and ensuring more sustainable business growth and job creation;
4) support for strengthening policies, institutions and investments for resilient, inclusive and sustainable recovery [12].

4.2 Regional Development Banks

The African Development Bank launched emergency assistance in March 2020 to support African countries. Among the most important steps are the following:

1) On March 27, 2020 the African Development Bank raised $3 billion from Fight COVID-19 social bonds with a bond maturity of 3 years and an interest rate of 0.75%, which was the largest social bond issue denominated in dollars ever issued on international capital markets. The proceeds from the 3-year bond issue will help mitigate the negative impact of the pandemic on Africa's livelihoods and economy.
2) On April 2, 2020 the African Development Bank in a joint event with the World Health Organization (WHO), provided $2 million in emergency assistance to strengthen the capacity of member countries to prevent, test and manage COVID-19 cases.
3) On April 8, 2020 the African Development Bank announced a COVID-19 response program that will provide governments and the private sector with up to $10 billion

to fight the disease and mitigate the problems caused by the economic downturn and job losses [13].

As for the Inter-American Development Bank, it is also actively involved in coordinated efforts to combat the COVID-19 outbreak with member countries and other international institutions, and stands ready to help Latin America and the Caribbean cope with the challenges posed by the pandemic. In order to finance companies in the countries of the region affected by the COVID-19 pandemic, the bank issued several bond issues during 2020. In January 2020, 5-year global bonds worth $3 billion were issued; in March - 5-year Sustainable Development bonds worth $2 billion, in April - 3-year Sustainable Development bonds worth $4.25 billion; in June - 5-year global bonds worth $4 billion. The Bank announced up to $12 billion in various programs for countries requesting financial support for disease monitoring, testing and public health services. The Inter-American Development Bank can also work with countries that have unspent loan balances to reallocate resources to efforts to combat the pandemic. In addition, the Inter-American Development Bank is studying the economic impact of the pandemic in its member countries and is considering resource allocation alternatives to help member countries withstand the medium and long-term negative impacts of the COVID-19 pandemic. Similar to other development banks, the Inter-American Development Bank is collaborating with WHO as well as the Pan American Health Organization and will continue to coordinate with its member countries and other multilateral agencies to ensure an appropriate response to the pandemic [14].

The Asian Development Bank has planned $20 billion aid package for its developing member countries to combat the impact of the COVID-19 pandemic [15]. The main activities of the bank are as follows:

1) budgetary support to developing member countries;
2) providing grants for food aid and health programs such as personal protective equipment, test kits and laboratories;
3) securing trade and value chain finance, thus supporting economic growth;
4) implementation of programmes to support the payment of wages for low-paid workers, which helps small firms to continue their activities and prevent job losses;
5) implementation of programmes of social cash payments and other social protection measures aimed at the poor and vulnerable groups of the population and households;
6) conducting economic analysis to help business decision makers plan the economic recovery process more effectively;
7) coordination with international financial organizations, WHO and UN agencies, as well as organizations dealing with development issues, which contributes to ensuring efficiency of joint activities;
8) in addition, the bank remains committed to supporting member countries on their path to sustainable development through projects and programs that go beyond the fight against the COVID-19 pandemic.

The Asian Development Bank has a Rapid Response and Cost Support Program in the fight against the COVID-19 pandemic. For instance, within this Program, an agreement between the Kyrgyz Republic and ADB was ratified (grant in the amount of $25 million and concessional loan of $25 million for a period of 24 years, including 8 years of grace period, with a rate of 1% per year during the grace period and 1.5% for each subsequent year).

The New Development Bank (NDB), established in 2014 by Brazil, Russia, India, China and South Africa to mobilize resources to finance infrastructure and sustainable development projects in the BRICS and other emerging economies, is also involved in measures to mitigate the impact of the COVID-19 outbreak in the BRICS countries. On June 16, 2020, the NBR placed its debut $1.5 billion bond issue on the international capital markets. The funds raised from the issue, as in similar cases with other international development banks, will be used to finance measures to combat COVID-19 and mitigate its negative consequences. By the end of July 2020, the NDB had already completed operations to provide funding for budgetary expenditures to four participating countries to combat COVID-19 and implement economic stimulus measures: China, India, South Africa and Brazil. The total amount of financial support from the NDB aimed at combating COVID-19 reached US $4 billion [16].

Nordic Investment Bank is also among the banks that have increased their bond issuance in 2020. This bank issued COVID-19 Response Bond to fund measures to prevent the social and economic impact of the coronavirus pandemic in the Nordic and Baltic countries. The funds will be directed to the healthcare sector. The total volume of COVID-19 Response Bond is 1 billion euros due in April 2023 [17].

An important sphere of work for international regional development banks is interaction with global financial institutions such as the IMF and the World Bank divisions. Actions by international financial institutions of various levels taken in terms of responding to COVID-19, include co-financing the upcoming 60 World Bank projects. As noted by D. Malpass, President of the World Bank Group, there has been a positive response to mutual calls from the World Bank Group and the IMF to temporarily suspend payments by poor countries to repay official bilateral debt [18]. The difficulty in the work of international regional development banks is due to the fact that an increase in allocated funds, along with a moratorium on debt repayment, will put the international financial institutions themselves in a difficult financial situation. The volatility of payments from middle-income borrowing countries can reduce aid to the poorest countries.

To sum up, since March 2020, international development banks have been providing funding to fight COVID-19. This was originally an emergency funding. It was organized as an operational, large-scale campaign around the world, since representatives of various development banks understood that the fight against COVID-19 must be waged everywhere [8]. Going forward, new support programs will be needed to help developing countries cope with the pandemic and re-focus on economic growth and sustainable development. Unfortunately, the pace of approaching the Sustainable Development Goals set in 2015 even in the pre-pandemic period was insufficient [19]. The COVID-19 crisis is likely to further impede sustainability challenges.

Under current conditions it can be recommended to pay attention to the opportunities that are provided by instruments of sustainable finance. Sustainable finance is defined by the IMF as the incorporation of environmental, social, and governance (ESG) principles into business decisions, economic development and investment strategies [20, p. 81]. Among sustainable finance instruments are the following:

- social bonds (bonds that raise funds for new and existing projects that create positive social outcomes);
- green bonds (specific bonds that are labeled green, with proceeds used for funding new and existing projects with environmental benefit);
- sustainability bonds (bonds with proceeds that are used to finance or refinance a combination of green and social project) [20, p. 86].

Good opportunities are provided by vaccine bonds that are used to fund vaccination programmes in the developing world. The first International Finance Facility for Immunisation was introduced in 2006 with participation of France, the UK and other European countries. Nowadays it's time to rapidly accelerate the availability and predictability of funds for immunization with such instrument as vaccine bonds. Advantages of the social responsible investments for the investors are as follows: alignment with social principles, attractive risk-adjusted returns, portfolio diversification.

As part of the long-term measures of the plan for post-crisis economic recovery, an increase in investments in infrastructure is required. However, such projects often require significant financial resources from development banks. One of the ways to solve the problem of insufficient financial resources of a separate development bank is syndication.

5 The Eurasian Development Bank's Activities Amid the Coronavirus Crisis

The Eurasian Development Bank (EDB) was founded in 2006 and is an international financial organization designed to promote the economic growth of the member states (the Republic of Armenia, the Republic of Belarus, the Republic of Kazakhstan, the Republic of Tajikistan, the Kyrgyz Republic, the Russian Federation), the expansion of economic ties between them and the development of integration [21].

In the context of the pandemic crisis, the EDB has developed a Program of Anti-Crisis Assistance to the member states. Taking into account the fact, that the situation in the respective countries differs, the support measures are also different: the revision of the loan payment schedules, introduction of a moratorium on interest payments, provision of short-term capital. For projects subject to currency risks, the EDB is considering the possibility of providing financing in the national currency. The bank plans to open targeted credit lines in the member countries using the Private-Public Partnership mechanism for the construction and modernization of medical institutions.

The EDB is working on the possibility of subsidizing the interest rate through government programs. Like other international financial institutions, the EDB needs additional liquidity. In this regard, the EDB plans to appeal to the central banks of the member states with a request to equate the issue of bonds denominated in national currencies with government securities. In addition, measures are being prepared that are not directly included in the special assistance programme, but also aimed at combating the coronavirus. For example, together with the Russian Export Center, the EDB is planning preferential lending for the supply of Russian-made products that are in demand to combat COVID-19.

The size and structure of the capital of banks, including development banks, are the main instrument for financing projects. The total paid-in capital of the EDB is $1.515 billion; other regional development banks are significantly ahead of the EDB in terms of capital. As it is noted by L.A. Badalov, with this amount of the EDB's capital, it is difficult to expect significant results from this bank in the field of international cooperation and integration [3, p. 14]. In the current conditions, it is advisable to increase the EDB's equity capital. There are opportunities for additional capitalization not in US dollars, but in the national currencies of the member countries. Another option for additional capitalization can be carried out with payment in gold [3, p. 15].

6 Conclusion

The revitalization of the work of international development banks is considered in the paper as a necessary condition for solving problems that have arisen in various sectors of the economy in the context of the coronavirus crisis. According to comparative analysis of the international development banks' activities in the context of COVID-19, we came to the conclusion that the volume of financial resources from them, forwarded to fight with pandemic, varies significantly: from $14.5 billion as fast track facilities and $160 billion in the next 15 months from the World Bank to significantly smaller amounts from some newly founded regional development banks. The amounts depend on comparative advantages of the financial institutions (their own capital, ability to mobilize and leverage financing, credit ratings, overall reputation and experience). International development banks acted rather similarly under the conditions of COVID-19 pandemic. As short-term and emergency measures, international regional development banks have implemented measures such as rapid response and support programs in the fight against the COVID-19 pandemic. Going forward, based on the initial anti-crisis measures, new assistance programmes are needed to ensure economic growth and sustainable development. International development banks should prioritize their work and plan both short-term (emergency) and long-term activities [22]. Transformation of the activities of international development banks in time of COVID-19 and beyond is presented in Table 1.

Table 1. Transformation of the International Development Banks' Activities in Time of COVID-19 and Beyond

Period	1st stage: short-term, emergency measures	2nd stage: medium- and long-term measures
Directions of crediting and investing	Preventing and limiting local transmission; building or expanding clinical care facilities, purchase of urgent medical equipment; supporting governments' short-term funding needs; urgent social response for protecting poor and vulnerable people; support for small and medium enterprises	Resilient, inclusive and sustainable recovery; economic growth, sustainable development, digitalization
Financial instruments and mechanisms	Grants; credits (including credits with highly concessional terms); trade financing, working-capital support; temporary suspension of debt payments	Credits for long-term projects; sustainable finance (sustainability bonds, social bonds, green bonds, vaccine bonds); Private-Public Partnership; co-investment in project finance, syndication; restructuring of the debts

In terms of bond issuance, international development banks play different roles, they can: 1) themselves act as issuers of obligations; 2) organize and help issue the bonds needed to raise funds for anti-crisis measures. Additionally to traditional, conventional bonds, it is advisable to use new types of obligations: sustainability bonds, green bonds, social bonds, vaccine bonds.

Due to the significant needs for financial resources international development banks will continue to increase the issue of their bonds. The level of demand for these obligations depends on the situation in the global financial market, the interest of investors in the strategies of the relevant international development banks in the field of sustainable development and support of the corporate sector in the fight against coronavirus. From the point of view of opportunity of attracting additional financial resources by borrowers, credit ratings assigned to them by international agencies play an important role.

It can be recommended for the international development banks to pay attention to the opportunities of ESG investments. As for necessary actions from the side of the policy makers, regulators should support the development of ESG instruments and corresponding markets by providing some incentives.

Measures taken by international development banks should be aimed at protecting the economies of member countries, rationally allocating financial resources, and effectively countering the negative consequences of the coronavirus crisis, both in the short and long term.

Acknowledgements. The article was written based on the results of the research carried out at the expense of budget funds, which were provided to the Financial University as part of the state contract.

References

1. Grishina, O.A., Zvonova, E.A.: Regulation of the world financial market: theory, practice, tools. INFRA-M, Moscow (2016). (in Russian)
2. Antropov, V.V.: Multilateral development banks in the world economy: features of activity and prospects for cooperation with Russia. Econ. Taxes Law **12**(1), 98–109 (2019). (in Russian)
3. Badalov, L.A.: Incentives for the development of the EDB as a regional financial institution. Bank. Serv. **5**, 12–17 (2020). (in Russian)
4. Bunich, G.A., Rovensky, Y.A.: Regional development banks: modern state and development prospects. Plekhanov Sci. Bull. **1**(15), 96–103 (2019). (in Russian)
5. Faure, R., Prizzon A., Rogerson A.: Multilateral development banks. A short guide (2015). https://www.odi.org/sites/odi.org.uk/files/resource-documents/10650.pdf. Accessed 25 June 2020
6. Nelson, R.M.: Multilateral development banks: overview and issues for congress. Updated February 11, 2020. https://fas.org/sgp/crs/row/R41170.pdf. Accessed 25 June 2020
7. Antipova, T.: Coronavirus pandemic as black swan event. In: Antipova, T. (ed.) Integrated Science in Digital Age 2020. Lecture Notes in Networks and Systems, vol. 136, pp. 356–366. Springer, Cham (2020). https://doi.org/10.1007/978-3-030-49264-9
8. World Bank Group President David Malpass: Remarks at High-Level Event on Financing for Development in the Era of COVID-19 and Beyond. https://www.worldbank.org/en/news/speech/2020/05/28/world-bank-group-president-david-malpass-remarks-at-high-level-event-on-financing-for-development-in-the-era-of-covid-19-and-beyond. Accessed 25 June 2020
9. EDB Macroreview: Outlook update/ April 2020. https://eabr.org/upload/iblock/881/EABR_Macroview_04_2020_EN_web_2.pdf. Accessed 25 June 2020
10. Griffith-Jones, S., Marodon, R., Ocampo, J.A.: It's time to mobilise development banks to fight coronavirus COVID-19. https://www.project-syndicate.org/commentary/mobilizing-development-banks-to-fight-covid19-by-stephany-griffith-jones-et-al-2020-04?barrier=accesspaylog. Accessed 25 June 2020
11. How the World Bank Group is helping countries with COVID-19 (coronavirus). https://www.worldbank.org/en/news/factsheet/2020/02/11/how-the-world-bank-group-is-helping-countries-with-covid-19-coronavirus. Accessed 19 June 2020
12. Saving Lives, Scaling-up Impact and Getting Back on Track. World Bank Group COVID-19 Crisis Response Approach Paper, June 2020. http://documents1.worldbank.org/curated/en/136631594937150795/pdf/World-Bank-Group-COVID-19-Crisis-Response-Approach-Paper-Saving-Lives-Scaling-up-Impact-and-Getting-Back-on-Track.pdf. Accessed 25 July 2020
13. African Development Bank COVID-19 response: moving from commitment to action. https://www.afdb.org/en/news-and-events/african-development-bank-covid-19-response-moving-commitment-action-36188. Accessed 25 June 2020
14. IDB launches $4 Billion 0.625% 5-year Fixed Rate Global. https://www.iadb.org/en/news/idb-launches-4-billion-0625-5-year-fixed-rate-global. Accessed 25 June 2020
15. ADB's Response to COVID-19: on the side of our developing members in their time of need. https://www.adb.org/news/videos/adbs-response-covid-19-side-our-developing-members. Accessed 25 June 2020
16. NDB approves USD 1 billion COVID-19 Emergency program loan to Brazil. https://www.ndb.int/press_release/ndb-approves-usd-1-billion-covid-19-emergency-program-loan-brazil/. Accessed 23 July 2020
17. NIB issues inaugural Response Bond. https://www.nib.int/who_we_are/news_and_media/news_press_releases/3476/nib_issues_inaugural_response_bond. Accessed 23 July 2020

18. Malpass, D.: April 3, 2020 end of week update: Forward progress. https://blogs.worldbank.org/voices/april-3-2020-end-week-update-forward-progress. Accessed 25 June 2020
19. Special edition: progress towards the Sustainable Development Goals. https://undocs.org/E/2019/68. Accessed 25 June 2020
20. Global Financial Stability Report: Lower for Longer. International Monetary Fund. October 2019. https://www.imf.org/en/Publications/GFSR/Issues/2019/10/01/global-financial-stability-report-october-2019. Accessed 15 June 2020
21. https://eabr.org/en/about/. Accessed 25 June 2020
22. Tsvirko, S.: Activities of international regional development banks in the context of the coronavirus crisis. Sci. Rev.: Theory Pract. (8) (2020). (in Russian)

A Preliminary Investigation into the Role of Virtual Sport Training Technology as Emotional Coping Mechanism During a National Pandemic Lockdown

Sunet Eybers[1(✉)] and Aurona Gerber[1,2]

[1] University of Pretoria, Private Bag X20, Hatfield, 0028 Pretoria, South Africa
sunet.eybers@up.ac.za
[2] Centre for AI Research (CAIR), Pretoria, South Africa

Abstract. The COVID-19 pandemic and the subsequent response by governments to introduce national lockdown regulations have confined individuals to their residential premises. As a result, no recreational or sport activities are allowed outside the (often small) boundaries of family homes, a situation often rapidly introducing social isolation. Research has proven that emotional coping mechanisms, such as sport, can lower the stressful and uncertainty burden on individuals. However, without the availability of this coping mechanism, many individuals have been forced to use virtual sport training technology to keep active. This preliminary quantitative study investigated the role of technology, in particular virtual sport training technology (if any) by cyclists as emotional coping mechanism during a period of national lockdown.

The results of an online survey indicated that sport, in general, has always been an emotional coping mechanism during normal challenging situations but that slightly more respondents used sport as mechanism during the lockdown period. Respondents indicated that virtual cycling training technology enabled them to continue with using their normal coping mechanism even in a period of national lockdown.

One of the benefits of a virtual training environment is the ability to socialize by riding with virtual team members. Surprisingly, the number of cyclists who preferred riding alone in the virtual cycling environment was slightly more than the cyclists who preferred to join scheduled rides with virtual team members.

The research is the first step towards an in-depth investigation into the adoption of technology as an emotional coping mechanism in stressful environments.

Keywords: Virtual sport training · Emotional coping · Pandemic

1 Introduction

The global COVID-19 pandemic has forced many countries to implement national lockdown regulations as part of their national disaster response programs in an attempt to curb the spread of the virus. An example is the lockdown regulations specified in the Disaster Management Act, 2002, Section 27(2) in South Africa [1]. In South Africa

T. Antipova (Ed.): ICCS 2020, LNNS 186, pp. 186–194, 2021.
https://doi.org/10.1007/978-3-030-66093-2_18

these regulations broadly restrict the movement of people and goods, prohibit public transport as well as the manufacturing and supply of goods and services for the initial period of 21 days extended by another 14 days. Failure to adhere to these regulations were punishable by 6 month imprisonment or a fine [1]. The restriction placed on personal movement is of particular interest during the national lockdown period and specifies, in South Africa, that no person is allowed to leave his/her place of residence for any other purpose than: "*performing an essential service, obtaining an essential good or service, collecting a social grant, or seeking emergency, life-saving, or chronic medical attention*" [2]. Furthermore, movement between geographical provinces or district areas were prohibited; no gatherings were allowed with special permission required to conduct funerals; and no shopping malls were allowed to trade. The implication of these restrictions were that no person is allowed to engage in any form of physical recreational activity, including sport, outside the boundaries of his/her own residence [2].

Physical movement restrictions enforced on individuals have globally introduced major uncertainties due to the inability to perform daily activities such as going to work (to earn a living) and engaging in social and sport activities. Not disregarding the financial uncertainties, the lack of social contact can introduce personal feelings of anxiety, depression, and stress as a result of loneliness and/or social isolation due to the restrictions [3]. Loneliness refer to the feeling of being alone, a negative feeling due to the discrepancy between actual and desired actual contact [4]. Clinical psychologists distinguish between reactive loneliness (experienced after life changing events such as the passing of a partner), or chronic loneliness (the loneliness becomes permanent due to the lack of mental, emotional or financial resources to satisfy social interaction needs). It should be noted that loneliness is not restricted to physical isolation but the feeling of personal dissatisfaction of social connectedness. It is not uncommon for individuals surrounded by people to experience feelings of loneliness. Research has suggested that these feelings can have a detrimental impact on individual physical and mental health and even increase the risk of premature mortality [5].

Social isolation refers to the lack of regular contact and involvement with fellow human beings, either for the purpose of meeting social or work related objectives [6]. Due to the social nature of human beings, research has, not surprisingly, indicated that social isolated people were more vulnerable and experienced more challenges in coping with stressful situations than social connectedly people [7].

Although a correlation between social isolation and loneliness exist, the two words are not synonyms. The main difference between loneliness and social isolation is that loneliness is often perceived as temporary and involuntary contact with other people, whilst social isolation, also often temporary in nature, refers to the complete lack of contact between an individual and society. Social isolation can lead to feelings of loneliness. On the other hand, social connectedness refers to the feeling of being valued through interpersonal relationships, in this particular instance using virtual sport training technology.

This study focusses on social isolation, the result of physical movement restrictions introduced as part of South Africa's attempt to limit the effects of the COVID-19 virus outbreak. Humans are social beings and without social interaction, consequences such as stress, anxiety and depression have been identified [8]. There are many stress coping mechanism humans adopt in dealing with stressful situations which include task-oriented, emotion-oriented and avoidance-oriented [9]. Since no tasks can be directed to lessen the impact of a national pandemic (task-oriented) and the situation can not be avoided (avoidance-oriented), this study consider emotion-oriented styles (for example, meditation, relaxation, distractions) to coping with the current status quo of the national lockdown. The research question under investigation is *did virtual sport training technology assisted cyclists to cope emotionally during the national pandemic lock-down?* The outline of the article is as follow – stress coping mechanisms are identified, followed by the adoption of virtual technology in various industries and subsequent for the purpose of sport and recreation. The research design is described, followed by the data analysis, discussion and a conclusion section.

2 Stress Coping Mechanisms

The term coping refers to the ability of an individual to consciously address any problem (such as personal or interpersonal problems) in order to reduce the problem into a smaller, manageable, tolerable input [10–14]. Successful coping strategies (also referred to as constructive strategies) can reduce individual stress levels whilst some strategies will increase stress (referred to as maladaptive). Reactive coping is a direct result of a stressor whilst proactive coping focus on releasing future stressors. The effectiveness of the coping strategy adopted by the individual depend on the type of stress (acute, episodic acute or chronic), the individual (personality type including habits) and circumstances (social environment) [8].

No consensus exist amongst scholars when the classification of stress coping mechanisms are presented [8]. The classification of coping strategies proved to be challenging due to the ability of individuals to adopt mechanisms from more than one category; combine coping mechanisms across categories; and the variation of the combination of mechanisms over time and context. Furthermore, Folkman and Moskowitz [15] attributed the variation in classification due to the rationale behind the classification and the underlying empirical techniques used (for example factor analysis).

Although the objective of this research is not to present a psychological founded study on the stress coping classification categories, some of the most agreed classification categories are discussed. Broad classification coping mechanisms identified in literature include adaptive cognitive (appraisal-focused) [13], problem-focused (also referred to as adaptive behavioral) [8, 10, 13, 16, 17], emotion-focused [8, 10, 13, 16–18], support seeking, in some instances classified as part of emotion-focused [8, 10, 16], meaning-making coping (Folkman and Lazarus, 1986, 1980), occupation-focused [13] and avoidance [8, 17, 18]. Adaptive cognitive coping mechanisms refer to strategies adopted as a result of the way individuals think, for example denial or the alteration of values and goals to cope with the stressor [13]. Problem focused coping mechanisms refer to addressing the cause of the stressor or input by means of applying tasks to either address the stressor or lesser the potential impact [8]. Emotion focused coping mechanisms refer to applying internal personal mechanisms such as emotions as response to the stressor. These responses can be positive, for example relaxation and sport, whilst some can be negative crying etc. Some authors [10, 16] classify support-seeking mechanisms as a stand-alone category, whilst others classify the mechanism as part of the emotion focused category. It should also be noted that problem and emotion focused strategies are often seen as complimentary [8]. Meaning-making focused mechanisms refers to the utilization of individual beliefs and values in order to re-classify life priorities [8, 10, 16]. Occupation-focused coping mechanisms, as introduced by Weiten and Lloyd (2008) refers to the changing of occupations in an attempt to deal with stressors. For example, an accountant might decide to opt for a lecturer job with less stress.

The most prevalent coping mechanism, and the focus of this study, is the emotional coping mechanism employed by respondents in an attempt to deal with extraordinary circumstances during the national lockdown period.

Many scholars have investigated the adoption of stress coping mechanisms to deal with stressful events. For example Garber [19] found that active coping, acceptance and planning were used by pharmaceutical students during stressful times, whilst venting and self-blame contributed to higher stress levels. The study found that the adoption of an exercise regime contributed to lower stress levels.

Under normal circumstances, the adoption of exercise as coping mechanism has been identified as the 8th most important stress coping mechanism in Canada. The national survey conducted during 2014 was completed by more than 35,000 respondents. The results of the study indicated that participants who exercised were able to adopt more positive coping strategies than their counterparts. Therefore, the study suggested that exercise should be promoted in stressful environments and situations as possible coping mechanism [20].

3 Virtual Environments

Virtual environments refers to the re-creation of an existing physical environment using digital platforms [21]. These constructed digital environments allow users to interact with the environment and fellow virtually present people. Literature make an important distinction between virtual reality (environments), augmented reality and mixed reality [21]. Virtual reality (VR), probably the most well-known of the three concepts, allow users to physically interact in a virtual world through the use of hardware that interacts with specific software. The user actively engages in the environment and can apply their senses to actually perform tasks, therefore being fully immersive. For example, in military training, soldiers are prepared for combat by physically engage in sessions. Augmented reality (AR) uses technology to create a digital overlay overall an existing, real world. As a result, augmented reality is not fully immersive as in the instance of virtual reality. Augmented reality (MR) use, for example glasses or smartphones, to engage in the environment. A popular example of an augmented reality environment is Pokémon GO. Mixed reality provides a blend of virtual reality and augmented reality. User can physically interact, in real-time, with virtual objects placed in a virtual environment. The objects will interact with the user, similar to objects in a real-world example.

Virtual environments, irrespective of the application of VR, AR or MR, is no novelty and has been extensively used in many industries, including medical (online consultations), education (virtual classrooms), gaming/video, medicine (medical procedure simulation) and advertising [22].

3.1 Virtual Environments in Sport

In sport, such as cycling, virtual environments are often used whereby the rider's actual performance (power and speed) is transferred into a virtually created world. Through the selection of pre-programmed workouts and routes, the rider, represented by an avatar, is able to ride specific predefined virtual routes. The software calculates speed and power depending on the rider's weight. For example, the profile of the environment, such as a steep uphill, will interact with the physical hardware trainer (smart trainer), adjusting the resistance accordingly to simulate an uphill. This will force the rider to increase pedal output (measured in watts) that will be shown on screen together with fellow riders in the virtual environment. For riders without a smart trainer, a non-smart trainer and power meter combination also allow for riding in a virtual environment but without the automatic adjustment of resistance according to the terrain.

There are many software applications available to online cyclists that typically offer monthly subscriptions. Examples include Zwift, TrainerRoad and Sufferfest (to name few). Below an example of the Zwift training environment (Fig. 1).

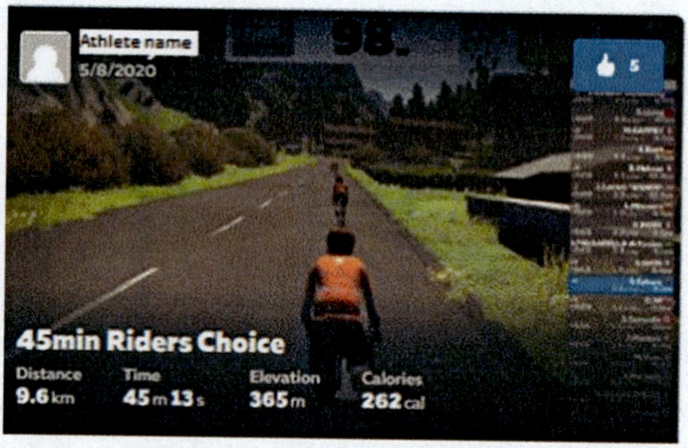

Fig. 1. Zwift cycling rider screen

The utilization of virtual reality technology in a sport training environment is a vibrant research topic [23] with varying objectives - from the improvement of physical motor skills (for example ball sports) [23, 24] to the psychological exposure of athletes to simulated environments to deal with competitive anxiety and stress during competitions (for example during a shooting range competitions) [24]. Although there is strong evidence of the successful application of virtual environments in these scenarios, researcher was not conclusive on the effect of, for example, a virtual audience on how an athlete adjusted behavior [24]. Also, [24] highlighted that the challenge often lies in reproducing an environment of "high pressure" [24], for example the pressure to score a goal in the last seconds of a football game.

4 Research Design

An online survey was developed that contain 10 questions mainly focusing on the utilization of virtual training technology by cycling enthusiasts during the national lockdown period. The survey was tested by three independent parties to ensure construct validity and reliability. After the testing phase adjustments were made to potential ambiguous questions. The target population was cyclists, irrespective of the age, gender or objective of their participation in the sport (recreational, serious or professional). The survey link was distributed by means of WhatsApp messages to cycling groups and it was therefore mostly completed on smart mobile devices. A total of 38 respondents started the survey whilst 29 completed the survey (a 76% completion rate) in an average time of 4 min.

5 Discussion of Results

The majority of the participants were residents of South Africa (97.4%) with a small minority (2.4%) based in the United Kingdom. Sixty one percent (61%) of respondents were male, whilst thirty nine percent (39%) were female. The majority of respondents belonged to an older generation with 43% of respondents older than forty five years, 25% older than 55 and 4% older than 64 years of age. Twenty nine percent were younger than 44 years of age (35–55 years 14%; 25–34 years 11% and 18–24 years 3%).

The study was completed by recreational cycles who ride more than three times a week. No professional cyclists completed the survey. The national lockdown had a profound impact on the respondents as the 79% of respondents prefer to cycle outdoors. These cyclists are now constrained to their own residential properties.

The survey allowed for respondents to complete the questions even if they do not have access to smart virtual trainer setup. The majority of respondents indicated that they have access (46%) whilst 36% used non-smart training equipment (for example cycling rollers). 18% of respondents didn't have access to either smart virtual trainers or non-smart training equipment. Although the overwhelming response was that respondents experienced tension or stress as a result of the current national lockdown situation (62%), surprisingly 31% didn't experience any tension, whilst 8% wasn't sure.

Not surprisingly, given the target population of the study, the majority of respondents indicated that they use any form of sport as emotional coping mechanism during normal stressful situations. During the national lockdown period, slightly more respondents indicated that they use sport as coping mechanism as this value increased from 26% to 28%. Given the values, a strong correlation of 0.803 indicated that cyclists use sport as coping mechanism under normal stressful conditions, and that the national lockdown situation didn't change that. Virtual cycling, however, did enable them to continue with their 'normal' coping mechanisms. Given the high overall correlation between normal coping mechanisms and COVID-19 coping mechanisms the individual coping mechanisms also indicated a strong correlation. A perfect correlation of 1 between the adoption of 'distancing themselves from the problem' as identified as normal coping mechanism and coping mechanism during COVID-19. Table 1 below display a summary of the individual coping mechanism correlations.

Table 1. Summary of coping mechanism correlation values

#	Coping mechanism	Correlation Value
1	Address the cause of the stressor	0.803
2	Learn new skills to address the cause of the stressor	0.818
3	Take control of the problem causing the stress	0.818
4	Distract myself	0.892
5	Relaxation techniques such as meditation	0.914
6	Avoid the cause of the stress	0.947
7	Seek social support	0.979
8	See the cause of the stressor in a positive light	0.978
9	Accept with no action	0.995
10	Distance myself from it	1
11	Use any form of sport to ease the stress	0.803

The standard deviation for the various survey questions are displayed in Table 2. In general the standard deviation is relatively low, except for question number 6, 7 and 8. This indicates the many different training software application used by cyclists (question 6) and the variety of coping mechanism adopted by participants (under normal and extra-ordinary circumstances, question 7 and 8).

Table 2. Summary of standard deviation per survey question

#	Survey question	Standard deviation
1	What cycling environment setup do you have at home?	1.10
2	How would you classify yourself as a cyclist?	0.74
3	Under normal circumstances, i.e. during times prior to the national lockout, how many times did you ride outdoors?	1.03
4	Do you prefer riding indoors or do you prefer riding outside?	0.88
5	If you have a smart trainer, do you ride alone or do you prefer to meet up with virtual team members?	1.05
6	Which virtual training software application do you prefer?	6.06
7	How do you normally cope with extra-ordinary or stressful situations? (you may choose more than 1)	3.71
8	How do you cope with the national COVID-19 lockdown situation? (you may choose more than 1)	3.59
9	If you have access to a smart cycling training environment, does the ability to ride in a virtual training environment with fellow team members assist you in dealing with the current situation during national COVID-19 lockdown?	1.45
10	Do you experience tension and/or stress as a result of the current COVID-19 lockdown?	0.65

The findings can be summarized as follow:

- Although the majority of respondents experienced tension or stress during the national lock down period, a very high percentage of respondents did not.
- Sport has always been an emotional coping mechanism, but this has slightly increased during the national lock down period.
- Virtual training technologies allowed cyclists to continue adopting normal coping mechanisms even during a period of national lockdown.
- Other emotional coping mechanism used during the lockdown period were the use of distractions, although no further information was available on the type of distraction adopted.
- In normal situations, even with a high local crime rate, the majority of respondents preferred cycling outdoors.
- The option to ride with fellow virtual team members wasn't preferred and used by respondents to the extent as one would expect.

6 Conclusion

The objective of the preliminary study was to investigate the role of technology, in particular virtual sport training technology, as emotional coping mechanism by cyclists during a global pandemic. Although the majority of results were not surprising and in line with limited literature published on the topic, what was surprising was that participants preferred to ride alone opting out of the opportunity to cycle with fellow riders. Also, virtual cycling technology allowed participants to continue adopting normal coping mechanisms even in a time where movement was restricted. It will be interesting, and recommended, that this study is expanded to include a bigger population, possibly participants from around the globe.

References

1. Disaster Management Act (2002)
2. Disaster Management Act: Amendment of Regulations (2020)
3. Zaharieva, R.: The dangers of social isolation during a pandemic. https://epha.org/the-dangers-of-social-isolation-during-a-pandemic/. Accessed 01 May 2020
4. Perlman, D., Peplau, L.A.: Toward a social psychology of loneliness. In: Personal Relationships in Disorder, pp. 31–54. Academic Press, London (1981)
5. Holwerda, T.J., Tilburg, T.G., Deeg, D.J., Schutter, N.: Impact of loneliness and depression on mortality: results from the longitudinal ageing study Amsterdam. Br. J. Psychiatry **209**, 127–134 (2016). https://doi.org/10.1192/bjp.bp.115.168005
6. Schrempft, S., Jackowska, M., Hamer, M., Steptoe, A.: Associations between social isolation, loneliness, and objective physical activity in older men and women (2019)
7. Robinson, S.: Isolation has profound effects on the human body and brain. Here's what happens. https://www.sciencealert.com/isolation-has-profound-effects-on-the-human-body-and-brain-here-s-what-happens

8. Carver, C.S., Connor-Smith, J.: Personality and coping. Annu. Rev. Psychol. **61**, 679–704 (2010)
9. Endler, N.S.: Stress, anxiety and coping: the multidimensional interaction model. Can. Psychol. Can. **38**, 136–153 (1997). https://doi.org/10.1037/0708-5591.38.3.136
10. Folkman, S., Lazarus, R.S.: Appraisal, coping, health status and psychological symptoms. J. Pers. Soc. Psychol. **50**, 517–579 (1986)
11. Cummings, E.M., Greene, A.L., Karraker, K.H.: Life-span development psychology: perspectives on stress and coping. Lawrence Erlbaum Associates, Inc., New York (1991)
12. Snyder, C.R.: Coping: The Psychlogy of What Works. Oxford University Press, Oxford (1999)
13. Weiten, W., Lloyd, M.A.: Psychology Applied to Modern Life. Wadsworth Cengage Learning, Boston (2008)
14. Zeidner, M., Endler, N.S.: Handbook of Coping: Theory, Research. Wiley, Hoboken (1996)
15. Folkman, S., Moskowitz, J.T.: Positive affect and the other side of coping. Am. Psychol. **55**, 647–654 (2000). https://doi.org/10.1037/0003-066X.55.6.647
16. Folkman, S., Lazarus, R.S.: An analysis of coping in a middle-age community sample. J. Health Soc. Behav. **120**, 219–239 (1980)
17. Higgins, J.E., Endler, N.S.: Coping, life stress, and psychological and somatic distress. Eur. J. Pers. **9**, 2530270 (1995)
18. Billings, A.G., Moos, R.H.: The role of coping responses and social resources in attenuating the stress of life events. J. Behav. Med. **4**, 139–157 (1981). https://doi.org/10.1007/bf00844267
19. Garber, M.C.: Exercise as a stress coping mechanism in a pharmacy student population. Am. J. Pharm. Educ. **81**, 50 (2017)
20. Cairney, J., Kwan, M., Veldhuizen, S., Faulkner, F.: Who uses exercise as a coping strategy for stress? Results from a national survey of Canadians. J. Phys. Act. Health. **11**, 908–916. https://journals.humankinetics.com/view/journals/jpah/11/5/article-p908.xml
21. Tokareva, J.: The difference between virtual reality, augmented reality and mixed reality. https://www.forbes.com/sites/quora/2018/02/02/the-difference-between-virtual-reality-augmented-reality-and-mixed-reality/#6ac262d02d07. Accessed 24 June 2020
22. Martín-Gutiérrez, J., Mora, C.E., Añorbe-Díaz, B., González-Marrero, A.: Virtual technologies trends in education. EURASIA J. Math. Sci. Technol. Educ. **13** (2017). https://doi.org/10.12973/eurasia.2017.00626a
23. Miles, H.C., Pop, S.R., Watt, S.J., Lawrence, G.P., John, N.W.: A review of virtual environments for training in ball sports. Comput. Graph. **36**, 714–726 (2012)
24. Sanz, F.A., Multon, F., Lecuyer, A.: A methodology for introducing competitive anxiety and pressure in VR sports training. Front. Robot. AI **2**, 1–11 (2015). https://doi.org/10.3389/frobt.2015/00010

Financial Market in the Period of the Pandemic: Trends and Forecasts

Irina Korostelkina(✉), Natalia Varaksa, Mikhail Korostelkin, and Marina Vasilyeva

Orel State University, Naugorskoe Highway, 40, Orel 302020, Russian Federation
cakyra_04@mail.ru

Abstract. The modern world at the end of 2019 faced an unprecedented shock-the coronavirus pandemic, which affected all spheres of society and the world economy. The global financial market, which has been significantly affected by the pandemic, is no exception. Many scientists and financiers around the world compare the current situation with the great depression and the consequences of World War II. The world is just beginning to calculate the real losses from the coronavirus epidemic. Most national economies forecast a severe recession in 2020, with global losses of about 3%. With a decrease in macroeconomic indicators, the stock market during the active phase of the pandemic shows growth. The specificity of stock markets is that they are not a mirror image of the state of the economy: index peaks usually occur just before the start of a recession, and index growth is usually an advance process that begins before economic growth begins. The article raises topical issues of COVID-19's influence on global financial markets. The aim of this article is to review the dynamics of the main stock market indices, determine the reasons for the deterioration of economic development with the application of the world's best management practices, and evaluate the measures of financial regulators for the growth of the financial market. The methodological tools of this research are General and particular research methods, methods and tools for graphical interpretation, comparative analysis, and related changes. The article gives a theoretical overview of the research of the problem, retrospective analysis of the impact of a pandemic on financial markets, examines the major stock market indices and forecasts for the global economy, the monetary actions of Central banks to stabilize the economic situation.

Keywords: Financial market · Pandemic · Global economy · Volatility · RGBi index · RTSi index · MICEX index · S&P500 index · Ruble exchange rate · Dollar exchange rate

JEL Classification: E44 · E69 · F37 · G10

T. Antipova (Ed.): ICCS 2020, LNNS 186, pp. 195–204, 2021.
https://doi.org/10.1007/978-3-030-66093-2_19

1 Introduction

The financial market as a system of economic relations that deals with the accumulation and distribution of monetary assets in the world practice operates on the basis of two main models: continental (based on Bank financing) and Anglo-American (based on a set of institutional investors). The 21st century is a century of scientific and technological progress and digital technologies, expanding country borders and globalizing the economy [1], so the convergence of financial market models determines their rapid development.

As a result of gradual convergence and interpenetration, the world's socio-economic and geopolitical processes are in a new phase of development. The shortcomings of the current financial system significantly affect the state of the world economy, which is characterized by the spread of global crisis trends [2].

The current global economic crisis is connected with the growing COVID-19 coronavirus pandemic, which presents a new historical challenge. This situation is not new for the world economy. Historically, there are several pandemics that affect the recession and the development of financial markets.

The first pandemic that caused significant damage to humanity was the Spanish flu pandemic (1918–1919), which caused damage both to society as a whole (the number of deaths according to official data was 29.5% of the world's population) and to the world economy (a decrease in GDP by 6.6%). During this period, a large-scale recession began in developed countries, which was extinguished by the development of exchange trading in the United States. However, the massive investment of free money in stocks of large companies led to the «inflating» of the stock bubble, which burst in 1929, which marked the beginning of the great depression [3]. However, the main Dow Jones stock market index was not affected by this situation.

It is not possible to fully assess the impact of the swine flu pandemic (1976 and 2009–2010) on the financial market due to the global financial crisis of that time. As for the economy as a whole, it lost about 1.5% of GDP in that period. In modern world history, an outbreak of SARS (SARS epidemic) was recorded in China (Guangdong province) in 2002–2003. However, the spread of this virus on the planet did not lead to serious economic damage (according to the world Bank, the damage from SARS amounted to about 59 billion rubles). [3], and the financial market was affected by this epidemic sporadically (oil prices fell by 31%, and then went up again). In 2014, the next epidemic was recorded – the Ebola epidemic, which also affected the world economy (damage amounted to $ 32.5 billion), but did not affect the stability of stock markets and the dynamics of major indexes.

The COVID-2019 coronavirus pandemic (beginning December 2019) is the first epidemic in the world's history that has caused significant damage not only to the economy, but also significantly affected financial markets. In February 2020, there was a collapse in the Chinese stock index SSE Composite index more than on 8%. At the same time, all Western stock markets rose and opened with a daily historical high. When the epidemic spread across countries (Italy, Spain, Germany, the United States) at the end of February 2020, the quotes of the leading indexes of almost all countries collapsed. Accordingly, we can say that the scale of the COVID-19 pandemic for the first time in history has significantly affected not only the world economy, but also financial markets. Experts predict a dropin GDP by more than 10% [4].

The modern economy creates new challenges and threats. Scientists and economists around the world are trying to solve the problem of ensuring financial stability by creating an optimal set of financial market tools. Therefore, the study of trends in the development of financial markets, the influence of objective (unmanageable) factors on them, as well as the assessment of the activities of state institutions is important and necessary at the present stage of economic development.

2 Materials and Methods

The methodological basis of the study was a set of General scientific approaches and methods, in particular, when writing the article, obtaining results and formulating conclusions, methods of comparison, analysis, comparison and classification were used. The research provisions are also reasoned using a systematic approach and private scientific methods. In the process of work, we used tools of graphical interpretation, comparative analysis, and the method of accompanying changes [5]. The study of the main indexes was based on the graphical method of technical analysis, comparison and deduction. The study of volatility indicators was conducted using the statistical method of volatility analysis. The identification of monetary policy measures is justified through the use of empirical and expert methods.

3 Theoretical Review

Considering the theoretical basis of the research, we note that for many years, domestic and foreign scientists have raised issues of adaptation of financial markets to new challenges and threats, analyzed structural and functional changes in the financial market as a whole and its individual segments. Global trends and financial crises have always had an impact on the financial market, which has been the subject of discussions among scientists. Such scientists as L. Grigoriev, M. Salikhov [6], M. F. Oztek, N. Ocal [7], K. A. Ilyina [8], B. S. Bernanke [9], and others have studied the emergence and development of financial crises. The result was a classic separation of financial crises into two rough types: occurs during the cyclic economic development and the tense international situation or the spread of the financial crisis in specific countries. At the same time, each of these types of financial crisis, although it poses a threat to the financial market, but their appearance and development can be corrected using the tools of state financial policy.

The COVID-2019 coronavirus pandemic is a new challenge for states as a whole and for financial markets. The severity of the problem is justified by the sudden unpredictable or poorly predictable appearance of a threat that cannot be attributed to any type of financial crisis. This circumstance contributed to a surge in the publication activity of scientists who studied the current situation in the financial markets and the direction of its development. So, L. M. Grigoriev, V. A. Pavlushina, E. E. Muzychenko [10] note that usually the initial stage of the crisis determines the variant of the subsequent development of events, but the coronovirus pandemic is a unique starting phenomenon of the world recession that has begun. Each country simultaneously found itself in this situation, but the way out of it will be different for everyone, both in the time interval and in the consequences. In this regard, the research of Christoph Stückelberger [11] on global cooperation and governance is noteworthy, while he focuses on the balance between international cooperation and national independence of the country. G. V. Vorontsova and D. I. Karlov [12] note the long period of countries return to the pre-crisis state, and they see the prospects for the development of the global financial system and financial markets in increasing their stability and financial globalization in the new conditions after the pandemic.

Each country has its own characteristics of financial markets development, which are reflected in the research of such scientists as A. D. Komarov [13], N. A. Mashkina, O. V. Aseev, N. B. Chernykh [14], E. A. Barinov [15], D. S. Salaev [16], B. Zeliba, F. Rahal [17] and others. In their works, scientists analyze measures to prevent the consequences of the crisis that has begun, assess losses by segments of the financial market, and characterize the methods and tools used for regulation. Cockroaches [18] studies the behavior of the main stock indices of the US and the Russian MICEX index and finds that increased volatility indexes as a result of instability in the financial markets. I. G. Gorlovskaya [19], exploring the tools of financial market regulation, priority is given to instruments of deterrence and prevention as the most promising in the management of financial risk. D. Fielding, J. Rewilak [20] consider the impact of the level of volatility and financial instability on the probability of banking crises and, as a result, the stability of the functioning of a particular segment of the financial market. M. F. Oztek, N. Ocal [7] describe the nature of the correlation between commodity and stock markets.

Despite the available research by scientists, further analysis of trends in the development of financial markets in the context of the ongoing phased situation of countries emerging from the coronavirus pandemic and assessing the consequences for the economy as a whole is required. In addition, the contribution of implemented support measures at the national and international levels to the stabilization and development of the financial market is not sufficiently evaluated, which requires the development of new effective tools aimed at increasing the stability of the financial market in the new conditions.

4 Results

The COVID-2019 coronavirus pandemic is a sudden stressful situation that leads to the development of an adverse cumulative effect on various elements of the financial market and the financial security of the state. The rapid spread of COVID-19 infection around the world has affected the revision of global economic growth forecasts. Thus, according to forecasts of the International monetary Fund (IMF), the decline in global GDP in 2020 will be almost 5% and will be the deepest since the great depression. Moreover, the IMF updated its forecast in June compared to April 2020. And it is less optimistic, since the COVID-19 pandemic has had a more negative impact on economic activity than previously thought. In 2021, global growth is projected at 5.4%. The same forecast indicates a possible decline in Russia's GDP by 6.6% this year [21].

Recovery in countries past the peak is expected to be slower than previously predicted, with an emphasis on providing assistance through international cooperation. A striking example in this case is the situation in Italy – a country where the economic situation has deteriorated sharply as a result of the spread of the pandemic. Italy, which has a high level of public debt and unemployment, did not receive support from EU countries, and support was provided by countries that did not expect it – China and Cuba. Further gradual development of countries requires more stringent measures to limit the spread of coronavirus and support economic recovery. The international monetary Fund has developed a baseline and alternative scenarios that will be implemented in the event of a re-outbreak of the pandemic in 2021 and affect key measures to provide financial support to countries in need.

Following the impact on the real sector of the world economy, the global epidemic led to an increase in risks in financial markets, whose participants reacted to the introduction of restrictive measures by outflow of funds from funds that invest in securities, which, in turn, led to a sharp drop in stock and bond prices. For example, the s&P500 broad market index lost 35% of its value from February 19 to March 23, 2020 (Fig. 1).

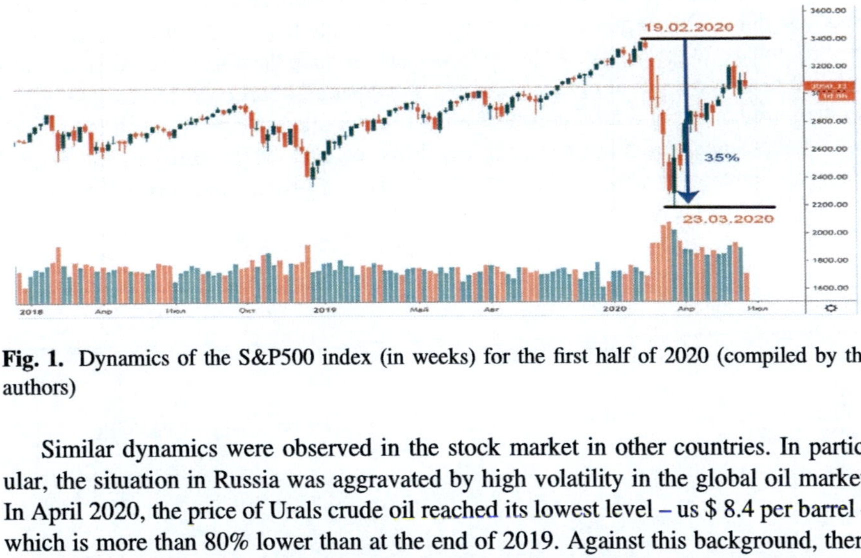

Fig. 1. Dynamics of the S&P500 index (in weeks) for the first half of 2020 (compiled by the authors)

Similar dynamics were observed in the stock market in other countries. In particular, the situation in Russia was aggravated by high volatility in the global oil market. In April 2020, the price of Urals crude oil reached its lowest level – us $ 8.4 per barrel – which is more than 80% lower than at the end of 2019. Against this background, there was an «explosion» of volatility in the domestic stock market, which is clearly confirmed by the graph of the IMOEX index itself (the Mosbirzhy index) and the graph of its volatility (the lower area) (Fig. 2).

The chart shows that volatility increased sharply in March 2020 and surpassed the short-term April surge in 2018, when sanctions were imposed on RUSAL.

Due to fears of a further collapse of stock markets, global investors reduced their investments in Russian assets in order to reduce risks. This was accompanied by the withdrawal of capital at the end of February-March 2020 and the purchase of foreign

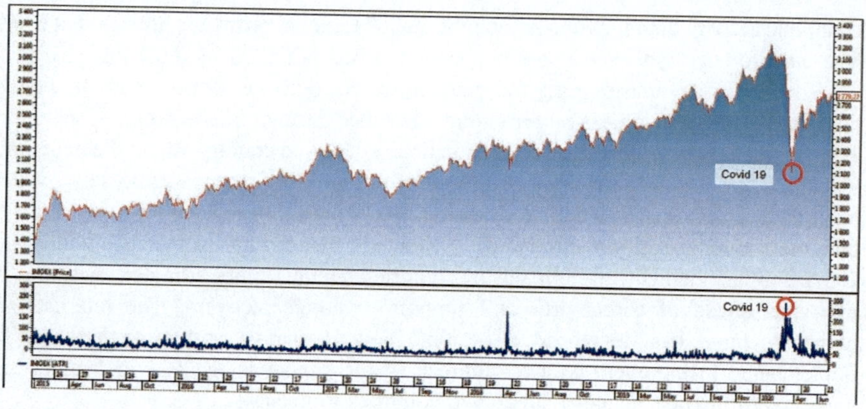

Fig. 2. Dynamics of the IMOEX index (Moscow Exchange index) and the level of its volatility over the past five years in days (compiled by the authors)

currency on the domestic market. In this regard, there was a sharp fall in the ruble exchange rate against the currencies of developed countries.

The loss of the ruble to the us dollar (March 2020) reached about 18% compared to 2019. However, it is noticeable that the fall was significantly less than in previous shock periods. This is also evidenced by the relatively low volatility of the exchange rate, compared to the peak of 2014. Here we can conclude that the elasticity of the ruble relative to the price of oil in recent years has decreased, it has become possible thanks to the introduction of a mechanism of «fiscal rules», the essence of which is financed with funds from the national welfare Fund lost oil and gas revenues of the Federal budget, if any, are formed in the case of reduction of prices on oil below benchmark (42,4 USD. per barrel) [22].

In this regard, it is worth paying attention to another main index of the Russian stock market-RTSi, which is the Mosbirzhy index, but expressed in USD (Fig. 3).

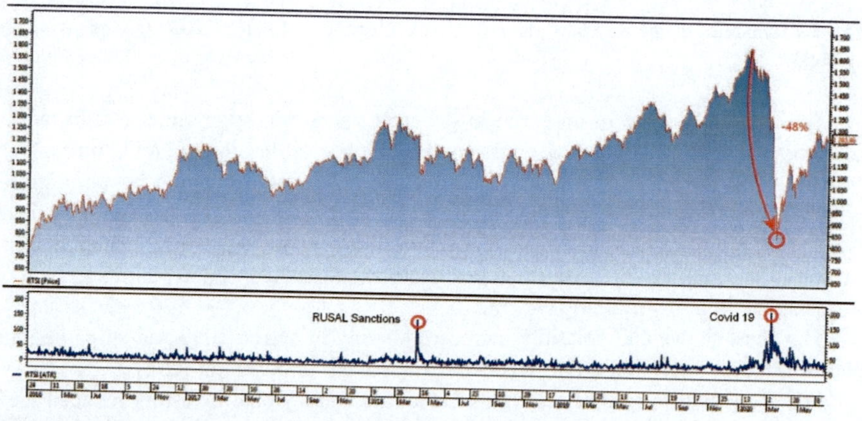

Fig. 3. Dynamics of the RTSi index (the Moscow Exchange index in USD) and the level of its volatility in days (compiled by the authors)

The fall of The Russian stock market index-RTSi was 48% from the top of the beginning of 2020. In comparison with the ruble index (Fig. 2), this is an order of magnitude higher, since this index was influenced by the ruble exchange rate.

In addition to the decline in the main industry indices and the weakening of the ruble, the closing of positions by foreign funds on the Russian stock market affected the growth of bond yields. Figure 4 shows the dynamics of the Russian government bond index (RGBI). This index reflects the dynamics of changes in the market price for a basket of OFZs (Federal loan bonds) with a fixed coupon.

Here, you can also see a sharp increase in profitability and a surge in volatility (March 2020). However, when compared with the shock period of 2014, these fluctuations were insignificant and short-lived. This was made possible, in part due to the actions of the Russian Ministry of Finance, which suspended OFZ auctions.

As the epidemiological situation in the world and in Russia recovered and measures were taken to support the economy, foreign investors gradually began to return to the Russian financial market, and OFZ yields returned to the levels observed before the period of increased volatility. Thanks to the ultra-soft monetary policy of the world's Central banks, the situation on the world financial markets has stabilized. «A soft» monetary policy supports lending and ensures stable dynamics of interest expenses in the banking sector. Regulatory easing and reduction of macro prudential allowances will allow banks to gradually cover losses and ensure the continuity of their operations» [23].

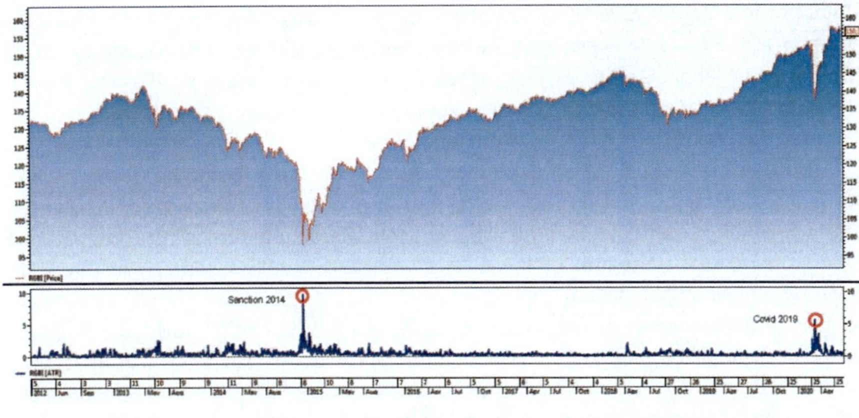

Fig. 4. Dynamics of the RGBi index and its level of volatility in days (compiled by the authors)

The current pandemic, the rapid spread of COVID-19 in the world has forced States to take large-scale measures to support the financial market and ensure a sufficient level of liquidity. Financial regulators of the world powers had to adapt their policies in the field of regulation and supervision. In particular, the following measures were taken:

- mitigating regulations (relief in respect of capital buffers, reducing counter-cyclical allowances, the establishment of preferential crediting);

- regulatory easing measures (waiving liability in case of violations, credit restructuring);
- easing of Supervisory requirements (cancellation of on-site inspections, extension of reporting deadlines, introduction of restrictions on short sales in the securities market);
- taking targeted measures to support the corporate sector (launching and expanding special refinancing programs, government support) [24]. According to the International monetary Fund, the volume of support measures in % of the country's GDP is: Japan – 21.1%, the United States – 14.3%, the European Union – 4.3%, China – 2.5%. Russia is no exception. In particular, the Bank of Russia, in the context of the coronavirus pandemic and the sharp decline in oil prices, has taken additional measures to provide the banking system with liquidity. Along with a significant reduction in the key rate from 6% in March to 4.5% in June 2020, the Bank of Russia is implementing measures to support:
- lending to small and medium-sized businesses, allowing them to maintain access to Bank lending and avoid insolvency in the face of reduced income as a result of the pandemic. The essence of measures in this area is reduced to reducing interest rates on loans and debt restructuring;
- mortgage lending, in the form of reducing premiums to risk factors, reducing the amount of the initial payment, including a number of mortgage bonds in the Pawnshop list [25].

In addition, the Bank of Russia has reduced regulatory and Supervisory burdens on financial organizations, in particular, increasing the reporting deadlines, reducing the number of non-urgent requests, and postponing verification deadlines. Coordination of measures at the level of international forums and organizations – the Group of 20, the IMF, the financial stability Board, as well as standard-setting bodies (BCBS, the international organization of securities commissions, the international Association of insurance supervisors) is of great importance [24].

The impact of COVID-19 on the financial system of the Russian Federation is huge and is realized mainly through the deterioration of foreign economic conditions, primarily the fall in demand and oil prices. Multilateral cross-country cooperation is essential to reduce the depth of the COVID-19 shock and its damage to the global economy and financial system.

The future dynamics of trends in the financial markets of the world and Russia in particular depends on how the situation with the COVID-19 pandemic will develop. In the event of new waves of infection, the ongoing prolonged economic downturn will lead to large-scale negative consequences. States will have to continue to increase their budget expenditures and increase their debt burden – this will lead to increased risks in the area of public finances. In this regard, financial markets may experience a more significant surge in volatility, which will negatively affect the banking sector.

5 Conclusion

In modern history, economic crises and pandemics have occurred simultaneously many times, but usually the financial market has not been affected as much as the impact of COVID-19. As a rule, financial markets slightly decreased their activity, and epidemics affected only the economy as a whole. The influence of COVID-19 was an exception. At the beginning of the pandemic, there was a collapse of financial markets and a decline in indices, but since the financial market does not mirror the economic situation, there is already a stable growth in the midst of the pandemic. The financial market usually reacts in advance to an unstable situation. The COVID-19 pandemic had a significant negative impact on the financial system, following its impact on the real sector, which required urgent measures to support the financial market and ensure sufficient liquidity, as well as adaptation of the regulatory and Supervisory policies of the world's financial regulators. Due to the easing of Supervisory requirements, regulatory instruments and the adoption of targeted support measures, a favorable trend has emerged in the development of many countries of the world and Russia.

By the summer of 2020, the situation in Russia had stabilized, which was facilitated by the Bank of Russia's measures to ensure the liquidity of the banking system. The improvement of the epidemiological situation in the country contributed to the return of foreign investors to the Russian market. OFZ yields have stabilized and reflect the situation before the period of increased volatility. Stability in the oil market also helped improve the situation.

It is very difficult to predict the occurrence of sudden emergencies similar to the coronavirus pandemic and assess its consequences, so the financial market must have a high level of stability to ensure security. In order to reduce the impact of COVID-19 and the damage to the world economy, global cooperation of countries is required. The situation with the coronavirus pandemic is not over yet and the dynamics of trends in the financial markets of the world and Russia depend on its further development.

References

1. Basnukaev, M.Sh., Popova, L.V., Korostelkina, I.A., Dedkova, E.G., Bisultanov, A.N.: Research on opportunities and development drivers for the digital economy. In: Conference «The European proceedings of social & behavioural sciences EPSBS» SCTCGM 2018 – Social and Cultural Transformations in the Context of Modern Globalism (Grozny, 2018, November 01–03). Conference Chair(s): Bataev Dena Karim-Sultanovich – Doctor of Engineering Sciences, professor, director of the Complex Scientific Research Institute n. a. H.I. Ibragimov of the Russian Academy of Sciences. – Grozny: Published by the Future Academy, pp. 193–200 (2018)
2. Pshenichnaya, V.P., Grebennikova, M.Yu.: Trends in the development of the world financial market. Bull. Ural Inst. Econ. Manag. Law **3**, 4–10 (2016)
3. Coronavirus - the first pandemic that brought down stock markets. https://dev.by/news/istoriya-vliyaniya-epidemii-i-pandemii-na-fondovye-rynki
4. Klepach, A.N.: Consequences of the pandemic for the Russian economy. https://zen.yandex.ru/media/id/5e274bc843863f00acd7ed97/posledstviia-pandemii-dlia-rossiiskoi-ekonomiki-5eda140901587e1eb3b15711

5. Korostelkina, I., Vasilyeva, M., Popova, L., Korostelkin, M.: Strengthening the state's economic security in the tax sphere: problems and prospects of the Russian Federation. In: International Conference on Integrated Science in Digital Age ICIS 2020: Integrated Science in Digital Age 2020, pp. 221–232 (2020)
6. Grigoriev, L., Salikhov, M.: Financial crisis-2008: entering a global recession. Econ. Issue **12**, 27–45 (2008)
7. Oztek, M.F., Ocal, N.: Financial crisis and the nature of correlation between commodity and stock markets. Int. Rev. Econ. Finan. **48**, 56–68 (2017)
8. Ilyina, K.A.: Analysis of the risk of financial crisis in Russia. ECO **3**, 146–154 (2015)
9. Bernanke, B.S.: Causes of the recent financial and economic crisis. Statement before the Financial Crisis Inquiry Commission, Washington, 2 (2010). https://www.federalreserve.gov/newsevents/testimony/bernanke20100902a.htm
10. Grigoriev, L.M., Pavlushina, V.A., Muzychenko, E.E.: Fall into global recession in 2020. Econ. Issue **5**, 5–24 (2020)
11. Stückelberger, C.: Post-corona world: balancing international cooperation and national sovereignty. J. Law Adm. **16**, 10–17 (2020)
12. Vorontsova, G.V., Karlov, D.I.: Prospects for the development of the world financial system in modern conditions. Fundam. Res. **5**, 45–53 (2020)
13. Komarov, A.D.: Financial markets of Russia and prospects for their development in modern conditions. Econ. Bus. Cans **1**, 185–193 (2018)
14. Mashkina, N.A., Aseev, O.V., Chernykh, N.B.: International financial markets. Features of the Russian financial market. Bull. Kursk State Agric. Acad. **9**, 272–279 (2018)
15. Barinov, E.A.: Coronavirus: impact on the economy and financial markets. Entrep. Guide **2**, 111–118 (2020)
16. Salaev, D.S.: Investments in 2020 and measures to prevent threats of uncertainty. ISJ Theor. Appl. Sci. **4**, 828–832 (2020)
17. Zheliba, B., Rahal, F.: Financial results of stock market of the Gulf State cooperation: a comparative study. Econ. Manag. Innov. **2**, 69–76 (2019)
18. Tarakanov, S.I.: Increased volatility of stock markets against the background of instability of financial markets. Entrep. Guide **44**, 193–201 (2019)
19. Gorlovskaya, I.G.: Development of tools for regulating professional activity in the securities market in the conditions of mega regulation of Russian financial markets. Bull. Omsk Univ. Ser. «Econ.», **4**, 23–33 (2018)
20. Fielding, D., Rewilak, J.: Credit booms, financial fragility and banking crises. Econ. Lett. **136**, 233–236 (2015)
21. Bulletin of the prospects for the development of the world economy of the International Monetary Fund, June 2020. https://www.imf.org/~/media/Files/Publications/WEO/2020/Update/June/Russian/WEORUS202006.ashx?la=ru
22. Information message of the Ministry of Finance of the Russian Federation «On the mechanism of the budget rule when oil prices fall below the base level», 9 March 2020. https://www.minfin.ru/ru/press-center/?id_4=36986-informatsionnoe_soobshchenie
23. Central Bank: The Russian economy faced an unprecedented shock due to the COVID-19 pandemic. https://1prime.ru/state_regulation/20200528/831532645.html
24. Information and analytical material of the Bank of Russia «Review of financial stability» (2020). https://cbr.ru/Collection/Collection/File/25489/OFS_19-02.pdf
25. The Bank of Russia approved measures to support citizens, the economy and the financial sector in the context of the coronavirus pandemic. https://cbr.ru/press/PR/?file=20032020_133645if2020-03-20T13_36_08.htm

Covid-19: Financial Impact and Disclosure

Olga Efimova and Olga Rozhnova(✉)

The Financial University under the Government of the Russian Federation, Moscow 125993, Russia
{Oefimova, ORozhnova}@fa.ru

Abstract. This paper identifies key areas of financial reporting that need to be considered when determining the impact of the pandemic on the companies' financial position, the performance and disclosure of material information. The article collected and analyzed experts' opinions among the representatives of the academic environment, the professional community and the reporting enterprises from the economic sectors which were influenced by the pandemic. In the practical part of our research we have analysed financial reports of companies from different sectors over the first quarter of the year 2020 with a view to identifying practical cases of the disclosure of COVID-19 impact consequences. For analysis we chose companies' reports from the sectors which were affected strongly by the pandemic: air transportation, online travel and related services, retail, metallurgy. Based on the analysis performed, we made recommendations on the disclosure of material information in the financial statements. These disclosures should be correlated with companies' social and environmental responsibility reporting.

Keywords: Coronavirus COVID-19 · Coronavirus impact · Financial consequences · Financial Statements · Disclosure

1 Introduction

The spread of Covid-19 in the world has impacted seriously on the global economy, financial markets, and business, regardless of the size of enterprises. The pandemic resulted in a fall in solvent demand, a decrease in production and sales volumes, an increase in additional costs, in particular, in order to maintain safe working environment, also difficulties in managing supply chains and inventories, high volatility of financial markets, growing financial risks, primarily liquidity and credit, rising unemployment and associated social risks and threats. The continuation of these circumstances could have a prolonged impact on an entity's financial results and cash flows. This situation makes it necessary to consider and monitor the financial and accounting consequences of Covid-19 carefully. While the pandemic has impacted on almost all businesses either directly or indirectly, some of the most hit sectors are air transportation, tourism, hospitality and retail. The aviation industry is facing massive disruption owing to travel restrictions, the tourism and hospitality sector has been affected with low occupancy in business and holiday travel. Also, we can expect significant consequences of the pandemic for production sectors, for example, for the

T. Antipova (Ed.): ICCS 2020, LNNS 186, pp. 205–213, 2021.
https://doi.org/10.1007/978-3-030-66093-2_20

steel and metal industry due to a sharp drop in demand, the closure of borders and the suspension of the supply chains across the world.

Despite the fact that some companies did not experience serious negative impact and some entities could benefit from the current situation, the activities of very few enterprises remained unchanged. Therefore, organizations should consider the Covid-19 financial impact on the business and the areas of financial reporting that will be affected in order to determine the necessary information.

Companies are required to test their assets for impairment when indicators of impairment are identified. Cases of forced shutdown of equipment or its use to solve urgent problems of the spread of COVID-19, as well as a decrease in sales and the recession in production demand, can be attributed to possible reasons for impairment.

It is known that goodwill should be tested annually for impairment. COVID-19 could affect goodwill due to significant changes in the business climate, outbreaks of infection among staff and key staff losses, and a serious decrease in the market capitalization of the enterprise. In addition, growing financial and social risks may require increasing of the discount rate, which may also lead to impairment of goodwill.

In accordance with IFRS 9 Financial Instruments, expected credit losses should be recognized for some types of financial assets. A negative economic outlook as result of COVID-19 could lead to credit risk increasing. Even if estimates do not require reevaluation, full disclosure of the circumstances should be provided in the financial statements.

The revenue of an entity may reduce due to the spread of the pandemic. Under pandemic conditions companies run a promotion in order to collect money or offer refunds or credits to customers for goods or services that cannot be used during this crisis period. Consequently, some contracts may become less profitable, or even loss-making.

Some entities may be facing supply chain disruptions. Companies would need to assess whether adjustments to the carrying value of their inventory are required to bring them to their net realizable value in accordance with IAS 2 'Inventories'.

Rapidly changing economic environment is subject to the new assessment of credit, liquidity and market risks. Management should determine if additional risk disclosures are required.

To remain a going concern companies must have the ability to meet their obligations as they come due within one year after the financial statements are issued. Management needs to determine whether there are conditions and events that raise 'substantial doubt' about a company's ability to continue as a going concern and to provide appropriate disclosure.

This article identifies key areas of financial reporting that need to be considered when determining the impact of a pandemic on the financial position, results, and disclosure of information in the financial statements. This is not a final list, and there may be other areas not included in this article that should be considered by preparers and interested users.

2 Research Method

The research intended to analyze the following issues. Since the impact of COVID-19 pandemic around the globe is unprecedented (Fig. 1), we have set a goal of collecting, generalizing and analyzing experts' opinions among the representatives of the academic environment, the professional community and the reporting enterprises from different sectors which are influenced by the pandemic greatly.

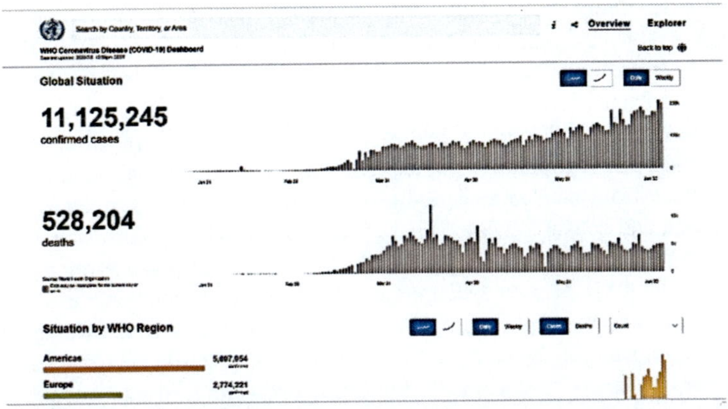

Fig. 1. COVID-19 aftermath for the world https://covid19.who.int/

Three most affected countries are presented in Fig. 2.

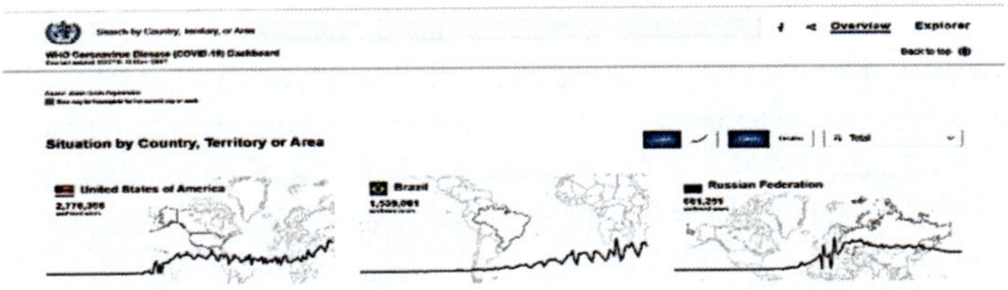

Fig. 2. Countries affected by the pandemic most of all according to World Health Organization https://covid19.who.int/

In accordance with Fig. 2 we can single out the USA and Russia among the countries that have suffered from COVID-19 badly. That is why we chose academic publications originated from these countries as objects of our study. We have analyzed Russian scientific publications that investigate new trends in accounting and financial reporting in connection with Covid-19 and came to conclusion that there are virtually no articles about the pandemic impact on accounting issues for the present. We have analyzed USA scientific journals as well. We have studied journal issues of American Accounting Association: Accounting Horizons, The Accounting Review, Issues in

Accounting Education[1] for the year 2020. We have not found any articles on the pandemic topical issue and accounting in these journals either. It is caused by the fact that scientists need time to conduct a comprehensive analysis of the events. Besides the publication process in serious scientific journals takes a long period of time in the current context. For instance, in Russia depending on the level of a scientific journal it can take from half a year to a year and a half from the moment of passing an article to an editorial office till the time of its publication. The period of considering, reviewing and publishing an article in the stated academic periodicals in the USA exceeds the abovementioned period of 6–18 months in Russian ones.

There is no doubt that considerable scientific papers based on in-depth research results and reliable information will be published soon. But at the same time there is also a concern that the events taken place in the spring and in the beginning of the summer of 2020 are not reflected in a timely manner in academic periodicals. It is obvious that new models of scientific studies in the field of accounting and the forms of their promotion should be found, and furthermore, digitization gives wide opportunities to implement it.

As distinct from academic periodicals, practical accounting publications [1–3], issues of audit and consulting companies [4–6] or international professional bodies [7–9] have a lot of articles on this problem. The authors of practice-oriented journals analyze COVID-19 impact on accounting and give recommendations how to act most properly in the current situation. For example, based on the analysis, Kulaga, E. offers recommendations to management and accountants on how to reflect the impact of the pandemic in the financial statements for 2019, taking into account events after the reporting date [10]; Lesnova, Y.V. studied the way in which enterprises would have to present economic aftermath of COVID-19 in interim and annual accounts for the year 2020 [11].

It should be noted that there is a link between climate change, environmental disasters and arising epidemics. This link, even if it is not expressed clearly at first sight, has become especially evident in 2020. The study of Brenner and Marwan (2018) indicated the existence of a direct link between climate change and the occurrence of epidemics seems extremely relevant [12]. The presented results focus on the effect that climate change has on spreading patterns of airborne infectious diseases. On the one hand, the pandemic exhibited one of the catastrophic scenario options resulting from climate change. On the other hand, it made it evident that even temporary reduction in the scale of activity had an impact on the environmental situation. Antipova (2020) recognizes the pandemic as "a black swan event" and considers its economic and social impact on all sectors of life [13]. Rozhnova and Efimova (2020) argue the necessity of considering financial implications of climate change risks as information important for the disclosure [14]. We believe that such disclosures are also appropriate for the pandemic.

We support the opinion of Kuter et. al (2018) who argue that study of the historical parallels is useful for the understanding of the current situation [15]. In continuation Peterson K. Ozili [16] discusses the accounting practices and financial reporting options for firms during a pandemic. The author concludes that accounting practices can help to dampen the effect of a pandemic on companies' financial reports.

---

1 https://meridian.allenpress.com/aaa.

Online academic debate and practical discussion concerning the issues of the pandemic impact on the accounting are very heated. New online platforms are still appearing. In our opinion these are the most widely used means of discussing the questions concerning accounting and pandemic:

a) Online publications and the placement of research results on the sites of international professional organizations. Such publications draw a wide response in both academic and practice-oriented accounting environment.
b) International scientific conferences where important issues on the current state of accounting and its evolution are discussed.
c) Organizing forums for discussing accounting issues of the pandemic, inviting to a broad discussion and publishing research results.

In the practical part of our research we have analyzed financial reports of companies from different branches over the first quarter of the year 2020 with a view to identifying practical cases of the disclosure of COVID-19 impact consequences. For analysis we chose companies' reports from the sectors which were affected strongly by the pandemic: air transportation, online travel and related services, retail, metallurgy.

We studied key issues concerning the assessment of the current situation, financial implications of the pandemic, perspectives and ongoing efforts to ensure sustainability made by the companies themselves.

Research results of the disclosure of the information in financial statements of the studied companies on key development issues in the context of the pandemic are presented in Table 1.

Table 1. Disclosing the impact of the pandemic in company reporting over the first quarter of the year 2020.

Analysis of the situation	Financial implications of the pandemic	Ongoing efforts	Perspectives
Air transportation PJSC "Aeroflot"			
Negative –Detailed explanations are given concerning both the branch and the company	Negative –Figures of revenue decline, passenger traffic costs are given	Considerable – Measures to effectively address threats of the pandemic and its negative impact on the company are disclosed	Negative – Negative expectations for the second quarter of the year 2020 are described
Retail Group "M.Video – Eldorado"			
Positive - Full transition to an online platform and transformation into online business	Positive - data of transaction growth in the first quarter of the year 2020 are shown as compared with the first quarter of the year 2019	Considerable – Several projects which allow providing customers with necessary domestic equipment and devices in a contactless, safe way promptly in conditions of self-quarantine are launched	Positive - Developing the ecosystem ONE RETAIL. Using RTD (Real Time Dealing); promoting strategies on entering telecom segment

(*continued*)

Table 1. (*continued*)

Analysis of the situation	Financial implications of the pandemic	Ongoing efforts	Perspectives
Metallurgy PJSC«MMK»			
Cautious - the Group has established a special committee under the leadership of the Parent Company's CEO and is promptly monitoring the situation	Negative, caused by the decline of product demand To date, the Group has not experienced any difficulties with the liquidity and supply of basic raw materials, equipment or spare parts	Considerable – The Group is taking active measures to prevent the spread and reduce the risk of coronavirus In order to reduce the risks to the Group's financial stability and optimize capital expenditures, the stress scenario provides for a reduction in CAPEX, with the active phase of construction of the coke oven battery complex being shifted to a later date	The Group and its largest debtors are included in the list of backbone enterprises established by the Russian Government. This allows the Group to take advantage of government support measures, if necessary
Online travel and related services Booking Holdings Inc.			
Negative- «The COVID-19 has resulted in a material decrease in consumer spending and an unprecedented decline in travel and restaurant activities and consumer demand for related services»	Negative -Negative financial implications for the finance results, cash flows, credit and other financial risks are disclosed in detail	Very considerable («elimination of non-essential business travel, cancelation of internal company events, reducing in marketing spend worldwide, a general company-wide hiring freeze, suspending share repurchases and other actions to provide additional financial flexibility»)	Negative- «Even if economic and operating conditions for the Company's business improve, the Company cannot predict the long-term effects of the pandemic on its business»

The analysis found that impact consequences for the financial position of the companies and cash flows as well as the completeness itself of the disclosure in the companies' reporting varied significantly. Such issues as going concern assessing; revenue recognition; goodwill and long-lived asset impairment testing; expected credit losses; measurement of financial instruments; inventory management and valuation; debt covenants; lease accounting; government support; forward-looking information; cash flow impacts and some others facts can be highlighted among the key topics related to COVID -19 impact. These issues are essential for a true and fair presentation of financial position, current and future financial results.

3 Results

Based on the analysis we have made the following recommendations:

- It is necessary to give coverage to topical accounting issues in academic periodicals during the period when such issues are in great demand. It is possible to publish mini-papers in a special column covering 'Opinions and Recommendations' even in serious scientific journals. This should be done to create relevant environment for the scientific debate and for the further in-depth researches on those topical issues.
- We had studied practical cases of reporting companies' disclosures and had analyzed them with a view to identifying redundant, insufficient, unsubstantiated information to discovering the best practical cases of the disclosure. We suppose that such studies should be transformed into a regular practice in the process of professional accountant education.
- The role of interim accounts is becoming more important in the rapidly changing world. Interim accounts allow companies to react promptly to significant events and to bring relevant information to interested users.
- Disclosures in financial reporting must be coordinated with links to non-financial reporting (for example, the number of cases, the measures to staff protection taken by a company, the cost estimation of these measures). Such studies were not carried out in the analyzed reporting documents; this reduces their value for users.
- Disclosures should be structured more clearly in financial reporting to make data more comparable.
- It is needed to encompass the linkage between the company's strategy and the results of reporting analysis.
- It is advisable to disclose information about the contribution of an enterprise to the fight against the spread of threats in the context of global crises comparable with the pandemic in reporting. This information should be correlated with social and environmental responsibility in non-financial reporting.
- It is useful to include such a section as "Risks, forecasts" in the company's disclosure of material accounting aspects.

4 Conclusion

The research has certain limitations. Although the paper presents the results of the analysis of only four companies from different sectors, we analyzed financial reports of more than 30 enterprises in the first research stage. Then we selected several companies which reflected the pandemic impact (positive and negative) to a greater extent. The first stage showed that many enterprises do not disclose issues of pandemic impact on their activity and its results at all. Due to available information limitation the research is also not sufficiently detailed. But now is the time to raise a range of issues and attract as many academics and practitioners in the field of accountancy as it is possible to the debate. We consider that management, accountants, auditors and standard-setters should make certain conclusions as early as 2020.

Further course of the study must be aimed at the digitization of accounting and reporting [17] since it allows collecting more extensive information, processing it quickly, detailing and aggregating it in the right directions, including users' opinion and making more accurate forecasts [18]. Furthermore, it is evident that accounting should help choose the right strategy and business model of a company, consider all risks allowing reducing a negative influence of different climate, social, environmental disasters on business.

References

1. IFRS in Focus - Accounting considerations related to the coronavirus 2019 Disease (2020). https://www.iasplus.com/en/publications/global/ifrs-in-focus/2020/coronavirus
2. EY. Applying IFRS - IFRS Accounting Considerations of the Coronavirus outbreak (2020). https://www.ey.com/en_gl/ifrs-technical-resources/applying-ifrs-accounting-considerations-of-the-coronavirus-outbreak
3. Grant Thornton. Reporting the impact of COVID-19 (2020). https://www.grantthornton.global/en/insights/supporting-you-to-navigate-the-impact-of-covid-19/reporting-the-impactof-covid-19/
4. IFRS Foundation. Covid-19 (2020). https://www.ifrs.org/news-and-events/2020/03/the-coronavirus-and-the-foundations-work/
5. IOSCO. Statement on Importance of Disclosure about COVID-19 (2020). https://www.iosco.org/library/pubdocs/pdf/IOSCOPD655.pdf
6. IPSASB. Covid-19 Relevant IPSASB Accounting Guidance (2020). https://www.ifac.org/system/files/publications/files/IPSASB-Staff-QA-COVID-19-Relevant-Accounting-Guidance_0.pdf
7. KPMG. COVID-19 Financial Reporting (2020). https://home.kpmg/xx/en/home/insights/2020/03/covid-19-financial-reporting-resource-centre.html
8. Moss Adams. COVID-19 Disrupts Financial Reporting (2020). https://mossadams.com/articles/2020/03/covid-19-disrupts-financial-reporting
9. PWC. Covid-19 Audit Committee financial reporting guidebook (2020). https://www.pwc.com/us/en/library/covid-19/audit-committee-financial-reporting-guidebook.html
10. Kulaga, E.V.: Impact of COVID-19 pandemic on financial reporting and its auditing for the year 2019/2020, №. 4 source: corporate financial reporting. Int. Stand. **4** (2020). https://finotchet.ru/library/94/349/
11. Lesnova, Y.V.: Impact of the current economic situation on the impairment of assets/corporate financial reporting, 2020, №. 5, source: corporate financial reporting. Int. Stand. **4** (2020). https://finotchet.ru/library/94/349/
12. Brenner, F., Marwan, N.: Change of influenza pandemics because of climate change: complex network simulations. Rev Epidemiol Sante Publique **66**, 424 (2018). https://doi.org/10.1016/j.respe.2018.05.513
13. Antipova, T.: Coronavirus pandemic as black swan event. In: Integrated Science in Digital Age, pp. 356–366. Springer (2020). https://doi.org/10.1007/978-3-030-49264-9_32
14. Efimova, O., Rozhnova, O.: Financial reporting and climate-related disclosures. J. Digit. Sci. **2**(1), 67–76 (2020). https://doi.org/10.33847/2686-8296.2.1_6. ISSN: 2686–8296. © Institute of Certified Specialists
15. Kuter, M., Gurskaya, M., Andreenkova, A., Bagdasaryan, R.: Asset impairment and depreciation before the 15th century. Account. Hist. J. **45**(1), 29–44 (2018)

16. Peterson, K.O.: Accounting and financial reporting during a pandemic. SSRN Electro. J. (2020). https://doi.org/10.2139/ssrn.3613459. https://www.researchgate.net/publication/341725351.
17. Lugovsky, D., Kuter, M.: Accounting policies, accounting estimates and its role in the preparation of fair financial statements in digital economy. In: Antipova, T. (ed.) ICIS 2019. LNNS, vol. 78, pp. 165–176. Springer Nature, Switzerland (2020). https://doi.org/10.1007/978-3-030-22493-6_15
18. Efimova, O., Rozhnova, O., Gorodetskaya, O.: XBRL as a tool for integrating financial and non-financial reporting. Advances in Intelligent Systems and Computing, AISC, vol. 114, pp. 135–147 (2020). https://doi.org/10.1007/978-3-030-37737-3_13

Digitization's Technology

Digital Technologies in Transport Industry

Albina Bilyalova[1(✉)], Irina Vaslavskaya[1], Marina Guskova[2], Larisa Egorova[3], and Roman Khalturin[3]

[1] Kazan Federal University, Naberezhnye Chelny 423800, Russia
abil171@mail.ru
[2] Russian University of Transport, Moscow 127994, Russia
[3] Moscow State Medical and Dental University A.I. Evdokimova, Moscow 127473, Russia

Abstract. At the beginning of the XXI century there was a significant change in the development of digital technologies and their rapid introduction into the economy. The transformation of innovative technologies in the field of information and communication has led to the concept "Digital economy". The digital economy is a modern type of economy characterized by the predominant role of data and methods of their managing as a main resource in the sphere of production, distribution, exchange and consumption. The transition to a digital economy has led to the emergence of the term "digital logistics". Digital transport logistics is based on the use of modern information technologies, methods and conceptual computer modeling of the supply chain, which virtualize transport resources in order to optimize the development of this field of activity. Current trends in transport development lead to the transformation of digital logistics, aimed at implementing new requirements for the level of service and quality of cargo transportation. The article aims to analyze the role of digitalization and the intellectual information use in the management of transport processes, it describes the advantages of implementing digital technologies and also examples of digital transformation of transport and logistics companies are given. It is shown that digitalization affects all transport and logistics companies and it is the subject of economic relations between the state and business. It is proved that digital transport logistics is one of the digital economy key tools, which allows not only to optimize the processes of cargo transportation in modern conditions, but also to significantly reduce costs in the transport and logistics industry.

Keywords: Transport industry · Transport science · Transportation · Digital economy · Digital technologies · Blockchain technology

1 Introduction

In modern world, transport and logistics systems (TLS) are an important industrial segment of the Russian economy and one of the most important elements of the development of a market economy of the Russian Federation. TLS plays a specific role in the process of cargo transportation, services for the delivery of goods, organization of flows in the production process, which ensures effective commodity exchange in the

T. Antipova (Ed.): ICCS 2020, LNNS 186, pp. 217–226, 2021.
https://doi.org/10.1007/978-3-030-66093-2_21

economic system of regions, creates conditions for the movement of people, goods, and services, and contributes to the strengthening of vertical-horizontal intersectoral and territorial networks. The structure of the transport and logistics system of enterprises, presented in Fig. 1, indicates the specificity of this industry. The Entrance to the transport and logistics system is labor, means of labor, objects of labor, necessary for the functioning of the TLS. The transport and logistics system Exit is the socio-economic result of functioning, expressed in terms of achieving the goal through obtaining a combination of effects (synergistic, integration, cooperation, coordination, unification). It should be noted that the elements of the TLS is variable, because they depend on a number of factors, such as the cyclical nature of the functioning of the system; parameters of the control object; characteristics of the subject of management; the difficulty of determining the boundaries of the TLS; the level of development of links of sustainable interaction on technological redistribution; the level of development of the organizational and economic mechanism as a whole and its relevant tools for effective management [1].

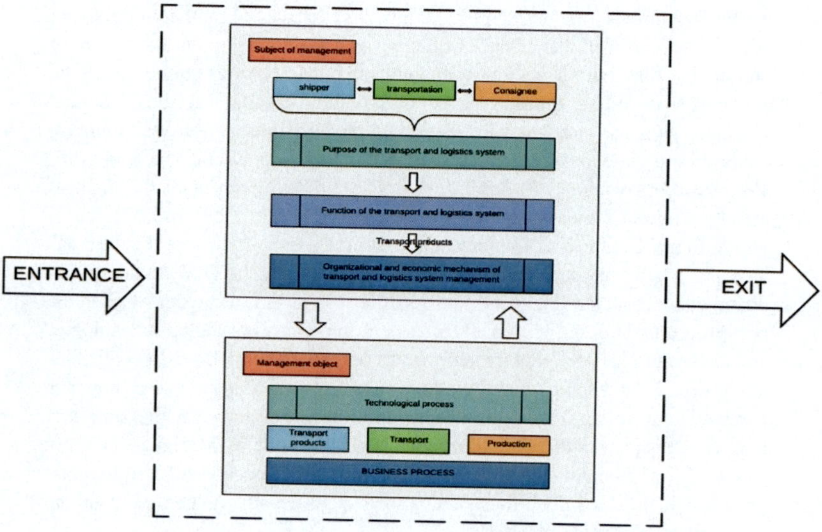

Fig. 1. The structure of the transport and logistics system

Thus, the development of the economy of the country's regions is inextricably linked with transport and logistics systems.

Analysis of the scientific literature has led to the conclusion that most researchers support the view that logistics is a specific business concept, the purpose of which is to optimize the resources of the enterprise associated with the management of material and related flows [2–8]. Accordingly, logistics is a strategic platform that affects the efficiency of the enterprise through the correct choice of logistics concepts, systems and technologies; the correct organization of the logistics process; solving problems of cross-functional and inter-organizational coordination and integration.

Innovative achievements of science and technology in the field of information and digital technologies, which lead to the possibility of creating a digital image of the real world, determine the need for theoretical and practical understanding of the processes of changing socio-economic life of society in accordance with the requirements of the new technological order.

The above has led to the concept of "digital economy". This term, first proposed in 1994 by the Canadian scientist Don Tapscott [9], has become widely used, and the study of problems and prospects for the development of the digital economy has become one of the topics considered in the framework of modern interdisciplinary research. Currently, special attention is paid to the issues of digital transformation of various sectors of the economy and spheres of activity, as evidenced by a large number of published works on this issue [10–17].

Digitalization of transport logistics systems is one of the priorities of the country's strategic development and requires a change in technology, which is one of the key factors for improving efficiency. The most popular technologies in the field of TLS are the Internet of things, big data, intelligent systems, Blockchain, wireless communication technologies, virtual and augmented reality technologies.

2 Methodology

As a theoretical and methodological basis for the work, we used scientific research and articles of domestic and foreign scientists on the development of the transport industry and the application of Blockchain technology to improve the quality of services in the field of freight forwarding service.

A significant number of scientific studies are devoted to information and intelligent technologies in logistics, digital logistics, application and organization of intelligent transport systems and traffic flows. Among them are the works of Murphy, P. R, Angeleanu, Andra, Blanchard, B. S, Daganzo, C. F, Kopylov O., Stewart P. D., Fedorova N. V., which are devoted to logistics and Economics of transport and forwarding activities, including issues of processes of digital transformation of global and national economies, of inter-transport interaction, how the transport companies can change business model according to digitalization [1–6, 8].

As the methodological base of the study, we used systemic, processual and integrated approaches to the study of the stated problem. The article used methods of generalization, analysis and synthesis, schematization, structuring and modeling, comparison and groupings, scientific abstraction, ranking, evaluation.

3 Results

Digital logistics is the management of material, information, financial and human flows in order to optimize them through the modern information technologies use. Logistics is one of the most digitalized industries, both in the world and in Russia. Digitalization of transport and logistics services largely determines the level of competitiveness of companies operating in the market. The main advantage of digital technologies is the ability to achieve significant success in optimizing costs and improving the transport and logistics services level.

The basic component of the digital transport logistics transformation is electronic document management. In the process of cargo transportation management, the rejection of paper transport invoices in favor of electronic transport invoices provides many advantages. First of all, it eliminates the cost of printing and delivering documents, document losses, and allows you to get information about the delivery and movement of goods online, as well as exchange electronic documents within a few minutes. The first cargo transportation with fully digital documents was carried out by Traft and L'etoile in August 2018. Traft delivered the cargo from the warehouse to a retail store in Moscow. Transport documents were issued using the electronic document management platform "Forma" [18]. As a result, in September 2018, the Fesco transport group delivered containers from Japan to Moscow by rail, which was fully digitalized, which reduced the time for container clearance in the port from 5 days to 21 h. By 2019, all railway documents have been converted to digital format, and electronic data exchange between Russian Railways and the Federal customs service has been introduced. Thus, the mass transition to electronic document management, in particular to electronic invoices, is necessary for the market of transport and logistics services, since it eliminates the problems associated with working with paper documentation.

Digitalization of the cargo transportation process sales is carried out using electronic services and trading platforms that operate on various types of transport. For example, the electronic trading platform "Cargo transportation" operates on railway transport. The owner and operator of the site is Digital logistics LLC, a subsidiary of Russian Railways. This platform is a fully digital service. The platform's clients are suppliers and buyers of transportation and related transport services. Due to the fact that communication between the platform's clients occurs only in digital format, maximum transparency and efficiency of work is ensured. Execution of the order can be tracked on line. The development plans for this site include expanding the range of services provided, in particular: loading and unloading, terminal operations, security services [19]. Also in the field of road cargo transportation, the TRAFFIC platform developed by BIA-Technologies serves as an aggregator connecting customers-shippers and carriers. The following services are available on the platform: monitoring current market prices for cargo transportation, online insurance under smart contracts, order tracking, and factoring financing (Fig. 2).

The analysis showed that using this platform can reduce downtime during loading/unloading by 50%, as well as reduce the load on staff by 30%. One of the platform's users is Delovye linii group, which places orders during peak periods (seasonal, pre-holiday, etc.) [20].

Digitalization not only affects individual transport and logistics companies, but also acts as a subject of dialogue between the state and business [7]. Thus, in 2018, it was announced that a single digital platform for the Russian transport complex would be created. The main objectives of the platform are to reduce transportation costs and unify transport and logistics solutions. The advantages of creating a digital platform for the transport system are shown in Fig. 3.

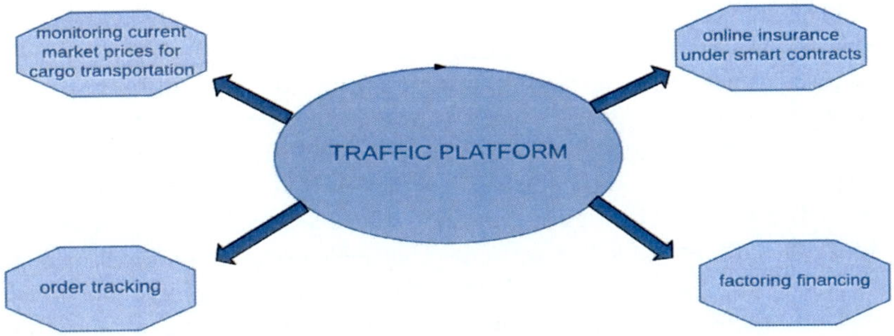

Fig. 2. Services on the traffic platform

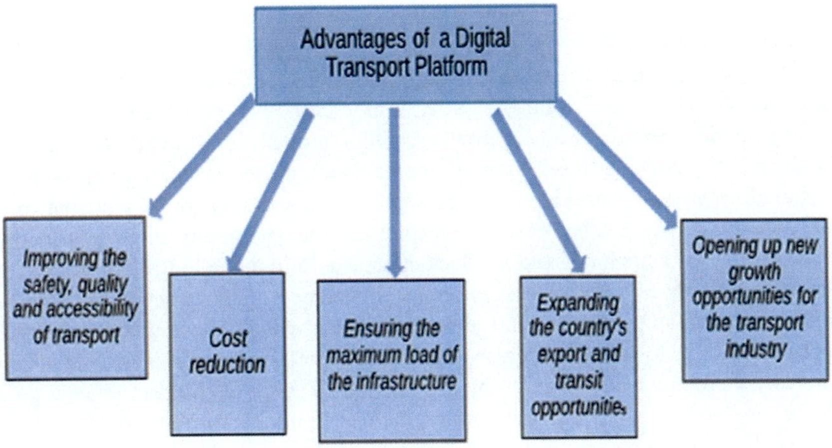

Fig. 3. Advantages of a digital transport platform

So, in modern transport logistics, when choosing optimal routes and transport, computer processing of initial data (orders, cargo parameters, fleet, etc.) is necessary. This is due to the constantly growing amount of data about the management objects state. Logistics is defined as the planning framework for the management of material, service and information flows. Logistics for physical goods usually involves integrating the flow of information, transportation, warehousing, and often security. Thus, logistics chains often cover multiple stages and hundreds of geographical locations. For this reason, it is often difficult to track events throughout the chain, check and verify transported goods, and quickly respond to unforeseen circumstances. In addition, due to the lack of transparency, it becomes extremely difficult to investigate illegal activities that occur anywhere along the route [2, 5]. A possible solution to this problem in logistics systems may be the introduction of Blockchain technology. A Blockchain is a continuous sequential chain of blocks (a linked list) that contains information in accordance with certain rules. Most often, copies of block chains are stored and processed independently on different computers. Computer algorithms written in a chain of blocks called smart contracts allow you to automate many logistics processes and, consequently, reduce their prices, as well as reduce the impact of the "human factor". As a transparent public register, the Blockchain is able to provide clients and auditors with simple and effective tools for tracking the entire route. One of the important aspects of Blockchain is that it can provide its advantages only if all members of the supply chain have access to the network. The implementation of Blockchain technology in logistics reduces risks and significantly increases stability, as it forms its practical orientation. The implementation of Blockchain technologies in industry is beneficial for the growth of document management, data storage, supply management, payment systems, e-Commerce and voting systems and public opinion research [21–24].

The advantages of implementing Blockchain technology for transport logistics in an enterprise are as follows: reducing the number of errors in the document flow; simplify work; ensuring transparency and reliability of information about manufacturers (suppliers) of goods and the process of commodity distribution, and others; reducing the duration of document flow; creating a unified infrastructure for managing product flows; reducing the length of the logistics cycle. Integration of transport logistics through the Blockchain system will allow the company to systematize the main processes, reduce labor costs for accounting for commodity flows and increase the security of valuable information. Blockchain is helpful in the fight against fraud and delivery errors, since one of the important advantages of the technology is instant simultaneous updating of information for all participants in the logitics chain. Also Blockchain is able to solve problems such as cargo theft, hidden damage, disputes over cargo ownership, as well as the problem of handling cargo to shippers to send goods quickly, and so on. Logistics chains often include hundreds of participants and geographical locations. Blockchain can make delivery more transparent, cheap, and secure.

Thus, when implementing Blockchain technology, even in the most basic its version, the technology will provide:

- ability to track the product from its place of production to the end user;
- protection of product buyers from unoriginal goods and fraud, which allows the buyer to be confident in the decision to purchase the product;
- dispersion, which allows all elements of the delivery process to participate in real time: tracking the movement of transport or purchased goods, making the necessary documents in the form of smart contracts, destroying illegal elements in the delivery process;
- security by using cryptography and not being able to make any changes to the delivery process.

So, we can conclude that the introduction of Blockchain logistics technology and intelligent digital technologies potentially provides real-time cargo tracking, reducing the workflow and increasing transparency. Blockchain turns out to be a much cheaper and safer infrastructure with a higher scale and ease of integration compared to other industries. The application of logistician Blockchain technology is currently not sufficiently developed and cannot be sufficiently consistent due to perceived risks and skepticism, but with each successful project, the future is approaching, and we can safely assume that in a few years logistician Blockchain technology will spread throughout the industry.

4 Discussions

Transport logistics is currently impossible to imagine without the use of innovative information technology. From the point of view of organizational conditions, it will be inefficient to form cargo delivery chains in the absence of intensive rapid exchange of information between the transport process participants, and the lack of opportunities for immediate response to needs transport market. Currently the speed of information exchange is quite high. In connection with the modern production and management technologies use, information communications are developing, which are the basis for analytical support of business processes in the transport industry at the national and international levels. Global transport corporations are aimed at using technological information systems which provide the customer with complete product information, in particular, the most important-where the product is currently located. One of these technologies is the Blockchain technology. Blockchain is so called a public and unchangeable "Internet Ledger" that records what things belong to whom. It is impossible to forge such a book. For example, Blockchain allows you to track the entire supply chain of a product from the manufacturer to the consumer. In this case, each transaction is recorded and added to the distributed database chain as a new fragment, which is manually assigned a unique multi-digit numeric cipher. This fragment stores data about the time, date, participants, transaction amount, and, importantly, information about the entire network. Here, any information transfer takes place in the form of a chain of blocks (block and chain), where each block always contains information about the previous block.

It is very important to realize that Blockchain is not just bitcoin, a cryptocurrency that has been much talked about and written about over the past few years. There are quite a lot of scenarios for possible use of this technology within an organization or within the interaction of individual companies. Thus, using Blockchain, you can organize financial transactions between banks around the world without a single center, develop a supply system of almost any type, create a fast and reliable system for concluding contracts between the buyer and seller or the recipient of the service and its supplier, organize a system of mailings, manage supply chains, etc. Blockchain can be extremely useful for organizations that exchange information, valuable data, but do not trust each other too much. In this case, the technology can simplify and improve the exchange and storage of information related to the interaction of these organizations. A distinctive feature and novelty of Blockchain technology is that it does not require a Central regulatory body (for example, a Bank) to approve and register any transactions. This allows, to get rid of intermediaries in the transaction system without losing its integrity. Blockchain technology theoretically provides the following features:

- replacing slow manual procedures with fast automatic;
- tracking the origin and properties of cargo (raw materials, semi-finished products, finished products) and passengers;
- information about cargo and passengers that ensures timely payment, insurance and payment of customs duties;
- information on the state of goods and passengers, allowing them to be tracked during transportation or during transfer between modes of transport;
- information for regulatory authorities about the goods and passengers involved participants (exporters, importers, carriers, etc.).
- Blockchain, being a relatively new area in the transport management system, is becoming a global technology that is projected to contribute to economic growth over the next several decades. This technology, as it was mentioned above, has many advantages, including the following:

  1. Process Integrity. Due to the security reasons, this program was made in such a way that any block or even a transaction that adds to the chain cannot be edited which ultimately provides a very high range of security.
  2. Traceability. Blockchain is designed in such a way that it can easily locate any problem and correct if there is any. It also creates an irreversible audit trail.
  3. Security. Blockchain technology is highly secure because of the reason each and every individual who enters into the Blockchain network is provided with a unique identity which is linked to his account.
  4. Transparency. Any data in the Blockchain can be viewable for any person, also if any changes were made in the Blockchain, those changes are publicly viewable. That's why Blockchain used in cryptocurrencies because, in cryptocurrency, every transaction is recorded and showed to the public.
  5. Faster and Cost Effective. If you used any cross-border transactions and/or between banks can take days and be quite expensive. But in the case of Blockchain-based cryptocurrency transactions, you can send money to anywhere and to anyone in the world because Blockchain is decentralized and no paper works needed.

Thus, it is obvious that the implementation of Blockchain Technology provided great convenience for the TLS, but, on the other hand, if the technology has advantages, it also has its drawbacks. So, Blockchain, like any technology, is not perfect, it has some obvious shortcomings, especially in terms of mass implementation of the technology. Among the shortcomings, the following should be noted:

1. High volatility of the most common Blockchain with the Proof-of-Work consensus algorithm due to the complexity of the transaction, which makes it an expensive technology.
2. The high cost of implementing this technology and the small number of specialists who are competent at it.
3. Uncertain regulatory status.

5 Conclusions

Digitalization is currently a factor that has a significant impact on transport and logistics. Companies that provide transport and logistics services have new opportunities that they can use to optimize business processes and improve the quality of service, thereby increasing their competitiveness.

In General, transport and logistics companies use systems such as TMS and WMS, conduct Analytics on Big Data arrays, use various sensors that track the state of the vehicle. In this paper, we focused on the use of Blockchain technology in the transport industry. Applications of Blockchain technology in various global transport and logistics companies were considered. It was revealed that Blockchain is used as a distributed registry model in the transport industry. In this case, all nodes of the system are kept signed and protected by cryptographic mechanisms from modifying the record of completed or completed operations. Such systems can store a wide range of information related to the transport industry. For example, this may be information about the operating conditions (modes of use) of the vehicle. The data collected in this way can be used later, for example, in insurance telematics applications. Another area is the use of distributed Ledger in logistics, especially in global (international) operations. Also, the characteristic of the Blockchain technology was given and the advantages and disadvantages of this technology were identified.

In conclusion, we would like to say that today the Blockchain technology does not stand still, and its capabilities have long gone beyond the cryptocurrency network. It is quite obvious that in the near future, the Blockchain will become one of the key tools of the digital society, which will not only optimize various processes, but will bring simplicity, transparency and efficiency not only to the transport and logistics industry, but also to the world economy. Thus, using Blockchain technology in combination with other technologies can significantly reduce the amount of paperwork, as well as reduce costs and increase efficiency. Also, we state that logistics is an industry in which the use of Blockchain technology can solve the security problem in this area, as well as simplify and reduce the cost of transporting goods through smart contracts.

References

1. Murphy, P.R.: Contemporary Logistics, 10th edn. Prentice Hall, Upper Saddle River (2010)
2. Angeleanu, A.: New technology trends and their transformative impact on logistics and supply chain processes. Int. J. Econ. Pract. Theories **5**(5), 413–419 (2015)
3. Blanchard, B.S.: Logistics Engineering & Management, 6th edn. Prentice Hall, Upper Saddle River (2003)
4. Daganzo, C.F.: Logistics Systems Analysis, 4th edn. Springer, New York (2005)
5. Kopylov, O., Mayboroda, V., Azarov, V.: An analysis of the use and difficulties in introducing information technology and information systems in transport and the transport infrastructure. In: Proceedings of the 2019 IEEE International Conference "Quality Management, Transport and Information Security, Information Technologies" (IT&QM&IS) (2019)

6. Stewart, P.D.: International Logistics: The Management of International Trade Operations, 3rd edn. Prentice Hall, New York (2010)
7. Vaslavskaya, I., Bilyalova, A.A., Tarasova, A.: Economic Education: the Role of the State in National Economy (2018). https://doi.org/10.22478/ufpb.2179-7137.2019v8n4.48381
8. Fedorova, N.V., Kukartsev, V.V., Tynchenko, V.S., Atluhanov, S.M., Gek, D.K., Zagudalova, E.A.: Problems of the digital economy development in the transport industry. In: IOP Conference Series: Earth and Environmental Science, vol. 315, p. 032047. IOP Publishing (2019)
9. Tapscott D.: The Digital economy. https://dontapscott.com/books/the-digital-economy/. Accessed 21 Apr 2020
10. Borremans, A.D., Zaychenko, I.M., Iliashenko, O.Y.: Digital economy. IT strategy of the company development. In: MATEC Web of Conferences, vol. 170, p. 01034 (2018). https://doi.org/10.1051/matecconf/201817001034
11. Brynjolfsson, E., Hu, Y., Smith, M.D.: Consumer surplus in the digital econmy: Estimating the value of increased product variety at online booksellers. Manage. Sci. **49**(11), 1580–1596 (2003). https://doi.org/10.1287/mnsc.49.11.1580.20580
12. Inshakova, A.O., Goncharov, A.I., Smirenskaya, E.V., Dolinskaya, V.V.: Modern comunication technologies in notification of notarial actions in Russia. J. Adv. Res. Law Econ. **VIII**(7(29)), 2144–2151 (2018)
13. Banalieva, E.R., Dhanaraj, C.: Internalization theory for the digital economy. J. Int. Bus. Stud. (2019). https://doi.org/10.1057/s41267-019-00243-7
14. David, H.W.: The future of human creative knowledge work within the digital economy. Futures (2018). https://doi.org/10.1016/j.futures.2018.10.002
15. Clifton, N., Fuzi, A., Loudon, G.: Coworking in the digital economy: context, motivations, and outcomes. Futures 102439. https://doi.org/10.1016/j.futures.2019.102439 (2019).
16. Vlasov, A.I., Shakhnov, V.A., Filin, S.S., Krivoshein, A.I.: Sustainable energy systems in the digital economy: Concept of smart machines. Entrep. Sustain. Issues **6**(4), 1975–1986 (2019). https://doi.org/10.9770/jesi.2019.6.4(30)
17. Tupchienko, V.A.: Digital Platforms for Lifecycle Management of Complex Systems. Nauchniy consultant, Moscow (2018)
18. You need to experiment with a digital bill of lading for at least a year. https://logirus.ru. Accessed 11 Sep 2019
19. Digital logistics. https://digitalrzd.ru. Accessed 11 Sep 2019
20. Logistics on the platform. https://plus.rbc.ru. Accessed 11 Sep 2019
21. Hsiao, W., Chang, T.: Exploring the opportunity of digital voice assistants in the logitics and transportation industry. J. Enterp. Inf. Manage. **32**(6), 1034–1050 (2019)
22. Larios-Hernández, G.J.: Blockchain entrepreneurship opportunity in the practices of the ubanked. Bus. Horizons **60**(6), 865–874 (2017)
23. Chen, Y.: Blockchain tokens and the potential democratization of entrepreneurship and innovation. Bus. Horizons **61**(4) (2018)
24. Bilyalova, A., Vaslavskaya, I., Gaifutdinova, R.: Digitalization of the transport industry: Technology of blockchain. Adv. Econ. Bus. Manage. Res. **138**, 152–156 (2020)

Electronic Banking and Financial Performance of the Russian Bank

Darya Rozhkova[1(✉)], Nadezhda Rozhkova[2], and Saidvalikhon Tozhihonov[1]

[1] Financial University under the Government of the Russian Federation, Moscow 125993, Russia
rodasha@mail.ru
[2] State University of Management, Moscow 109542, Russia

Abstract. The rapid development of the Internet has contributed to the emergence of electronic banking (e-banking). The introduction of new technologies makes it unnecessary to visit bank branches since to perform operations, it is enough to have a computer (laptop, smartphone etc.) and access to the Internet. Only after its widespread in developed countries, electronic banking began to demonstrate its noticeable growth in Russia, became one of the most dynamically upcoming services. Banks try to provide maximum comfort of service and to get closer to the client both in space and in time. Electronic (Internet) banking completely changes the form of interaction between banks and their customers, brings customer satisfaction, however what happens with performance in the banking industry? The question of a link between electronic banking and performance (ROE and ROA) was raised in the article. We found no significant relationship between offering of Internet banking and bank profitability. The findings of this study demonstrated that electronic banking is an essential part of industry development; however, bank management and shareholders should find other ways in order to increase performance.

Keywords: e-banking · Digitalization · Bank performance · Russian banks

1 Introduction

New technologies find their application in all areas of the economy and people's lives. They always have a specific goal: to improve, simplify, reduce the cost of different processes [1]. In conditions of high dynamism in the development of the banking sector, where advanced technologies are deploying extremely quickly, electronic banking services are important factor for banking sector development. Exposure to increased competition makes banks look for new, more promising ways to increase intensive and high-quality interaction with customers. The use of advanced technologies in servicing, allow banks to distribute their resources, minimize costs, improve quality of customer service.

Traditionally, banks try to provide maximum comfort of service and get closer to the client both in space and in time. This is achieved by increasing the network of branches. However, nowadays to achieve more effective results banks use modern technologies to provide remote banking services through ATMs, telephones, Internet etc.

T. Antipova (Ed.): ICCS 2020, LNNS 186, pp. 227–237, 2021.
https://doi.org/10.1007/978-3-030-66093-2_22

In order to get as close to customers as possible, banks use all existing technologies [2, 3]. Electronic (internet, digital, online) banking does not imply removing clients from the bank; on the contrary, banking services become more accessible to customers; they are always at their fingertips [4]. Electronic banking is a general term, reflecting the usage of electronic and telecommunication networks to provide bank services to clients [5]. In other words, e-banking includes all services to customers through Internet (for example, when a customer uses its personal computer or mobile-banking when he utilizes a Smartphone). The integration of unified theory of acceptance and use of technology (UTAUT) is widely used by researchers [6, 7].

The essence of electronic banking is customer self-service. A self-service technology is a technological form of the bank's interaction with customers, which allows to be serviced independently of a banking service employee. The bank's remote customer service system is a multifunctional software and hardware package that allows the bank's customers to carry out various operations, manage their accounts online and receive a wide range of relevant financial information without visiting the bank.

An important event related to the development of electronic banking in Russia is the end of the 2008 crisis. The reason was the general financial decline, as well as the decline of the banking sector in particular. Since there was a need to reduce costs, as well as the need for effective management of financial and human resources of the bank.

A suitable solution to this problem was Internet banking, which made it possible to optimize the work of banks. A positive effect of the integration of Internet banking systems is a reduction in the burden on the banking operations operators. All these effects have made the development of Internet banking services a necessity for banks.

The situation in Russia at the moment is extremely interesting, as the Russian Federation is in first place in many indicators in the field of Internet Banking. Currently, there are various ways to make a service unique and highlight its features. The vast majority of banks provide banking services, but the number and types of services are sometimes dramatically different.

The use of banking, in combination with traditional banking services in Russia, allows banking institutions to expand their market and increase their competitiveness.

The Internet helps ensure that the connection between the bank and the client becomes more efficient, which also makes it possible to work with clients depending on their preferences, willingness to take risks. Banks transfer some of the services to the Internet, as it is beneficial for the banks themselves (in terms of performance) and convenient for customers. Firstly, online service saves time, as bank customers do not need to visit a bank branch; in addition, all transactions take place within one banking day. Secondly, the cost of services provided via the Internet is much lower for customers, since their cost for the bank is also lower than traditional services. Thus, the question about the link between using e-banking and potential benefits seems very important.

In this article we target to address this issue, taking in mind that when we talk about using Internet technologies intuitively, we perceive some positive impacts: increasing market share, expanding product range, customized services.

Previous researches have examined the relative performance of banks offering Internet banking services. The results are quite controversial. For example, the investigation of Italian banks shows high dependence between offering Internet banking and profitability [8]. Multichannel banks in Spain are better in terms of ROE [9]. At the same time, analysis of Australian banks evidences no significant impact on performance and risk profile of banks [10].

Following the research path, the article addresses issue of main tendencies of Internet banking development in Russia and performance of the bank (in terms of ROE and ROA). According to the results of the analysis of the state of the Internet banking market in the Russian Federation, it can be said that Russian banks have recently shown interest in this area (approximately 2012). This interest has rapidly spread among most of the country's major banks. The Internet banking service, which was considered a unique service by which banks demonstrated their advantage, today is a necessity. The reason for this change can be considered the availability of the Internet in almost all regions, the availability of modern technology, which has become part of the everyday lives of most of the country's inhabitants. Increased competition in the market of remote banking services (remote banking services) will encourage providers to supplement their systems with additional functions that are not directly related to making payments. At the same time, providing the user with complete integrated solutions for managing personal finances will remain a key priority. We advance the body of knowledge on this subject by considering Russian experience and particularities.

2 Literature Review

In the modern world, there are two options for organizing banking: a traditional bank, which has its own offices and operating rooms and provides customers with the opportunity to manage their accounts via the Internet and a virtual bank, which operates only on the World Wide Web. The Internet creates an entirely new situation for the traditional economy, which is called differently: market cyberspace or the global market. Any client can "go to the bank in Internet" as if he comes to the branch to receive any banking service.

To use all the services of the bank, a web browser is used. A considerable advantage of this system is the lack of the need to install additional software. The Internet banking system is often available on a bank-client system using thin client technology. An essential aspect of security is the confirmation of transactions using one-time passwords, so that data interception does not allow attackers to gain access to clients' money.

However, traditional banking services still prevail in the domestic banking services market and combined is gaining momentum, which allows banking institutions to provide financial services not only directly through branches but also through various electronic and non-electronic sales channels. The combination of traditional and remote services allows capturing the market and satisfy the financial needs of various segments of the population, since a significant part of customers, especially the elderly, prefer traditional services, while young people often choose remote one.

At the present stage of economic development, Internet banking has become an everyday tool that is actively used in daily work, it is considered to be a transformation factor [11, 12].

The literature review indicated, that majority of Russian scientists study main trends and principles of digital banking evolving in Russian Federation.

The main benefits of such digitalization include a costs reduction and shorter length of service providing [13]. As well as internet banking is considered as a radical innovation, changing the way of interaction with clients and improves the quality of banking services [14]. The development of technologies makes it possible to reduce significantly the distance between the bank and the client. Besides, this trend leads to increased interbank competition, contributes to the qualitative and quantitative development of services in banks.

Currently, customers have the opportunity to study and evaluate the terms of the offered banking products, as well as to determine the attractiveness and functionality of their remote access systems.

3 Methodology of Research

3.1 General Background of Research

The method that we will use will be the case study of PJSC "Sberbank" (hereafter Sberbank). The case study method is not only beneficial because it allows combination of qualitative and quantitative methods, but as well because we get a better understanding of the whole by focusing on a key part [15]. Sberbank is one of the largest banks in the Russian Federation and the CIS countries. Almost 76% of the population of the Russian Federation uses the services of this organization. Sberbank includes a large sector of investment banking services. Its share in the Russian private deposit market is 46.6%. As well as almost every third loan from individuals and legal entities (34.7% and 33.9%). This credit institution also deals with corporate clients, about 36% of the loan portfolio accounted for by medium and small companies.

Being the largest and oldest bank in Russia, it is actively competing in the banking services market and focuses on rapidly changing customer preferences and the newest technologies. This organization is the owner of one of the largest networks of ATMs and self-service terminals (over 91 thousand) [16].

3.2 Digital Services of Sberbank

Digitalization of banking services is an important step in the economic development of modern society. The range of «traditional» services is very wide including making deposits, all kinds of loans, issuance of bank cards, implementation of money transfers, brokerage services, bank insurance.

In order to provide the best and most modern service, following the digitalization of economy, the bank follows the overall trend of increase a quality of services [17, 18, 19]. It tries to improve the channels of remote access for users. These remote access channels include:

- online banking ("Sberbabk Online");
- SMS service;
- mobile application.

"Sberbank Online" is an example of electronic (Internet) banking; a program designed in 2008 to remotely serve its customers. In this program, the user account is associated with his plastic card accounts. In personal account clients perform various operations with money. "Sberbank Online" may provide full information about accounts, deposits and loans of the user.

One of the main advantages of this service by the user is that a client does not need to visit the bank to perform any operation with his own funds. Internet banking as a remote banking service channel it is very famous in various habitations, regardless of their size. Among residents living in small towns, urban-type towns or villages, and using the Internet, almost 60% use Sberbank Online.

The opportunities of "Sberbank online" incorporate a few spaces. Firstly, it is a protected work process. All data that is given to Internet banking clients consistently has insurance, as encryption.

Secondly, this is customer identification. When a user enters the system and signs documents with an electronic digital signature, he is identified. This action helps reduce the cost of various identification tasks (for example, a "control call" from a call center). But it is also worth paying attention to cases of unauthorized access to information. As a result, it is necessary to conduct meetings with the client.

Thirdly, this is the availability of customer information in the system. This feature helps to reduce the time for entering user information and also helps to reduce the risk of errors—there is no need to constantly enter information such as a mobile phone or email.

Fourth, this is the opportunity to develop specialized forms for input. Almost all actions for Internet banking in various forms support the development technology of 47 different forms for entering documents. This can be applications, forms, orders. From accepted technology, this work can be performed either by the system supplier or by the customer bank itself.

Fifth, this is the ability to integrate with internal systems. An Internet banking system cannot exist without integration with the internal system of a bank, primarily with an automated banking system. Therefore, the main task for the integration of this system is to connect internal and external circuits; it has already been solved, as, among other things, it is the technology for exchanging information between systems.

In 2019 "Sberbank Online" is dominated by payments for housing and communal services, the share of which amounted to 29%, loan repayments are in second place with a share of 16%, and in third place - the payment of taxes with a share of 15%. Thus, among all payments made in the "Sberbank Online" system, housing and communal services payment is most popular among bank customers.

"Sberbank Online" can also be used as a mobile application, which is a complete replacement and supplement. This feature has been provided to clients since 2011. Mobile application "Sberbank Online" is one of the most downloadable free iPhone apps in Russia according to Apple, which indicates its popularity and effectiveness in providing digital banking services. Moreover, in 2018–2019, it entered the top 5 most popular Fintech applications of the world.

Monthly audience of "Sberbank Online", including the web version, mobile application and SMS service "Mobile Bank", in 2018 reached 58.5 million active users. Annual increase in an audience of the "Sberbank Online" mobile application in absolute terms, it was higher than in all previous periods, and amounted to + 12.6 million active users in 2018. In 2018, the Sberbank Online mobile application for iPhone and Android began to be updated 1 time in 3 weeks. During the year, 28 releases were introduced on the platforms iOS, Android, WinPhone.

In 2017–2018, significant changes were made to the functions of "Sberbank Online" and a number of innovations were made. In "Sberbank Online" mobile application, the «Tips service» was launched, which allows analyzing user data (spending, accumulation, profile data) and gives recommendations to the client based on mathematical analysis.

In 2018, the bank worked on the implementation of services that were most in demand by customers. The work was based on user suggestions, including reviews in app stores. As a result, the transaction history was updated in the application, a chat with the operator appeared, a call to the bank via the Internet, ordering a debit card and details on its issuance, payment of traffic police fines and searching the application.

An important milestone in the development of the application was the appearance opportunities for obtaining a consumer loan in a few minutes. From the launch in November until the end of 2018, 120 thousand customers received credit in the mobile application.

The number of "Sberbank Online" users in the Russian Federation is 81 million (2019). More than 75 million transactions pass through their online Banking system every day.

In 2018, more than 1 million people used the service and in "Sberbank Online" mobile application, it became possible to manage SMS notifications from the bank – the Mobile Bank service. Now customers can enable and disable notifications, as well as change their phone number to receive notifications without a visit to the office and without a self-service device. For example, 91% of active customers with Sberbank cards are connected to the Mobile Bank [20].

We underline that in 2018, for the first time, a decrease in the flow of customers in physical service channels was recorded. The Bank's clients began to conduct most of their main operations in digital channels (see Fig. 1).

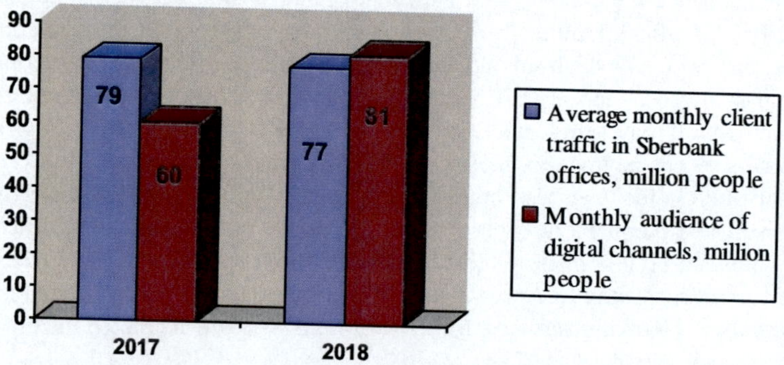

Fig. 1. Average monthly customer flow at Sberbank offices and monthly audience of digital channels, million people

Let's look at the statistics of connecting customers to electronic banking services for 2016–2019 (Table 1).

Table 1. Statistical data of users of Sberbank.

	2016	2017	2018	2019
Sberbank online (total users), millions	30	43,2	60	81
Web version, millions	17,2	17,5	17,9	31
Mobile App., millions	12,8	25,7	42,1	50

On average, the percentage increase is 23.4 percent each year. The electronic banking service "Mobile app", as well as the web version of "Sberbank Online", has an annual influx of users. The average inflow is 35 percent each year. Indicators of the dynamics of user growth in both electronic banking services are at a fairly high level, but if you go deeper into the figures, you can see that the popularity of "Sberbank Online" mobile app is growing faster both in quantitative comparison with the web version, and in percentage terms when comparing the annual indicators of the mobile app.

Such statistics may say that all these years, Sberbank is doing everything to ensure that the bank is the most comfortable and safe for the end user and, therefore, in the era of digitalization the bank gives its customers the possibility to choose between convenient ways to access the management of financial assets, while retaining security and speed of interaction with e-banking service, which is a clear priority in choosing a bank for the end user of the product.

3.3 Research Methodology

Following the methodology, proposed by the group of authors [21], we used secondary data extracted from official web-site of Sberbank in the audited financial reports (consolidated financial statements). With the secondary data collected, returns on assets and equity for the relevant years were calculated.

The data collected cover the period 2003–2019. Since the objective of the study has been to find a connection between e-banking and the profitability performance of banks in Russia with regards to the returns on equity (ROE) and returns on assets (ROA), the data was divided into two parts: pre- and post-adoption of electronic banking.

The years 2003–2007 cover the pre-adoption period and 2008–2012 and 2013–2019 covers the post full adoption period of electronic banking ("Sberbabk Online") in Russian Federation. For the greatest comparability, we identified 5 years before implementation and 5 years after, and then showed the subsequent dynamics.

4 Data Presentation and Analysis

Bank performance was measured by return on assets and return on equity. We calculated return on assets by dividing banks' net profit after taxation by the total assets held by the bank, return on equity was computed by dividing banks' net profit after taxation by the total equity capital. We computed means for 2003–2007, 2008–2012, 2013–2019 (Table 2 and 3).

Table 2. ROE and ROA in the pre-adoption period.

Indicators	2003	2004	2005	2006	2007	**Mean 2003–2007**
Net income after taxes, in billions of Russian Roubles	14	19	66	83	106	58
Total equity capital, in billions of Russian Rub	135	151	231	309	750	315
Total assets, in billions of Russian Rub	1 478	1 920	2 513	3 467	4 929	2 861
ROE	10,33	12,73	28,48	26,84	14,20	**18,52**
ROA	0,94	1,00	2,62	2,39	2,16	**1,82**

The next step was to perform the test for difference between means (for ROE and ROA) at 5% of significance.

Table 3. ROE and ROA in the post-adoption period.

Indicators	2008	2009	2010	2011	2012	**Mean 2008–2012**	**Mean 2013–2019**
Net income after taxes, in billions of Russian Rub	98	24	182	316	348	194	549
Total equity capital, in billions of Russian Rub	637	779	987	1 268	1 624	1 059	3 042
Total assets, in billions of Russian Rub	6 736	7 105	8 629	10 835	15 097	9 681	26 340
ROE	15,34	3,13	18,40	24,92	21,43	**16,64**	**17,43**
ROA	1,45	0,34	2,11	2,92	2,30	**1,82**	**2,05**

5 Summary and Conclusions

As presented in Table 2 and 3, on average ROE of Sberbank tends to decline (from 18.52 to 16.64); ROA of Sberbank tends to be the same (1.82), however we observe an increase of ROA during 2013–2019.

In terms of ROE, that means that a capacity of the bank to generate sustainable profitability goes down. The test for difference between means (at 5 percent levels of significance) showed that there is no significant difference between pre- and post return on equity and return on assets on adoption of electronic banking. The results are similar to the results of Egland et al. (1998), Carlson et al. (2001) [22, 23].

On average, every year the dynamics of growth of Internet banking customers increases by 32%. Our research confirms and explains the reasons and factors that influence this growth. Every year, the Internet banking functionality allows the client to interact with five new features by connecting them to the services of partners and government websites. Nevertheless, overbanking in Russia is often put forward as a reason for low profitability of banks. As well as along with numerous advantages, the use of remote banking is not without its drawbacks. The main factors and problems that hinder development remote banking services via Internet banking and mobile applications are: increased competition from non-Bank structures and payment systems services; low qualification of clients in terms of the use of all capabilities of these systems; lack of trust on the part of customers to remote operations and a lack of understanding of the advantages remote banking; insufficient qualification of bank employees; insufficient security of communication channels and information security; documentation in need of improvement functioning of the remote banking services tools.

As seen in Table 2 and 3, the dynamics of the profitability of bank' assets was unstable with an upward trend. These changes were due to the impact of a number of factors, among which the main one was fluctuations in bank profits. Moreover, we have observed stability during 5 years after adoption of Internet baking and then a growth. Electronic banking includes high initial capital spending and an increase in maintenance cost. Thus, the adoption of digital banking in the short run perspective shows no positive effect in terms of ROA.

Given the state of the Internet in Russia, we can say that it began to develop a little later, compared with the leading countries in this field, this, undoubtedly, influenced the state of the Internet in Russia, as well as the use of the latest technology.

The situation in Russia at the moment is extremely interesting, as the Russian Federation is in first place in many indicators in the field of Internet Banking. However, in modern Russia no efforts are made for assessment of the potential needs for digital personnel [24]. Currently, there are various ways to make bank service unique and highlight its features. An important area of improving information security is the improvement of the software interface used to make payments.

The vast majority of banks provide banking services, but the number and types of services are sometimes dramatically different. So, in one bank, the user has access to an extensive list of actions, including opening deposits, getting a loan, in another only information services can be available.

Summing up, we would like to note the ever-increasing role of Internet banking. Despite the fact there is no significant impact for banks' performance, data from recent years indicate a constant increase in interest in these services, and we assume that the constant improving of electronic banking services is a necessary requirement in the strategic planning of banking activities.

References

1. Antipova, T.: Digital view on the financial statements' consolidation in Russian public sector. In: Antipova, T., Rocha Á. (eds.) Information Technology Science. MOSITS 2017. Advances in Intelligent Systems and Computing, vol. 724, pp 125–136. Springer, Cham (2018)
2. Agumbayeva, A.Y., Chmyshenko, E.G., Pulyaev, N.N., Bunkovsky, D.V., Kolesov, K.I., Amirova, E.F.: Industrial transformation of Kazakhstan in digitalization's era. J. Adv. Res. Law Econ. **10**(6(44)), 1861–1867 (2019)
3. Rozhkova, N., Blinova, U., Rozhkova, D.: The concept of management accounting based on the information technologies application. In: Antipova T., Rocha Á. (eds.) Information Technology Science. MOSITS 2017. Advances in Intelligent Systems and Computing, vol 724. Springer, Cham (2018). https://doi.org/10.1007/978-3-319-74980-8_8
4. Jiménez, J.R.Z., Díaz, I.A.: Educational level and internet banking. J. Behav. Exp. Financ. **22**, 31–40 (2019)
5. Steven, A.: Information System: The lnformation of E-Business, pp. 11–36. Natalie Anderson, New Jersey (2002)
6. Blut, M., Chowdhry, N., Mittal, V., Brock, C.: E-service quality: A meta-analytic review. J. Retail. **91**(4), 679–700 (2015)
7. Rahi, S., Mansour, M.M.O., Alghizzawi, M., Alnaser, F.M.: Integration of UTAUT model in internet banking adoption context. J. Res. Interact, Market (2019)
8. Hasan, I., Maccario, A., Zazzara, C.: Do internet activities add value? The Italian bank experience. Rietumu bank group (2003), Consolidated and Bank Financial Statements and Auditor's Report for the year ended, T. 31 (2002)
9. Hernando, I., Nieto, M.J.: Is the Internet delivery channel changing banks' performance? The case of Spanish banks. J. Bank. Financ. **31**(4), 1083–1099 (2007)
10. Sathye, M.: The impact of internet banking on performance and risk profile: evidence from Australian credit unions. J. Bank. Regul. **6**(2), 163–174 (2005)
11. Янова, Е.А.: и др. Развитие интернет-банкинга как трансформационный фактор развития современной экономики России. Фундаментальные исследования **3**(9) (2009)
12. Sigova, M.V., Khon, O.D.: Digital banking in Russia: The mainstream of FinTech. Ученые записки Международного банковского института **20**, 44–55 (2017)
13. Rozhkova, D.: Main problems of creating an innovation ecosystem in Russia. In: The International Scientific and Practical Congress of Economists and Lawyers "The genesis of genius", pp. 161–165 (2014)
14. Ikhsanova, L., Protsko, E.: Electronic banking: tool of transforming the interaction between banks and clients and improving the service quality of the Russian banks. J. Econ. Econ. Educ. Res. **17**, 73 (2016)
15. Zahle, J.: Data, epistemic values, and multiple methods in case study research. Stud. History Philos. Sci. Part A **78**, 32–39 (2019)
16. Sberbank. https://www.sberbank.com. Accessed 06 May 2020

17. Antipova, T.; Shikina, I.: Informatic indicators of efficacy cancer treatment. In: 12th Iberian Conference on Information Systems and Technologies (CISTI), Lisbon, Portugal, 21–24 June 2017, pp. 1– 5 (2017). https://doi.org/10.23919/CISTI.2017.7976049.
18. Voskanyan, Y., Shikina, I., Kidalov, F., Andreeva, O., Makhovskaya, T.: Impact of macro factors on effectiveness of implementation of medical care safety management system. In: Antipova, T. (eds.) Integrated Science in Digital Age 2020. ICIS 2020. Lecture Notes in Networks and Systems, vol. 136, pp 346–355. Springer, Cham (2020). https://doi.org/10.1007/978-3-030-49264-9_31
19. Rozhkova, D., Rozhkova, N., Blinova, U.: Digital universities in Russia: prospects and problems. In: Antipova T., Rocha Á. (eds.) Digital Science 2019. DSIC 2019. Advances in Intelligent Systems and Computing, vol. 1114. Springer, Cham (2020). https://doi.org/10.1007/978-3-030-37737-3_23
20. Internet Banking Rank 2019 (Markswebb). https://markswebb.ru/report/internet-banking-rank-2019/. Accessed 06 May 2020
21. Abaenewe, Z.C., Ogbulu, O.M., Ndugbu, M.O.: Electronic banking and bank performance in Nigeria. West Afr. J. Ind. Acad. Res. **6**(1), 171–187 (2013)
22. Egland, K.L., Furst, K., Nolle, D.E., Robertson, D.: Banking over the internet. Q. J. Office Comptroller Curr. **17**(4), (2008)
23. Carlson, J., et al.: Internet banking: market developments and regulatory issues. Manuscript, the Society of Government Economists, Washington DC (2001)
24. Rogulenko, T.M., Ponomareva, S.V., Krishtaleva, T.I.: Competition between intelligent machines and digital personnel: the coming crisis in the labor market during the transition to the cyber economy. In: The Cyber Economy, pp. 185–194. Springer, Cham (2019)

Overview of Digital Solutions for Agriculture

Olga A. Frolova[1(✉)], Julia A. Yukhlina[1], Alexander V. Efremcev[2], Tatyana A. Dozorova[3], and Yevgeny V. Voronov[1]

[1] Nizhny Novgorod State University of Engineering and Economics, Knyaginino, Russia
ekfakngiei@yandex.ru
[2] AFG-National LLC, Nizhny Novgorod, Russia
[3] Ulyanovsk State Agrarian University named after P. A. Stolypin, Ulyanovsk, Russia

Abstract. The Purpose of this paper is to identify opportunities for digital transformation in the agro-industrial complex. The research is based on the processes of introduction of digital technologies in agriculture. Russian agricultural organizations were selected for the research object. The subject of the research is the process of digitalization, as well as digital solutions. The introduction of modern digital technologies into production allows us to create optimal soil-agrotechnical and organizational-territorial conditions that ensure a significant increase in productivity and labor productivity, reducing material costs, preserving soil fertility and protecting the environment throughout the life cycle of agricultural products.

Keywords: Agriculture · Digital agriculture · Digitalization · Transformation · Information technology · Innovation

1 Introduction

Today, the issue of solving the problems of agriculture, namely its development and modernization, with regard to the introduction of new and innovative technologies, remains relevant. Agriculture always remains in the first place, with regard to the aid allocated for its development.

In the modern world, everything moves and does not stand still, the goal is to improve the processes of further development of agricultural production. The agricultural sector is one of the most important, represented in almost all countries. We know a lot about new technologies, about new programs of development of agriculture, one of the most attractive and in trend is the program of digitalization of agriculture. The introduction of digital technologies in agriculture involves not only pilot technologies and other technologies of industrial modernization, but also integration and multi-layered forms of cooperation.

T. Antipova (Ed.): ICCS 2020, LNNS 186, pp. 238–250, 2021.
https://doi.org/10.1007/978-3-030-66093-2_23

2 Materials and Methods

The problem of analytical assessment of the agricultural sector of the Russian economy in today's economic conditions has determined the relevance of this study.

This issue is very relevant at the moment. In the course of previous research, works were published on the selected topic, namely: «The system of state support for novice farmers in the form of grants»; «Development of recommendations for granting grants to novice farmers», «The system of state support for agricultural producers in the digital agriculture. Russian and foreign experience of digitalization of agriculture».

Agriculture is a branch of economy aimed at providing the population with food and obtaining raw materials for a number of industries. Russian agriculture is the base complexing industry's largest cross – sectoral Association of the agroindustrial complex (hereinafter – AIC), in which, in addition to agriculture, there are food and processing industry, industry and processes, supporting and coordinating agriculture, such as agricultural machinery, fertilizers and means of protection, logistics, Finance and others.

Dmitry Patrushev, the Minister of agriculture, called the stability of the domestic food market a strategic reference point for development. At the same time, it is necessary to strengthen the position of Russian products abroad. This requires a systematic increase in the production of high-quality agricultural products.

According to operational data of the Federal customs service, 0.15 million tons of grain were exported in the current 2020/2021 agricultural year. The volume of wheat exports for the season was 0.12 million tons, barley-0.01 million tons, corn-0.02 million tons. As of June 5, Russian farmers purchased about 20 thousand units of agricultural machinery, including 5.1 thousand tractors, 1.7 thousand combines and 12.9 thousand units of other equipment, which is almost twice the figures for 2019. The average milk yield from one cow per day was 18.64 kg, which is 1.09 kg more than a year earlier.

The opportunities for modernizing the industry are huge. The need for food security and development of export potential transform agriculture in high-tech industry, capable not only to provide food for themselves, but also many countries of the world, and create opportunities for introducing new innovative technologies that did not exist previously, to encourage management decisions, is able to provide the population with quality and safe products.

In General, the Russian Federation is experiencing an increase in agricultural production, but since 2016, the growth rate of this indicator is slowing down, for example, in 2015, production increased by 11.2%, in 2016-by 6.6%, in 2017-by only 0.01%, and in 2018, the rate of production decreases by 0.002%.

At the same time, Russia should be ready for domestic agricultural producers to enter the foreign market, especially with products of high added value. The lack of processes compatible with the high production requirements adopted in foreign markets can lead to crisis phenomena in domestic industries with high potential and rapid growth dynamics in dairy farming, pig farming, poultry farming, sugar production, fat and oil industry, as the domestic market becomes saturated.

The development of agriculture in Russia in recent years, under the conditions of an embargo on the import of a number of products, has already identified some industry problems that need to be addressed. Agriculture in Russia is an integral part of the agro-industrial complex, which needs to be constantly developed and modernized.

In 2017, the goal was launched to make agriculture more technologically advanced, innovative, digital, taking a leading position in the economy. In Russia, digitalization of the agro-industrial complex has already been made a priority at the national level. The program "Digitalization of agriculture" should provide participants with the opportunity to use broadband, mobile, LPWAN communications, information technologies (small and big data, AI, management platforms) of domestic instrumentation (tags, controllers, sensors, controls) to significantly improve the efficiency of agriculture.

The current level of digitalization of domestic agriculture is of serious concern:

1. Lack of scientific and practical knowledge on innovative modern agricultural technologies and methodology;
2. lack of a global forecast for agricultural prices;
3. lack of proper amount of information technology tools and equipment;
4. poor logistics, storage and delivery systems lead to high production costs.

Only a small number of agricultural producers have the financial capacity to purchase new equipment, use it equipment and platforms.

In recent years, due to the automation of agricultural production and the replacement of heavy manual labor with machinery, the number of employed rural population directly in the organizations of the agro-industrial complex has been steadily decreasing. Accordingly, there is a problem of their employment, employment, ensuring profitability, which is addressed in the Republic by a whole set of interrelated measures. The program of support for novice farmers is aimed at improving the standard of living of rural residents, providing employment for the rural population, increasing the business activity of young people in rural areas, and preserving the rural way of life. These people want to devote themselves completely to the rural lifestyle in the future, we also believe in them, and we wish them successful joint work in the implementation of the selected projects", said Rishat Khabipov, Deputy Minister of agriculture and food of the Republic of Tatarstan.

A number of scientists were engaged in digitalization acutality: A. E. Shamin, I. V. Shavandina, T. N. kutaeva, O. A. Frolova in the article «Smart village. Problems and prospects in Russia» [21].

The aim of the digital transformation of agriculture is the growing contribution to the economy in 2024 to 5.9 trillion. RUB. the increase in export sales to 45 billion rubles, as well as increasing both management effectiveness and the efficiency of agricultural production and sales, reduce the cost of production processes, in turn leading to higher rural incomes. Russia has adopted normative documents on digitalization of agriculture, which regulate the main provisions.

To achieve this goal, it is necessary to solve the problems of digital transformation presented in Fig. 1.

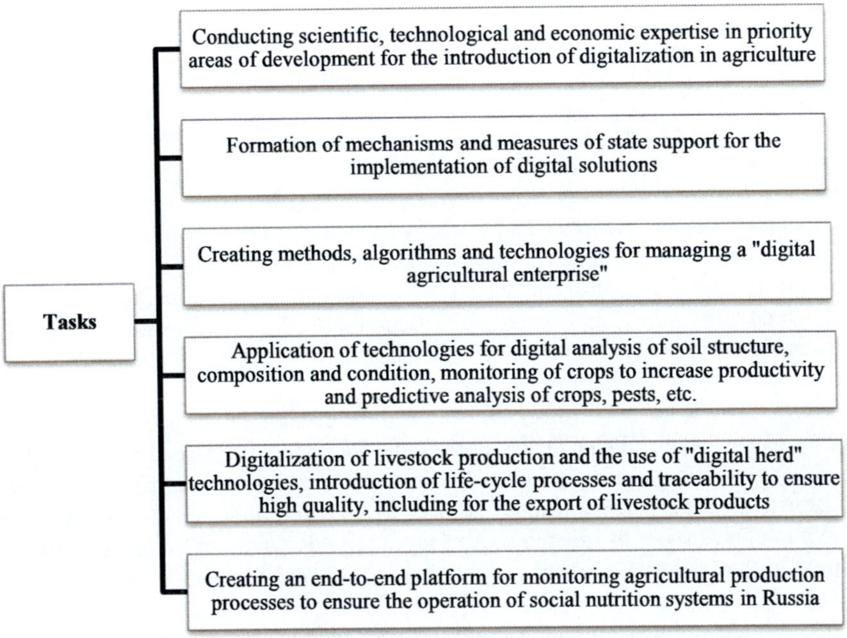

Fig. 1. Challenges of digital transformation of Russian agriculture.

As noted in the project «Digital agriculture», this platform will build work and provide a system of access to information about the counterparty, and this, in turn, will allow you to quickly check enterprises when solving serious issues, such as the financing of this organization, the implementation of lending and insurance [20].

As O. Frolova, V. Makarychev, N. Yashkova, L. Kornilova, and A. Akimov note in their works: "the Digital economy plays an important role in assessing the food security of the region. We believe that digital technologies will allow:

- optimization of logistics of raw materials and supplies;
- optimization of the composition of the raw material zone of the region according to pre-set criteria;
- calculate the necessary internal volume of food production, taking into account the population of the region and its gender composition;
- outstrip the required volume of raw material production in the region, as well as the need for resources for its production (acreage, livestock, livestock premises, agricultural machinery and labor), based on planned labor productivity.

We agree with the opinion of these authors and believe that in order to implement all plans for the future, starting with the digital transformation of Russian agriculture, it is necessary to create a sustainable business model that would promote the widespread use of digital technologies at all levels of the agricultural sector [18, p. 8].

Vinay Kellengere Shankarnarayan in the course of the study, notes that an emerging technology like Big Data, there are common issues that every industry must address in order to realize the benefits of digital transformation. For the agricultural sector, it is valuable to learn from other sectors that are further along in the adoption of Big Data analytics. A heavy investment in erudite agricultural equipment does not address the food-scarcity issue, and this is where the Big Data concept is required [2].

According to the Russian Ministry of agriculture, «the use of digital technologies in the agro-industrial complex can increase the profitability of agricultural production through point-to-point optimization of costs and more efficient allocation of funds. The introduction of the digital economy is calculated to reduce costs by at least 23% when implementing a comprehensive approach».

Agriculture is the most vulnerable sector of the economy from natural phenomena, largely dependent on climate factors. Over time, the impact of climate change on global food security will increase. The intensity, seasonality and amount of precipitation will become increasingly unpredictable, which will significantly reduce the ability of agricultural businesses to adapt to such changes. To this can be added the huge damage caused to the country's economy as a result of droughts or floods caused by climate change.

As a branch of economy agriculture has the following main features:

1) the Economic process of reproduction is intertwined with the natural process of growth and development of living organisms that develop on the basis of biological laws.
2) the Cyclical process of natural growth and development of plants and animals caused the seasonality of agricultural labor.
3) in contrast to industry, the technological process in agriculture is closely linked to nature, where the land acts as the main means of production.

The main limiting factor is the peculiarities of agricultural production. But a number of current trends suggest that this situation will change radically in the coming years. Smart agricultural technologies are being replaced by machine learning and neural networks, digital platforms, 3D printing, robotics, biosensors and Big Data. The opportunities for modernizing the industry are huge.

Agriculture in the world is turning from a traditional to a high-tech industry that can create new markets for innovative solutions and developments that did not exist before to solve a large number of practical problems. The time has come when smart digital solutions should help agriculture cope with the challenges of increasing productivity in the industry and its sustainable development [19].

In General, there are 3 stages of development and implementation of digital technologies in agriculture (Fig. 2).

Pilot technologies

- Airplane type or Flying wing
- Computer drones or drones

Market saturation

- use of GEODATA for crop forecasting,
- optimization of agricultural work, logistics management
- the arrival of a new generation of agricultural technologies – the Internet of things and blockchain.

Integration

- the creation of integrated cloud services
- remote sensing of the earth
- hyperspectral aerial photography
- weather forecast data

Fig. 2. Main components of the process of implementing digital technologies in agriculture.

Agriculture, by definition, has a number of features that determine the active use of ICT in it:

1. Multiplicity of factors that determine the results of the production process: natural-climatic, soil, biological, economic and social. And most of them are highly variable in time and space. This leads to significant management costs at the level of a particular economy.
2. the Multiplicity and territorial dispersion of economic entities, which significantly complicates management decisions on the scale of the industry.
3. Intensive and multi-sided intersectoral relations of agriculture with the enterprises of the I and III spheres of the agro-industrial complex, the number of partners of farms-suppliers of resources and buyers of products.

3 Results

There are six main directions of digital transformation of agriculture and scientific and technological development in the field of "smart agriculture", which involve the introduction in the subjects of the Russian Federation, concerning the introduction of information technologies for evaluating the effectiveness of agricultural policy, forecasting and regulating agri-food markets at the Federal and regional levels of agribusiness management. We have classified the main projects of the innovation cycle of cross-cutting digital systems (Fig. 3).

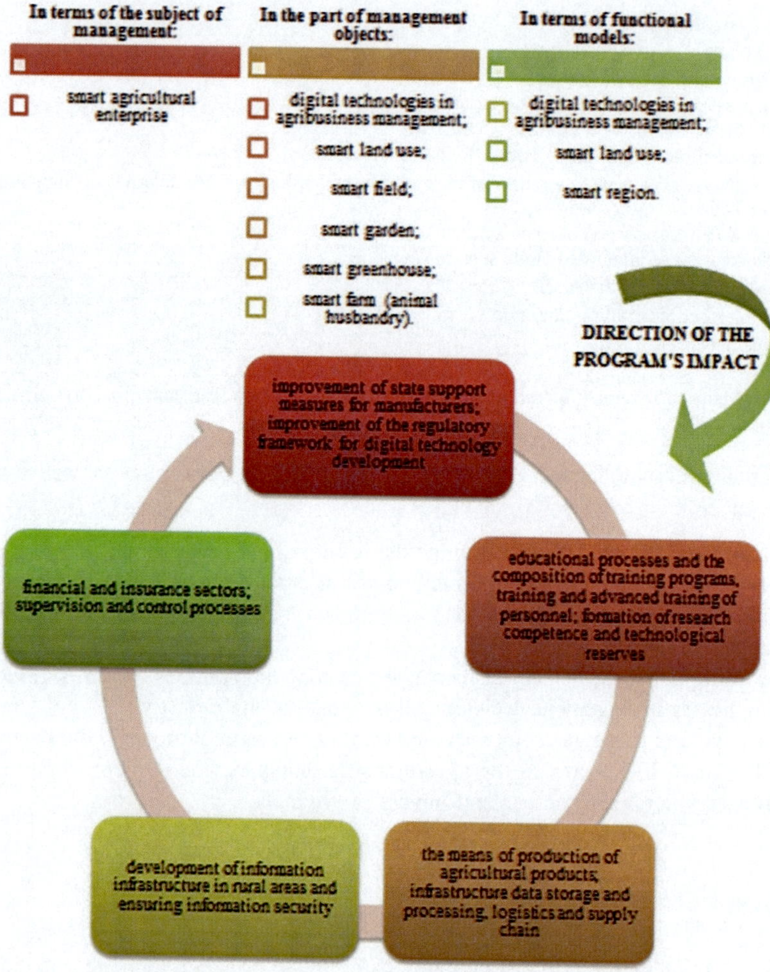

Fig. 3. Digitalization implementation Process in the agro-industrial complex.

The digital economy is now on everyone's ear, it gives a lot of opportunities to implement, modernize, improve, and actively develop your business. The Digital agriculture program will be gradually added to agriculture, gaining momentum, but at the same time actively influencing the production of one or more enterprises. In our opinion, the introduction of digital technologies will reduce costs for many needs of the organization. In the course of the study, it was found that the program can affect many aspects of agricultural production (Fig. 3).

In Russia, the digital economy is regulated by a number of regulations [3–13] (Fig. 4).

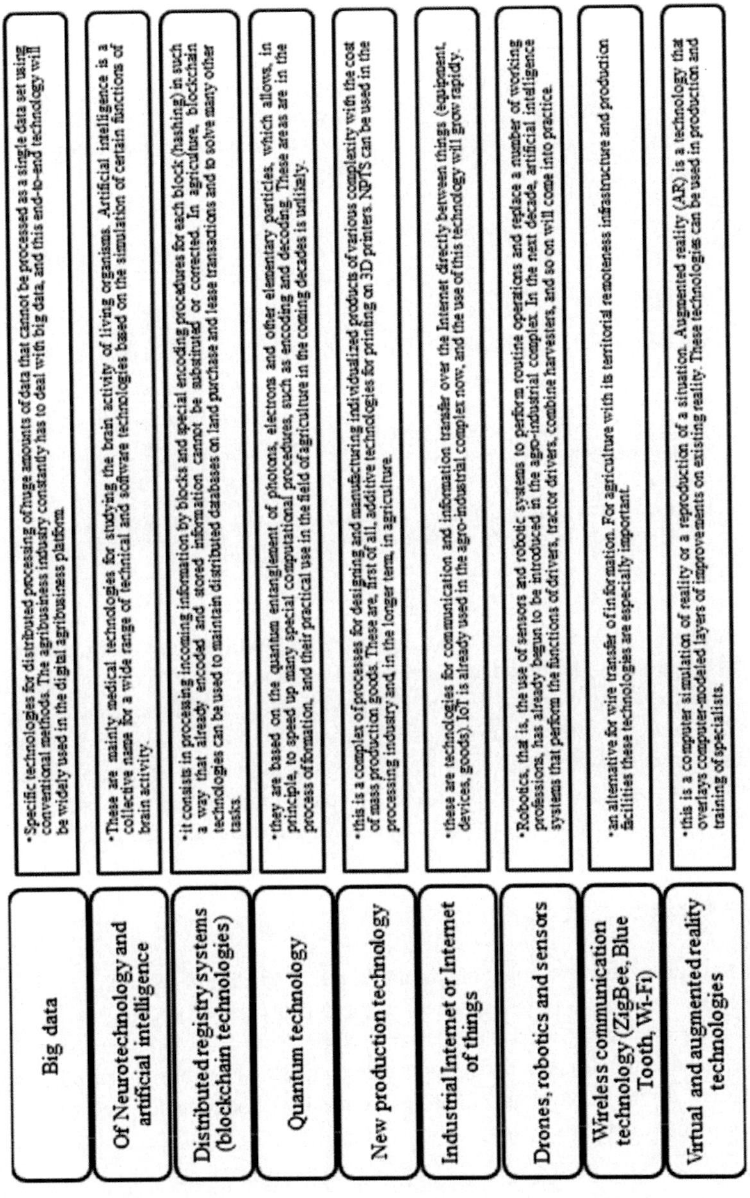

Fig. 4. Trends in digital technologies in agriculture.

If we draw a line between a traditional and digital enterprise, we can conclude that the difference will be very large. Figure 5 shows the main capabilities of a single enterprise before and after digitalization.

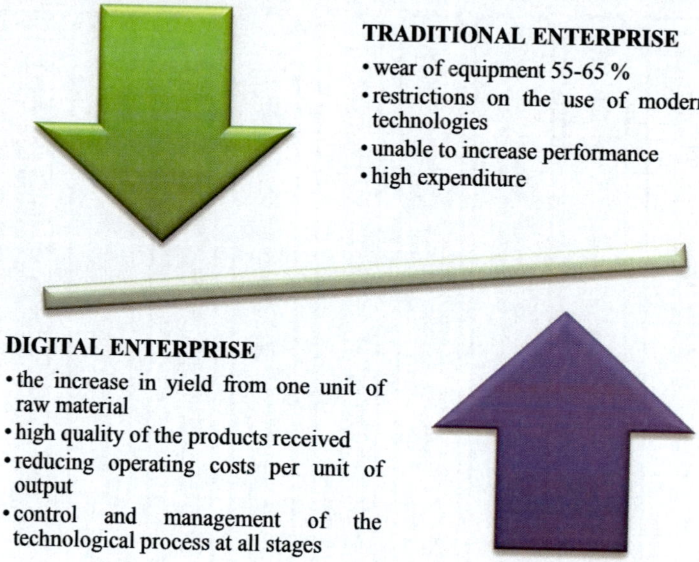

Fig. 5. Trends in digital technologies in agriculture.

Regardless of the exact term used, digitalization implies that management tasks on-farm and off-farm (in the broader value chain and food system) focus on different sorts of data (on location, weather, behaviour, phytosanitary status, consumption, energy use, prices and economic information, etc.), using sensors, machines, drones, and satellites to monitor animals, soil, water, plants and humans. The data obtained is used to interpret the past and predict the future, to make more timely or accurate decisions, through constant monitoring or specific big data science enquiries [14, 15].

4 Discussion

One of the successful examples of the introduction of digital technologies in agricultural production is the Viasna Agrofirm in the Nizhny Novgorod region.

The company has developed the project «Agriculture 4.0». its Essence is to build an effective management system for agricultural production based on modern management methods and digital technologies that ensure the best performance in the industry.

Main elements of this project:

- Process approach to business organization, modeling and design of ideal business processes with «0»;
- computing and telecommunications infrastructure (server, sensors, communication channels, iot) is built around the business process, integrated and integrated into a single complex;
- comprehensive automation of business processes (100% coverage of the business process by means of continuous monitoring and control, automatic generation of accounting documentation, robotic response to deviations from the standard);
- mobile apps for business process participants;

- training and organization of work of "field" and other employees of the company in accordance with the business process standard;
- formation of a knowledge base based on historical data, agro-chemical analysis data and other sources to optimize cultivation technology (point farming) and increase the intensity and efficiency of land and other resources use;
- integration with 1C-automated formation of accounting and management information.

The investment volume is more than 45 million rubles.

The project provides for measures that will allow for more effective implementation of agricultural policy both across the country, through the formation of appropriate information and analytical support on the basis of combining (currently disparate) various databases (Rosstat, Federal customs service, Rosreestr, Roscosmos, Roshydromet, etc.), and within a specific agricultural enterprise. Implementation of these solutions will make it possible to plan more effectively the development of the industry as a whole, the placement of crops taking into account the transport shoulder to the consumer and the provision of rolling stock for Russian Railways (which is especially important), and to monitor the state of crops in real time. In addition, it is planned to simplify the interaction of agricultural enterprises with the Ministry, banks and insurance companies through the «Intelligent system of state support measures» and «personal account of the recipient of subsidies». A complex of "digital agricultural solutions" for business ("Smart farm", "Smart field", etc.) and a training program that covers 55 thousand agricultural workers, create conditions for the widespread introduction of "point farming", which we discussed earlier, reducing agricultural risks and increasing crop yields and profitability of enterprises that have entered the «digital world».

Of course, the practical implementation of new methods of work, and especially such large-scale changes in the work of an entire industry, is fraught with many difficulties and you need to be prepared for this. And it is necessary to plan changes starting from the field – how the nature of the work of a simple worker will change, whether it is a machine operator, agronomist or engineer, what new skills and knowledge he should have, what and how to teach and learn. And on the basis of this, build the work of other departments, managers. Only in this way, by changing the basic principles of business process organization, you can get a real result from the implementation of IT solutions. And integrated software products will allow specific businesses to reduce the cost of product ownership and receive centralized support.

In the course of our research, we identified the main problems that slow down and hinder the process of digitalization of agriculture:

1. Low level of financial security for the introduction of ICT in most agricultural producers.
   As already noted, in the agricultural sector, it is necessary to stimulate the implementation of the process of digitalization through state support, because most farms are working on the verge of payback using outdated technologies.
2. Currently, there are no educational technologies for training specialists for digital agriculture, which leads to a shortage of qualified personnel.
   According to the Ministry of agriculture of Russia, today there are half as many it specialists working in agriculture in Russia as in countries with a traditionally developed agro-industrial sector. At the moment, the Russian agricultural sector is provided with training of highly qualified personnel for digital agriculture.

3. Digital inequality between the city and the countryside persists, which is why there is a barrier to the introduction of digital technologies in rural hinterlands. The digital economy will be a new technological basis for the social and economic sphere of the Russian Federation.
4. Shortcomings of legislative and regulatory regulation.

Issues of development of the state information support system in the field of agriculture are regulated by the same article 17 of the Federal law «On the development of agriculture», adopted in 2016. However, this article needs to be changed.

The scientific novelty of the work is the development of a digital enterprise model for agricultural organizations, which describes business processes and software products (platforms) for their automation and optimization (Fig. 6).

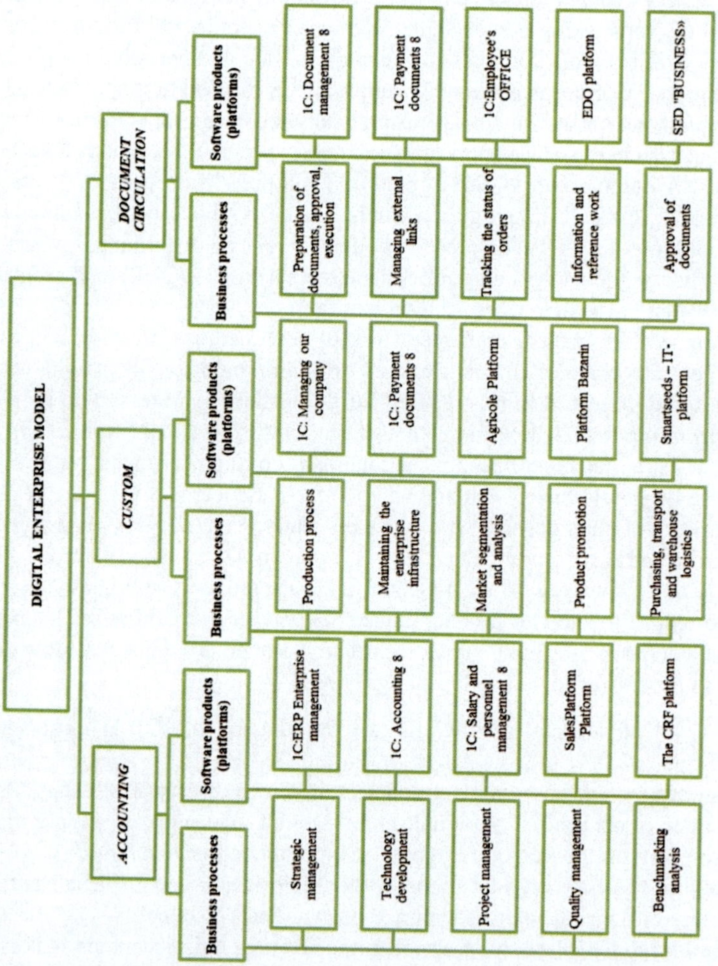

Fig. 6. Digital enterprise model for agricultural organizations.

Thus, digitalization contributes to the automation of agricultural activities, which leads to an increase in both productivity and product quality, has a beneficial effect on the growth of labor productivity at enterprises, and pays attention to the high technological breakthrough of the enterprise. Do not forget that the growth in demand for IT specialists who will be able to work after graduation is increasing. The output of products from the raw material unit increases, the costs per unit of output are reduced, while the control and management of the output of products and the entire technological process is automated at all stages.

We see that the future of agricultural enterprises is being built out of outdated traditions thanks to the breakthrough of a new digital solution.

References

1. Patnaik S., Sen S., Mahmoud M. (eds): Big Data for smart agriculture. Smart Village Technology. Modeling and Optimization in Science and Technologies, vol. 17, pp. 69–80. Springer, Cham (2020)
2. Shankarnarayana, V.K., Ramakrishnab, H.: Paradigm change in Indian agricultural practices using big data: challenges and opportunities from field to plate. https://doi.org/10.1016/j.inpa.2020.01.001. Accessed 01 Apr 2020
3. Decree of the President of the Russian Federation of 7 May 2018 «On national goals and strategic development goals of the Russian Federation for the period up to 2024»
4. Decree of the President of the Russian Federation № 642 of 1 December 2016 «On the strategy of scientific and technological development of the Russian Federation»
5. Decree of the President of the Russian Federation № 350 of 21 July 2016 «On measures to implement the state scientific and technical policy in the interests of agricultural development»
6. Program from 28.07.2018 № 1632-R, approved by the order of the Government of the Russian Federation «Digital economy of the Russian Federation»
7. Resolution № 528 of 29 March 2018 «On budget allocations for the implementation of priority measures of the program» Digital economy of the Russian Federation»
8. Resolution of the Government of the Russian Federation of 25 August 2017 № 996 «On approval of the Federal scientific and technical program for agricultural development for 2017–2025»
9. Decree of the Government of the Russian Federation dated 07 July 2017 № 1455 «Strategy for the development of agricultural machinery in Russia for the period up to 2030»
10. Decree Of the Government of the Russian Federation dated 8 December 2011. strategy for innovative development of the Russian Federation until 2020
11. Forecast of scientific and technological development of the Russian Federation for the period up to 2030, approved by the Government of the Russian Federation on 10 July 2018
12. Order of the Ministry of agriculture of the Russian Federation dated 12 January 2017 № 3 «On approval of the Forecast of scientific and technological development of the agro-industrial complex of the Russian Federation for the period up to 2030»
13. Government Decree «On the procedure for developing and implementing comprehensive scientific and technical programs and projects of the full innovation cycle in order to implement the priorities of scientific and technological development of the Russian Federation» (draft)

14. Managing socio-ethical challenges in the development of smart farming: from a fragmented to a comprehensive approach for responsible research and innovation. https://doi.org/10.1007/s10806-017-9704-5. Accessed 21 Apr 2020
15. Eastwood, C., Klerkx, L., Nettle, R.: Dynamics and distribution of public and private research and extension roles for technological innovation and diffusion: case studies of the implementation and adaptation of precision farming technologies. J. Rural Stud. **49**, 1–12 (2017)
16. In the starting blocks for smart agriculture: the internet as a source of knowledge in transitional agriculture, https://doi.org/10.1016/j.njas.2019.100309. Accessed 01 May 2020
17. Janssen, S.J.C., Porter, C.H., Moore, A.D., Athanasiadis, I.N., Foster, I., Jones, J.W., Antle, J.M.: Towards a new generation of agricultural system data, models and knowledge products: information and communication technology. J. Agric. Syst. **155**, 200–212 (2017)
18. Frolova, O., Makarychev, V., Yashkova, N., Kornilova, L., Akimov, A.: Digital transformation of agro-industrial complex. In: 5th International Conference on Agricultural and Biological Sciences (ABS) WGD Conference Series: Earth and Environment Science, vol. 346 (2019)
19. AgroXXI.ru agro-industrial portal. https://www.agroxxi.ru/mirovye-agronovosti. Accessed 21 Mar 2020
20. Vaganova, O.V., Solovjeva, N.E., Aulov, Y.L., Prokopova, L.I.: Transformation of agriculture through digitalization, innovative solutions, and information technologies. In: III International Scientific and Practical Conference «Digital Economy and Finances». Advances in Economics, Business and Management Research, vol. 137, pp. 65–68 (2020)
21. Frolova, O.A., Shamin, A.E., Shavandina, I.V., Kutaeva, T.N.: Smart village. Problems and prospects in Russia. In: Digital Science. Scopus Database, pp. 400–486 (2019)

Educational Technologies

The Processes of Informatization and Digitalization in the Linguistic Education

Angelina Bagateeva[1(✉)], Chulpan Ziganshina[1], Albina Islamova[2], and Anifa Akhmetshina[2]

[1] Naberezhnye Chelny Institute, Kazan Federal University, Naberezhnye Chelny, Russia
angel803@yandex.ru

[2] Naberezhnye Chelny State Pedagogical University, Naberezhnye Chelny, Russia

Abstract. The article considers the main processes in modern education – informatization and digitalization as one of the directions of modern education modernization. Digitalization, as one of the dominant areas of development of the modern economy, has also been reflected in the field of higher education. Being one of the most important part of society, the digital environment of higher education leads to significant changes in the educational process organization as well as in its content. The description of information resources of the digital system of foreign language education (computer training programs, electronic educational resources, Internet capabilities) is given. The main digitalized instruments for linguists are identified: databases, archives, Internet resources, dictionaries, encyclopedias, etc. Their main feature is full plainness and free of use. The information is being constantly updated in all electronic resources as well as the main information communication technologies (ICT) are being updated. The practical significance of the article consists in the ability of ICT and Internet resources can speed up and simplify the research work of linguists.

Keywords: Informatization · Digitalization · Computer information technologies · Glossary · Education · Linguistics

1 Introduction

Digitalization is one of the dominant areas of development of the modern world. Nowadays informatization and digitalization have also been reflected in the field of higher education. Being one of the most important part of society, the digital environment of higher education leads to significant changes in the educational process organization as well as in its content. Digitalization acquires special significance in a foreign language training of students in the system of higher education. The digital transformation in the higher education system allows to organize and systematize the educational process, change the content and format of education [3], involve students in an interactive environment and increase motivation to learn a foreign language.

T. Antipova (Ed.): ICCS 2020, LNNS 186, pp. 253–259, 2021.
https://doi.org/10.1007/978-3-030-66093-2_24

Various types of information technologies are widely used, radically updating the education not only in new forms, but also in new ways to achieve learning goals.

A foreign language course, supported by ICT is the most effective way of building a foreign communicative competence among university students. At the same time, the most common ways of application of ICT in education are the application of a number of ICT tools based on different platforms or the application of a uniform academic platform which enables to integrate a range of ICT applications based on it [1, 2].

Various types of information technologies are widely used, radically updating the education not only in new forms, but also in new ways to achieve learning goals [3, 4].

Linguists began to use computer technology somewhere about 70 years ago. That led to appearing massive dictionary databases, dictionary collections, etc. That was the period of creating new methodology of morphological, syntactic, lexical, semantic types of the linguistic analysis of the dictionary and structural analysis [5]. Various programmes were created for linguistic aims of study to produce quick and effective analysis of the vocabulary of the text. This managed to accelerate the process of making a detailed report on the web of the text.

2 Methods

Methodology functions, related to a foreign language teaching, are methodology variants of these technologies in the development of speaking language skills and shaping social, cultural and intercultural competence of the students.

The efficiency of the academic process depends on a number of factors, the most critical of which involve the way of representation of the course structure and contents. Contemporary foreign educators resort to the term "teacher training design" to denote methods, principles and approaches to the formation, structuring and content-building of the academic course based on ICT.

In connection with the rapid development of mobile technologies and ICT, there is a growing attention to the teaching techniques aids of e-learning format. Meanwhile, it's vital to attend not only to the teaching techniques but to the academic process systematization. The traditional method of teaching a foreign language is a communicative one. The communicative cycle rests on the notion "building a situation, characterized by the information deficiency" (building the information gap). Thus, communicative approach to the structure and content-building of the course reflects the following stages: crucial environment for communication (building this environment), finding out information and defining a peculiar information gap, the acquisition of language material, relevant to the solution of the communicative task, communicative training, a comeback to the initial situations in this context.

The most efficient methods of using ICT in the educational process of linguists are the following:

- a method of passing on new knowledge. This model is based on digital education Internet resources which facilitate a massive access to learning materials irrespective of temporal and spatial boundaries. The downfall of this method is that the student is a passive recipient of information.

- a method of building skills. ICT application enables to create a reusable database of educational and teacher training resources and have a regular feedback, which enhances the acquisition of the material under study. The systematic application of this method taps ICT potential to the full extent and reconsiders the role of a teacher.
- a method of building new knowledge based on the experience of ICT application of the above mentioned models. Due to this method, students are supposed to build the strategy to solve academic tasks on their own, have a critical appraisal of the information and analyze approaches to the solution of the tasks. The use of ICT in the process of learning a foreign language has the following advantages: appeal for the students to evoke interest in learning a foreign language, interactivity of educational materials; the possibility of adapting the educational process to the needs of each student. The demerits of this method in connection with ICT implementation are high demands to the information and communication culture of the university teaching staff.

3 Results and Discussion

The continuously increasing role of information technologies and a significant number of qualitatively new scientific achievements that have led to transformation in all spheres of human activity in general, and the field of foreign language teaching in particular.

The authors pinpoint the definition of the term "mobile education" from their own perspective and offer the checklist of the most efficient mobile technologies and ICT used in teaching techniques (e-mail, blog-technologies, wiki-technologies, podcasts, web-forums, e-dictionaries, information and reference internet resources, video communication means, etc.) which aim at the development of foreign communicative competence of university students. The introduction of ICT into the education process while teaching a foreign language has a beneficial influence on the achievement of the main goal of modernization in education, which is supposed to create quality education, that contributes to the development of harmonious personality who will fit in the information domain and be knowledgeable in ITC as long as there is a surge in social demand of information community.

The main components of the digital education system aimed at solving different scientific research of a linguist, in our opinion, information resources and means of monitoring the educational process occupy a special place [6, 7]. To form all the necessary competencies of students of a linguistic university, we use:

1. Computer-based training programs to intensify professional foreign languages preparation. The computer program "SANACO" used by us is a language software complex, which provides all the main aspects of language learning - auditing, pronunciation and working with written text; vocal exercises improve intonation and pronunciation; It is possible to organize individual work and in groups (up to 6 students); The use of multimedia training materials helps to diversify the lesson, while practicing oral speech.

2. Electronic educational resources (EER) that enable the reproduction of educational materials using electronic devices. The development of information technologies funded to the development of a new form of education - electronic education, which requires new means of learning - an electronic educational resource. The EER developed for the areas of preparation “Linguistics” has the following structure:
    - glossary - a small dictionary in which words for a specific module are collected;
    - chat rooms - for instant messaging between the student and the teacher;
    - news forum - a teacher in a news forum creates topics with explanations of certain sections, tests and others;
    - theoretical material - used for self-study and offered in the form of a file for download (pdf-document for self-study by the student offline), either in the form of an HTML document for self-study by the student, or in the form of a link to electronic resources on the Internet (texts, audio or video, electronic library resources).
    - electronic textbooks - designed to study each of the modules;
    - dictionaries - they are structured by modules, have a menu, by hyperlinks from the positions of which you can access the desired module;
    - tests - contain tasks of open and closed types, can be used for preliminary, current, repeated, periodic control.
    - electronic tasks - they are dedicated to the independent work of students, being equipped with an assistance function and a feedback mechanism;
    - interactive presentations - the main task of such presentations consists in mastering language by students themselves. The interactive feed makes such a study visual and interesting.
    - videos - a selection of educational videos corresponding to the lecture and practical course of each of the modules, and allowing you to organize various forms of training work in an interactive format.
    - test topics - a list of topics that must be disclosed to the student in the MOODLE system, a file located in the ESM, or implemented outside the system.
    - final test - tests with instructions on their implementation. The question bank for tests contains questions of two or more types (multiple choice, correspondence, short answer, etc.).
3. Internet resources (dictionaries, text and video glosses). Being an information system, the Internet offers students a variety of information and resources: e-mail, the ability to publish personal information, the ability to communicate with native speakers, conduct webinars with the participation of other teachers, create their own home page, access to information resources, reference catalogs and search systems [8].

Nowadays the Internet provides linguists with various resources in the form of arrays (databases) of information among which we can define the following:

1. Electronic encyclopedias (feb-web.sh).
2. word Association Networks (wordassociations.net).
3. Electronic libraries (eLIBRARY.RU, Cyberleninka, ARBICON).
4. University Databases.
5. Author’s sites.

6. Forums.
7. Word processors (MicrosoftWord, LotusWordPro OpenOffice Writer)

There is a significant part of the Internet resources devoted to the educational and scientific spheres. It can be presented in different forms - in the form of online libraries, dictionaries, encyclopedias, or even databases and archives. The availability and mobility of the information provided allows them to be used by various representatives of the educational community - scientists, teachers, students and so on. Due to the complete openness of electronic information resources, there is a huge variety of their types. Using the electronic resources, linguists sufficiently reduce the time of preparation the research papers and scientific studies [9].

The linguist due to the specific sphere of his work studies various encyclopedic sources in order to obtain the necessary information and/or compare it with other sources. In the age of informatization and digitalization, electronic sources for scientific work provide mobile and fast information needed.

To analyze the dynamics of linguistic processes occurring in the modern language, linguists use electronic databases which contain different verbal associations for the purpose of a visual representation. These databases become the subject of study for philologists engaged in psycholinguistics. The Word Association databases are regularly updated. This gives philologists the opportunity to track the associations arise with a particular word in different context (in the context of a noun; in the context of the adjective; in the context of a verb; in the context of the dialect) [10].

The existing electronic information resource "network of verbal associations" (wordassociations.net) makes it possible to study verbal associations that occur in the head of the addressee or sender with a specific word. Creation process and compiling a list of associations is based on a software module. This module analyzes classic (diachronic) and modern (synchronous) works of Russian writers and masterpieces of foreign literature. In this case, the principles of a systematic phased (research) approach are used. Combination of algorithms created by the programmer allow you to display a set of associations associated with required word. According to the author of the project, when developing algorithms, mental and nervous processes occurring in the human brain were taken into account. However, it is important to note that the output list of associations can be considered solely as an average value result [11]. Various types of electronic libraries provide quick access to the necessary fiction and non-fiction texts.

Author resources allow adding various documents that are open to the public, without prior notice moderation (i.e. content is controlled by a network moderator before publication). Example of this author's site is durov.com, Russia's largest website for Philology students. Various forums of literary and linguistic orientation allow philologists to exchange experience and research results, coordinate their actions with native speakers of other languages, and receive from these carriers the necessary information useful for scientific research.

An example of such forums is Linguoforum (lingvoforum.net this resource is dedicated to discussion of topics such as: questions about the Russian language (the norm of modern literary language, literacy, etc.)), etymology of proper names, General

Linguistics, Applied Linguistics, etc. Various types of word processors (for example, MS Word) are almost the main object interaction of the philologist and computer technologies. Such programs are widely used for creating, editing, and preparing various text documents. Text editors of various types allow you to format the text (for further sending), and check the text for typos and descriptions.

The program in word processors is based on dictionary and grammatical databases compiled by Philologists together with programmers. This allows real-time tracking of the degree of literacy of the created text: from punctuation and spelling to its stylistic features.

Electronic mental maps can be of great help in the work of literary scholars. With their Help, you can reconstruct any story structure of a work of art and form it. This function is provided by the ultra liner program developed by Russian programmer Yuri Blokhin. This allows you to decompose the story elements into a form of mental maps between which you can form a semantic and logical connection. This allows the literary critic to trace the structure of the plot and the sequence of elements of the story: the set of "values", the relationship of characters to each other and in combination with the described events, to determine the points and vertices of the plot, to trace the balance the intensity of the character, the charge and the arc. As a result, we get the "skeleton" of the story, completely abstracting from the text layer [12].

4 Conclusion

The basis of the fundamental set of information technologies is communication, which it is inextricably linked with language and, consequently, with linguistics.

The use of information technologies significantly facilitates and accelerates accumulation the latest knowledge (in the context of the linguistic sphere), their search and editing. However, Information technologies can become for a philologist as well as for a linguist not only a source of dynamic digital content information, but also directly the object of research as a communicative linguistic phenomenon [11, 12].

To sum it up, it may be stated that research in this area is designed to become a catalyst for the development of modern means of teaching foreign languages based on ICT, to promote the study and implementation of ICT for linguists. The content and philosophy of the study should be aimed at promoting advanced technologies in the field of teaching and learning foreign languages, the educational process and the training of highly qualified specialists.

References

1. Ptushchenko, E.B.: Innovative approach in teaching computer science as the basis for the formation of professional information and technological competence of an expert. Bull. Adyghe State Univ. Ser. Pedagogy Psychol. **5**, 168–171 (2008)
2. Khokhlova, Yu.E.: Glossary on the Information Society. Institute for the Development of the Information Society (2009)

3. Vigursky, K.V.: Philology and modern information technology. http://feb-web.ru/feben/feb/media/philo-info.htm. Accessed 10 June 2020
4. Khrolenko, A.T.: Fundamentals of Modern Philology: A Textbook (2013)
5. Electronic libraries, dictionaries, encyclopedias. http://edu-top.ru/katalog/?cat=6. Accessed 10 June 2020
6. Fudamental electronic library. Russian literature and folklore. «The Network of Verbal Associations» project. https://wordassociations.net/en/about. Accessed 10 July 2020
7. Gorny, E.N.: Internet and philology. Electronic resource of the «RZh» jounal. http://old.russ.ru/netcult/20001201_gorny.html. Accessed 10 June 2020
8. Durov.com. http://www.durov.com/durov.htm. Accessed 10 June 2020
9. Zakharova, I.G.: Information technologies in education: textbook for higher educational institutions, Academy, p. 338 (2008)
10. Syisoev, P.V.: Directions and prospects of informatization of language education. High. Educ. Russia **10**, 90–97 (2013)
11. Robert, I.: Modern information technologies in education: didactic problems, and use perspectives: Russian Academy of Education, Moscow (2010)
12. Amazon Prime The Screenwriter's Bible. 6th Edition: A Complete Guide to Writing, Formatting, and Selling Your Script. https://www.amazon.com/Screenwriters-Bible-6th-Complete-Formatting/dp/l935247107. Accessed 10 June 2020

Academic Freedom for the University as an Intellectual and Creative Organization

Alexandra Ponomareva[1,2](✉) and Maxim Ponomarev[1,2]

[1] Department of Marketing and Advertising, Rostov State Uiversity of Economics and South Federal University, B. Sadovaya. 69, 344002 Rostov on Don, Russia
alexandra22003@rambler.ru

[2] Department of Economic Theory and Enterpreneurship, Russian Academy of National Economy and Public Administration, Pushkinskaya Street, 70, 344011 Rostov-on-Don, Russia

Abstract. The article reveals the essence of the "academic freedom" concept and phenomenon, develops the directions for the development of universities as an intellectual and creative organizations based on the implementation of the academic freedoms' principle, and offers a matrix for the realization of academic freedoms and autonomy of the university. The practice-oriented matrix presents the infrastructural components, the development of which creates the transformational basis of the university, the principle of academic freedom autonomy.

The author considers the features of the university as an intellectual and creative organization, determining the specifics of its management based on the implementation of the principle of academic freedom. This approach involves the transformation of the marketing paradigm of the university as a management tool that activates the intellectual and creative potential of its employees based on the implementation of the principle of academic freedoms in the context of the digitalization of the scientific and educational process.

The main function of the "academic freedom" concept is revealed, which consists in the formation of a theoretical and ideological paradigm, which serves as the basis for creating sociocultural, economic and managerial mechanisms to counter the most powerful negative impact vector on the university environment, which has a destructive effect on the essence of university science and education.

The specifics of the concept's contents and the essence of the managerial tool "academic freedom" are studied in relation to the current stage of the university as a social institution development, which allows them to be used as the basis for transformational processes aimed at de-bureaucratizing the university and building a future university self-government system.

Keywords: Academic freedom · Creative organization · University

T. Antipova (Ed.): ICCS 2020, LNNS 186, pp. 260–268, 2021.
https://doi.org/10.1007/978-3-030-66093-2_25

1 Introduction

The socio-economic institution "university" has a long history and traditions. The purpose of the university is to be the center of advanced science, education, culture, the focus of the intellectual and creative elite, to form and develop the intellectual, scientific, creative potential of society. Historically, an important component of the university's concept, the foundation for building its corporate culture, scientific and educational relations between scientists, teachers and graduate students, students were academic freedoms that allow teachers, researchers, students and graduate students to be independent in their scientific and educational activities. Massification of science, reduction of state financing of higher education, strengthening of state control in relation to universities, quantitative methods for assessing the effectiveness of its activities, the destruction of the university's self-government system, the development of a managerial approach to the organization and planning of science and education in higher educational institutions and research institutes, including control of the content component of the educational and scientific process, total digitalization [1], including in its chaotic forms during the coronavirus pandemic, practically led to the destruction of the institute of academic freedom, which contradicts the essence of the university and the activities of scientists, teachers, students and graduate students within its walls. Thereby, the main research problem is the contradiction between the listed dominant trends in higher education and the essence of a university based on the principles of academic freedom and autonomy. The scientific and educational product of the university is the result of the intellectual, scientific and creative work of scientists and the cooperation of the faculty with students and post-graduate students. The formation of a high-quality scientific and educational product is based on human capital development, the intellectual and creative potential of the university, regard for the personality of a teacher-researcher, post-graduate student, student. Our hypothesis is that the university is an intellectual and creative organization, and the leading principle of university management marketing as an intellectual and creative organization is the principle of academic freedom, which activates the intellectual and creative potential of scientists, teachers, graduate students, students on the basis of self-management of scientific and educational activities.

The materials of our research are the publications of a scientific and practical nature devoted to the problems of modern educational systems, practical cases representing fragments of the educational process at the university, academic freedom program documents, publications in professional communities, social media posts. Observation, monitoring were used as research methods in order to identify the features of the implementation of academic freedom principles in the modern scientific and educational process; the descriptive and analytical method made it possible to present approaches to the "academic freedom" and "university autonomy" concepts in program documents; Semantic analysis of the content of publics in social networks devoted to educational problems has revealed negative phenomena in the modern academic environment associated with the restriction of academic freedoms.

2 Results

Studying the concept of "academic freedom" in diachronic and synchronous aspects led to the following conclusion: its essence is changing, depending on which of the socio-economic forces exerted on the university, striving for autonomy, independent and objective scientific knowledge and truth, the greatest negative impact, prevented the realization of his social mission. In different historical periods, the church, the state, the fundraising and financial management system of the university, its administrative apparatus, and the public became such a force. In the Middle Ages, the university and the scientists developed their scientific ideas to oppose churches and declarations, and then to realize academic freedoms in order to create a university space of science that is safe for scientists who could not influence the pro-clerical forces. Thus, scientific thought and scientific discoveries, opposing and contradicting clerical dogmas, gain a right to exist, and their authors should be in relative safety, conduct their research and express their ideas within the walls of the universities. It is believed that the institutional consolidation of the "academic freedom" concept, that arose in an autonomous medieval universities, occurred in 1811 under the influence of the ideas of J. Fichte. H. Helmholtz at the end of the 19th century noted in his speech that the professor is free to choose the direction of scientific research, teaching methods and lifestyle [2]. Especially relevant in the period of transformation of the management of scientific and educational activities of universities, an example of which is the Russian system of higher education, in the light of the discussion about modern rating systems and effective contracts as tools for evaluating the work of scientists and teachers is the idea of G. Helmholtz that they can only be evaluated by peers - scientists with authority in a certain field of scientific knowledge. The main basic value of the classical German University is truth, the search for which is carried out by scientists in compliance with the principle of academic freedoms - freedom to learn and freedom to study based on the principle of the unity of research and education [2]. The modern European approach to the concept of "academic freedoms" was laid down by the Unesco Recommendations (1974), Magna Charta Universitatum (1988) [3], the Declaration on Academic Freedom and Autonomy of Universities (1990), and the Unesco World Declaration on Higher Education (1998), The Bologna Declaration (1998) and other documents [3].

The analysis of Russian documents and social processes associated with the phenomenon of "academic freedom" shows that the latter are the point of application of forces in the struggle for democratic values and part of the liberal picture of the world. Thereby, one of the first documents in Russia, declaring the academic freedom were so-called the Privilege on the Academy in Moscow in 1682 [3]. In addition to academic freedoms, the Privilege declares class equality, the right of its own charter and diploma. This document was never adopted, and only under Peter I, in 1706, a decree on academic freedom was signed at the Moscow Academy [3]. The St. Petersburg Academy of Sciences, founded in 1724, in fact, lacked autonomy and academic freedom, the Academy was under the control of government officials of the Academic Chancellery [3]. Under Alexander I, two important documents were adopted that developed academic freedom in Russia: the Decree Act for the Imperial University of Dörpt (1802), which subsequently extended to other Russian universities, and the First

General University Charter (1804), which declared university autonomy [3]. University freedom in Russia in the 19–20 centuries become the object of struggle of various political forces. So, in 1835, the adoption of the University Charter abolished most academic freedoms, in 1863, the new Charter, as part of the liberal reforms of Alexander II, returned part of the academic freedoms to the profession, in 1884, Alexander III, during the "counter-reforms", essentially eliminated academic freedoms and the autonomy of universities [3], and in 1905, Nicholas II, yielding to the requirements of all-Russian student strikes, returns the rectors' election, which later was canceled in the Soviet period in 1929 and returned in 1987 [3]. In 2012, the Federal Law "On Education in the Russian Federation" was adopted in Russia, in which the academic rights and freedoms of teachers and students and the autonomy of educational organizations are declared as principles of state policy [4].

The content of the "academic freedom" concept was transformed when it was transferred to the US academic environment (in 1885 the first publication on academic freedom was published in the USA [2]). The source of negative influence that can deform the essence of the scientific process has become part of the internal environment of the university. The university's fundraising system, designed to provide the financial and economic foundations of its activities, began to directly influence the thematic and essential direction of university scientific research. The possibility of commercializing scientific research is becoming an important motive for supporting various areas of scientific activity. The fundraising system negatively affects one of the main academic freedoms - the freedom of informed choice of topic and direction of scientific research. In addition, the university's management structures began to step on the civil liberties of the professors. The reason for the creation of a special document - the Declaration on Academic Freedoms [3] - was the dismissal of 11 professors. The document can be conditionally divided into two parts - conceptual (declaration of principles) and practical. In the conceptual part, three elements of academic freedoms are highlighted - freedom of research, of learning, and freedom of expression and action. The declaration emphasizes that the problem at the moment is precisely the freedom of expression and action as part of the professor's civil and political position both within the university and outside. The existence of the problem is due to pressure from the board of trustees, which require the university to implement the policies of their owners, investors, and dissent can lead to dismissal. In the discussion of this issue, the Declaration contains an important idea for our study that the material motivation of scientists as workers does not dominate the university. Much more important elements of a scientist's motivation are the possibility of self-realization, research, an independent position, the ability to share the results of one's research, scientific and civic position with students and the general public [4]. This kind of motivation is a striking feature of the creative class [5] to which university professors belong. The Declaration traces the idea that if a scientist is not endowed with academic freedoms, society will not be able to get the benefits and knowledge necessary for its development from science, and the university is considered in its civilization mission to provide society with the knowledge it needs for development, to teach new things and provide society with trained specialists, labor resources. Students' respect for the teacher is based on the belief that the professor is independent and free in expressing his ideas and views. The authors of the Declaration emphasize that respect for the professor and the educational process are destroyed if students have a suspicion that Teacher is intimidated, repressed, does not dare to truthfully and sincerely express his position. The Declaration points to the importance of board of trustees, however, the principle of equality of board of trustees and faculties should underlie a balanced university system. The morale of the university, which is based on academic freedom, is a factor that ensures the effectiveness of the university in relation to public interests. It is interesting that the authors of the Declaration consider the university as a refuge from the tyranny of the ruler in a totalitarian society, as well as a salvation from the tyranny of public opinion in a democratic society. The authors of the declaration offer a number of practical recommendations that can be used in conflict situations related to the attack on academic freedoms at the university.

An analysis of the processes taking place in the modern system of higher education in Russia demonstrates that disregard for the principle of academic freedom has a negative effect on the educational and scientific process. Despite the fact that academic freedom and universities' autonomy are declared as the main principles of state policy in Russia, the realization of academic freedom faces certain institutional and social barriers. So, for example, the existing system of research funding absolutely deprives of chances or significantly reduces the opportunity to receive financial support from existing funds [6]. The stated policy of the Russian Science Foundation and the analysis of the grants provided by the fund show that the bulk of its financial resources are directed to the technological and natural science spheres of scientific research, while humanitarian, socio-economic areas of research are less likely to receive financial support. This does not exclude the possibility that scientists can carry out these studies on an initiative basis, which is quite possible in some scientific fields, but absolutely impossible in those areas in which it is necessary to attract significant material and financial resources that make it possible to conduct research. In the field of education, academic freedom and universities' autonomy are largely limited by existing federal state educational standards, the system of accreditation of universities, and the bureaucratization of their management.

The principle of academic freedom is closely connected with another principle - the autonomy of universities, their joint implementation creates the foundation for a high-quality scientific and educational process. The main contradiction of the current stage of universities' development in Russia is that they are not able to be autonomous in economic terms, and society through state mechanisms ensures their financing [6]. Together with state funding, the state comes to the university, going far beyond the financial and economic activities of universities, permeating all aspects of their activities, destroying their autonomy and making it impossible to observe the principle of academic freedom. An alternative to public funding is a concept in which "universities finances themselves" through business contracts, grants, and student fees. This concept cannot be realized in Russian modern conditions, because it does not have a sufficient social base - a well-off one, capable of paying for a higher education of the middle class and a developed system of grant-giving funds, market participants who could take on the functions of creating financially sound trustee boards universities. Processes of this kind are typical for most post-Soviet countries [7]. In this situation, a need has arisen in Russia for a new Declaration of Academic Freedoms and the Autonomy of Universities, followed by the formation and conclusion of a Social Contract between universities, society and the state. The initiation of such an agreement would be possible "grass roots": in the university teaching environment, the prevailing situation is negatively assessed, but this is currently unlikely, since the processes of consolidation of the university community are weakly expressed [8]. A possible development of events is the initiation of the Russian Declaration of Academic Freedoms and Autonomy of Universities and the Social Contract "from above" as a result of the desire of political elites to propose a solution to existing problems in university science and education [9, 10].

The following directions for the development of the university based on the implementation of the principle of academic freedom and autonomy of the university as an intellectual and creative organization are proposed:

- the recognition of the university as a special social institution that performs the functions of preserving, transmitting, searching for and creating the scientific knowledge necessary for the development of society, the formation of a new generation and the provision of society with qualified labor resources;
- the recognition of the university as independent of state ideology, religions, political movements and decisions of authorities;
- the introduction of the presumption of the profession integrity principle;
- the revival of the institution of higher education reputation and the development of the principle of participants' responsibility in the scientific and educational process for the results of scientific and educational activities;
- the formation of a system of self-government by scientific and educational activities;
- digitalization of management processes based on the principle of academic freedom;
- creating the conditions for the integration between the scientific and educational process at the university;
- freedom to choose the topic of scientific research and the direction of its development;
- freedom in presenting one's scientific position on the content of scientific research, in putting forward scientific hypotheses;
- freedom of access to resources for research;
- freedom in academic mobility;
- freedom in expressing one's civil and political position regarding socio-economic events and decisions in society;
- freedom to criticize scientific points of view;
- freedom in the organization and contents of the educational process of the university;
- recognition of the professor's free time as a resource for the formation of his scientific, human and intellectual-creative capital;
- the transition to a qualitative assessment of the results of scientific and educational activities of the teaching staff by members of the professional community;
- election of faculty for 5 or more years, an increase in the number of perpetual contracts;
- transformation of the scientific and educational timing system of the faculty on the basis of changes in approaches to the formation of the load in the framework of the educational process;
- modernization of approaches to the formation of state educational standards and the system of accreditation of universities, the development of the autonomy of universities in the formation of standards for scientific and educational activities;

- transformation of the licensing system and accreditation of the university on de-bureaucratization basis;
- de-bureaucratization of educational and methodological support of the educational process;
- creation of institutional conditions for the university fundraising system development.

The solutions we developed are displayed in the form of the Matrix, the implementation of the academic freedom and autonomy of the university (Fig. 1).

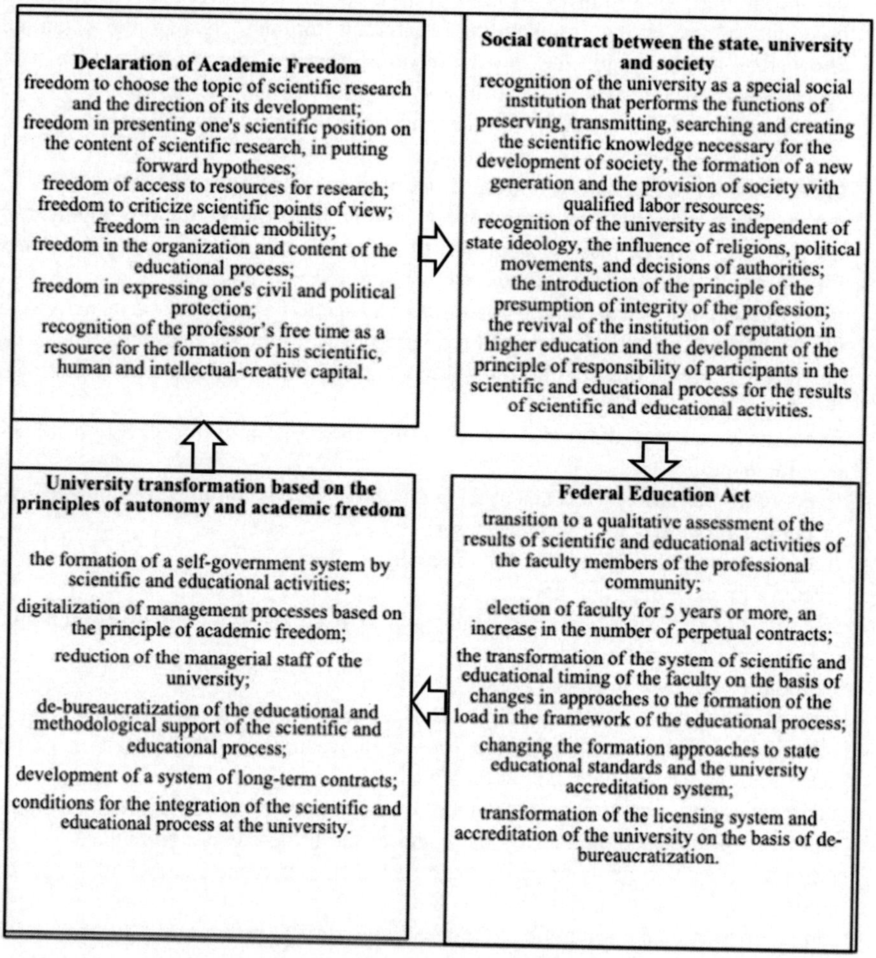

Fig. 1. Matrix implementation of academic freedom and autonomy of the university (includes the institutionalization of the "academic freedom" and "autonomy of the university" concepts in three documents - the Declaration of Academic Freedoms, the Social Contract and the Educational Law and shows the ways of transforming the university based on the principles of autonomy and academic freedom)

3 Conclusion

The academic freedom of the university is its important competitive advantage both for students and post-graduate students in the market of scientific and educational services, and for scientists and teachers in the labor market. The academic freedom of the university is a factor that forms a special scientific and educational space, the environment most favorable for the formation of the intellectual and creative potential of professors, graduate students and students, the development of human capital and the scientific potential of society as a whole. Academic freedom is an important tool to counter the negative influence of the external and internal environment on the scientific and educational process, and as sources and vectors of negative influence change over time, the underlying document postulating the content of academic freedom must change. The urgent task facing the university environment is the development of a modern Declaration of academic freedom that can withstand the challenges of modern society, preserve the ability of the university's scientific and educational space to stimulate the intellectual and creative activity of scientists, teachers and students, and produce high-quality scientific and educational products. The Declaration of academic freedom should conceptually reflect their essence at the present stage of development of science and education, form the basic principles of the organization of managerial and communication processes of the university, the implementation of which in practice can free the scientific and educational community from negative external and internal influences, give a qualitatively new impetus to the development of scientific educational process and its results.

References

1. Mavrina, I., Mingaleva, A.: The digitization of the russian higher education. In: Antipova, T., Rocha, A. (eds.) Digital Science. DSIC18 2018. Advances in Intelligent Systems and Computing, vol. 850, pp. 134–143. Springer, Cham (2019). [digital access]. https://www.hse.ru/data/2019/01/15/1145439993/Черногорский%20сборник.pdf
2. Golovko, N.: Declaration of principles on academic freedom and academic tenure: historical background and the principles of academic freedom. Scholarly notes of Transbaikal State University. Soc. Sci. Ser. **12**(3), 60–70 (2017). https://doi.org/10.21209/2500-171X-2017-12-3-60-70
3. Skripnik, K.: Magna Charta Universitatum and historical background for its prinicples. SFU Pedagog. Sci. Bull. **1**, 107–116 (2015)
4. Novikov, M., Perfilova, T.: University Charter of 1884: illusion of academic freedom. Yaroslavl Pedagog. Bull. **1**(1), 89–101 (2015)
5. Florida, R.: Creative class: people who change the future, 421 p. Publishing House «Classics XXI», Moscow (2007)
6. Lobova, S.: Precarious work of scientific and pedagogical workers: contents and consequences. Regional Econ. Dev.: Govern. Innov. Staff Train. **6**, 243–259 (2019)
7. Gorin, S.: Transformation of the Russian educational system: the problem of academic freedoms and the choice of individual educational technologies. J. Prof. Educ. **10**(2), 3682–3687 (2020). Siberian University of Consumer Cooperation, Novosibirsk

8. Soloveva, T., Solovyev, D.: To The issue of realization of academic freedom principle in modern university education. Bull. Omsk State Pedagog. Univ. Hum. Res. **2**(15), 110–114 (2017)
9. Iskakova, G.: The role of the university as an institute of academic freedom and production of knowledge and innovation agent in the development of civil society. In: Materials of the Second International Scientifical Conference "Traditions and Innovations of the Civic Education for the Modern Society", pp. 33–38 (2019)
10. Kurbatova, M., Kagan, E.: Opportunism of university lecturers as a way to adaptate the external control activities strengthening. J. Inst. Stud. **8**(3), 116–136 (2016)

Hierarchical Clustering to Identify Emotional Human Behavior in Online Classes: The Teacher's Point of View

Susana A. Arias[1(✉)], Pablo Moreno-Ger[2], and Elena Verdu[2]

[1] Facultad de Ciencias de la Salud, Universidad Tecnica de Ambato, Ambato, Ecuador
sa.arias@uta.edu.ec

[2] Escuela de Ingeniería y Tecnología, Universidad Internacional de la Rioja, Ave. de la Paz, 137, 26006 Logroño, Spain
{pablo.moreno, elena.verdu}@unir.net

Abstract. Teacher and student emotions are a fundamental basis in the development of the teaching-learning process. In this paper, we aim to verify whether it is possible for the emotions that are registered on a teacher's face to constitute emotional vectors and therefore be grouped hierarchically in order to obtain the teacher's emotional behavior during a virtual class. The experimental process demonstrated that it is possible to obtain an emotional funnel whose result is reflected in a valid hierarchical cluster to identify the set of emotions of a teacher when dealing with a specific topic or in the course of a time window. The work is in progress, but the conclusions it offers are valid enough to be proposed as a basis for recommendations in the teaching-learning process.

Keywords: Teaching · Learning · Emotions · Hierarchical cluster

1 Introduction

At the present time, learning processes are being developed in online models, thereby also allowing flexibility of the process. In turn, this helps to establish personalized intelligent systems adapted to the educational model [1]. During the teaching-learning process, the registering of emotions is very important because the level of learning is primarily based on this factor [2]. Identifying the moments or topics in which there are emotional changes is especially attractive for recommending the way in which the process should be approached and it is also the best way in which the teacher can effectively transmit their knowledge to students [3]. Emotional learning is recorded in face-to-face classes as a fundamental element of this process [4]. However, this type of learning has not been studied for online classes and its operation is currently being pilot-studied [5]. The statistical interpretation of the data and its handling using mining algorithms are likewise attractive when identifying how emotions are registered in the teaching-learning process. For example, with the use of intelligent tutorials interaction with cognitive processes is achieved. Therefore, the emotions produced whilst being in the learning environment are reflected in them. This allows data to be obtained, which

T. Antipova (Ed.): ICCS 2020, LNNS 186, pp. 269–275, 2021.
https://doi.org/10.1007/978-3-030-66093-2_26

in turn, help to predict and elucidate adaptive data mining mechanisms – the conclusion of which is the self-regulation of student learning [6].

Asynchronous and synchronous learning must be focused on the student since it constitutes timeless and non-spatial learning. This enables us to work with experimental data in order to predict student performance based on symmetry and hierarchical cluster algorithms – the result of which helps us to recommend learning strategies for high-risk groups [7]. In fact, even using robots for classes has allowed students to obtain positive emotions and improvements in the concentration process, which are higher than 80%, i.e. whose values could be studied in the faces of the participants and which proved to be reliable when determining the introduction of new technologies to help to better understand the emotional process of students during the development of a class [8]. Hierarchical linear models have been used to evaluate the emotional effects towards the different learning groups. This allows us to verify the response capacity and the teaching influence of the instructors while the process lasts, and it can even be verified via grouping processes to see if the learning routes are valid or not for the quality improvement of the process [9]. The data processing allows us to analyze the reflections of novice teachers in the teaching-learning environment through grouping processes, i.e. whose conclusion showed that it is possible to achieve a positive emotional impact and propose teaching strategies towards teacher educators from the groups increasing with the quality of the process [10]. In this work we intend to obtain the set of emotions that the teacher registers during the course of a virtual class and to identify whether it is possible to obtain emotional vectors (a set of emotions) whose emotional unity is recorded in each frame of the teacher's face while the recording is being carried out. To record emotions by frame we used the emotion software[1] on the face which allows us to obtain the emotion and its emotional percentage. See Fig. 1.

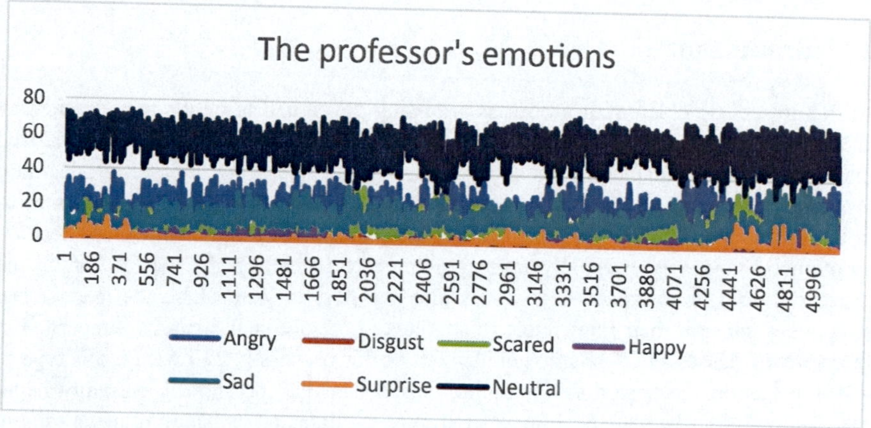

Fig. 1. Emotions on a teacher's face during the teaching-learning process

[1] https://github.com/omar178/Emotion-recognition.

Figure 1 shows the emotions that the teacher registers during a virtual class. The set of emotions is recorded per frame and their union forms the emotional vector, which are recorded in a single matrix and will serve as input for the emotional cluster algorithm – the result of which is the maximum Euclidian distance that will allow us to verify the number of emotional clusters that can be obtained for each class. The point to stress is that since the vectors are registered by frame, it is very useful to identify at what moment the grouping of emotions is shown since the teacher's treatment of a specific topic and what emotional vectors it is producing can be analyzed. To verify the effect of the clusters, the process was applied in a teacher's class and the appearance of the groups was verified.[2] The conclusions showed that the set of vectors can be grouped into a cluster creating a kind of emotional funnel that facilitates the identification of the groups and the valid emotional behavior to record recommendations in the teaching-learning process. The following sections describe in greater detail the process mentioned above, as well as the experiment and the conclusions obtained from the methodological application.

2 Methodology

In order to verify whether it is possible to group the teacher's emotions into hierarchical clusters in this work, the following steps were followed:

- A video was obtained from a teacher during the course of an online class (10 min long)
- The emotions that the professor displayed were recorded during class by means of the emotion detection software.
- Emotions and their percentage were inputted for the emotional cluster algorithm.
- The maximum Euclidian distance was verified in order to improve the process, which was done according to the number of emotional clusters obtained during this stage.
- Clusters were assigned to each emotional vector.
- The following section describes the methodological application in the experiment and the results obtained.

3 Experimentation

The emotions of the teacher's face during the development of a virtual class. Table 1 shows the emotion records:

[2] https://github.com/rana-othman/Huam-Affective-States-and-their-influences-Ontology-HASIO.

Table 1. Teacher emotions

Angry	Disgust	Scared	Happy	Sad	Surprise	Neutral
14.0402407	0.11996793	2.79443283	0.12591766	12.1250264	0.27545181	70.519
14.6925688	0.10669181	3.3781264	0.15289683	13.9707342	0.35269132	67.3463
15.2073145	0.10132749	1.89444721	0.15539704	9.2721045	0.19167637	73.1777
17.1839014	0.58863652	2.07637902	0.14946956	11.9800881	0.19307568	67.8285
12.9969999	0.36724636	1.73648205	0.18941454	12.0724775	0.16710624	72.4703
14.7960231	0.47395094	2.31586676	0.25176611	12.1891052	0.2537376	69.7195
19.0575928	0.38461972	2.96046399	0.12769676	12.7564535	0.40908195	64.3041

Table 1 shows the numerical registration of the emotion vectors: each recording time corresponds to a video frame. A total of 9,308 vectors were processed for a 10-min video. The data in Table 1 were entered into the Hierarchical Cluster algorithm:

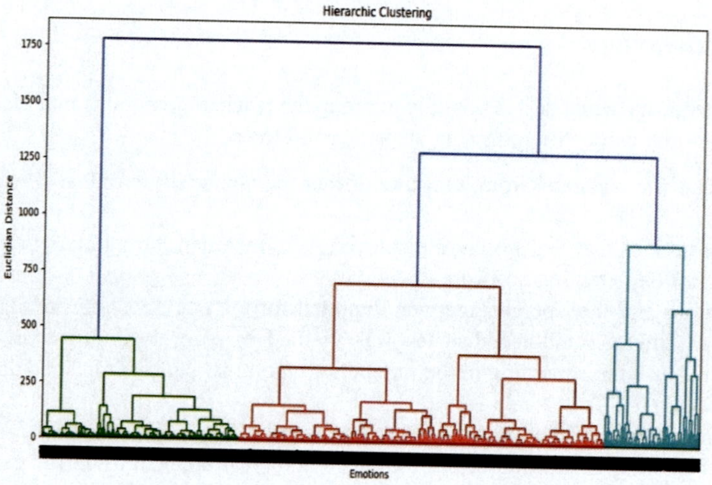

Fig. 2. A dendrogram of a teacher's emotions

Figure 2 shows the teacher's emotions in a hierarchical cluster. The maximum distance (blue line) shows that 2 emotional clusters are formed. However, there are two clusters (red and light blue) that can form a single maximum distance cluster:

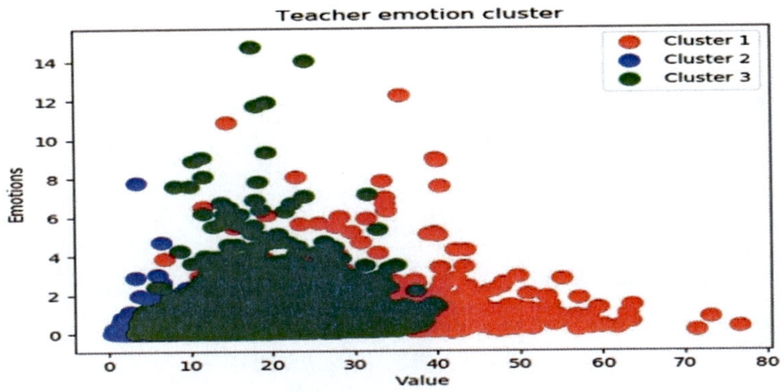

Fig. 3. Emotional clusters

In Fig. 3, the clusters are observed. Indeed, cluster 1 is separated from the combination of cluster 2 and 3. The following table shows an example of the assigning of vectors to clusters:

Table 2. Assigning the vectors to the hierarchical cluster

Angry	Disgust	Scared	Happy	Sad	Surprise	Neutral	ClusterJ
14.0402407	0.11996793	2.79443283	0.12591766	12.1250264	0.27545181	70.519	3
14.6925688	0.10669181	3.3781264	0.15289683	13.9707342	0.35269132	67.3463	3
15.2073145	0.10132749	1.89444721	0.15539704	9.2721045	0.19167637	73.1777	two
17.1839014	0.58863652	2.07637902	0.14946956	11.9800881	0.19307568	67.8285	3
12.9969999	0.36724636	1.73648205	0.18941454	12.0724775	0.16710624	72.4703	3
14.7960231	0.47395094	2.31586676	0.25176611	12.1891052	0.2537376	69.7195	3
19.0575928	0.38461972	2.96046399	0.12769676	12.7564535	0.40908195	64.3041	3
25.7531405	0.42532343	3.66705656	0.07578548	15.6549722	0.32367643	54.1	3
18.695797	0.35221099	2.83385776	0.14869279	12.9973486	0.3765709	64.5955	3
13.8354883	0.20205483	2.22563855	0.06138291	11.8289858	0.09812451	71.7483	3
14.881292	0.26700334	2.57337838	0.14506866	6.51338026	0.39386661	75.226	two
11.7139429	0.22257618	1.83929931	0.1903706	11.0048987	0.18053913	74.8484	two
15.5197114	0.34933933	4.84564863	0.1376551	19.8164329	0.52083135	58.8104	3
15.3472424	0.35366656	4.80443798	0.13780966	17.8539857	0.55300021	60.9499	3
18.6558679	0.55179163	2.97201425	0.17050438	15.6483173	0.26209087	61.7394	3
16.0976604	0.11841914	2.52950471	0.21400708	9.12730321	0.33777498	71.5753	two
16.9832707	0.23630667	2.61152256	0.09076577	11.919225	0.18161545	67.9773	3
20.2393502	0.34154691	4.97516505	0.10750336	12.5454664	0.4592767	61.3317	3
3.27282548	0.26092234	10.3739366	14.9593607	14.3721506	27.089268	29.6715	one

Table 2 shows an example of the assigning of vectors to clusters. In total, 1,347 vectors were assigned to Cluster 1: Cluster 2: 2,800 vectors, and Cluster 3: 5,161 vectors, which corresponded to 14.47%, 30.08% and 55. 44%, respectively.

4 Conclusions

If Hierarchical Cluster 3 is analyzed because it is the majority in the emotional grouping, the most prevalent emotion is neutral and then angry:

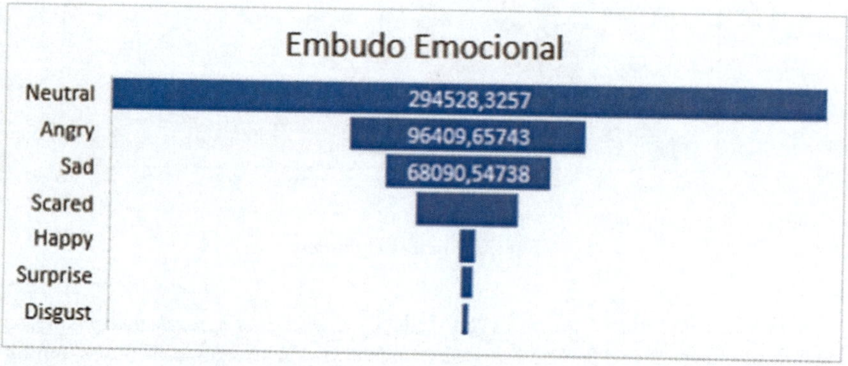

Fig. 4. Cluster 3 emotions

Figure 4 shows the emotions that prevail for cluster 3. These emotions produce an emotional behavior, which, when assigned to the ontology, produces emotional behavior: Agonistic (angry) and Submitting (sad). It is very important to describe the fact that since the vectors are recorded by frames, it is possible to detect when the Cluster 3 vectors assigned at a specific time occur, and to detect which topic was being dealt with in the course of the time window, and to be able to recommend improvements towards the teacher. In the case of agonistic, for example, this type of behavior can occur because the teacher tries to show anger on his face to control the class or because he cannot clearly express his ideas to make himself understood by the students. In the case of submitting, the teacher tries to keep their attention (pressure from the class) under control by means of humility and resignation, which at the time can be similar to sadness. The important thing is that it is possible to determine the emotional behavior of the teacher by the emotional vectors assigned to a time window, and to know what recommendations to provide to deal with a specific topic during the teaching-learning process and thereby feed a recommender system.

References

1. Cui, Q., Gou, J.: Review of online learning behavior analysis. In: 4th International Conference on Economy, Judicature, Administration and Humanitarian Projects (2019)
2. Hokka, P., Vahasantane, K., Paloniemi, S.: Emotions in learning at work: a literature review. Vocat. Learn. **13**, 1–26 (2019)
3. Deldjoo, Y., Schedl, M., Cremonesi, P., Pasi, G.: Recommend systems leveraging multimedia content. ACM Comput. Surv. **53**, 1–38 (2020)

4. Meyers, D., Domitrovich, C., Dissi, E., Trejo, J., Greenberg, M.T.: Supporting systemic social and emotional learning with a schoolwide implementation model. Eval. Program Plann. **73**, 53–61 (2019)
5. Humphry, D., Hampden-Thompson, G.: Primary school pupil's emotional experiences of synchronous audio-led online communication during online one-to-one tuition. Comput. Educ. **135**, 100–112 (2019)
6. Azevedo, R., Gasevic, D.: Analyzing multimodal multichannel data about self-regulated learning with advanced learning technologies: issues and challenges (2019)
7. Hung, H.C., Liu, I.F., Liang, C.T., Su, Y.S.: Applying educational data mining to explore student's learning patterns in the flipped learning approach for coding education. Symetry **12** (2), 213 (2020)
8. Gomez, H.F., Arias, S.A., Lozada, T.E., Martinez, C., Robalino, F., Castillo, D., Aguirre, P.: Emotional strategy in the classroom based on the application of new technologies: an initial contribution. In: de Information and Communication Technology for Intelligent Systems, Singapore (2019)
9. Hennessey, A., Humphrey, N.: Can social and emotional learning improve children's academic progress? Findings from a randomized controlled trial of the Promoting Alternative Thinking Strategies (PATHS) curriculum. Eur. J. Psychol. Educ. **35**, 1–24 (2019)
10. Ledger, S., Ersozlu, Z., Fischetti, J.: Preservice teacher's confidence and preferred teaching strategies using TeachLivE virtual learning environment: a two-step cluster analysis. EURASIA J. Math. Sci. Technol. Educ. **15**(3), em1674 (2019)

Should e-Learning Providers Encourage Users to Set a Target Score?

Takumi Kato(✉)

Saitama University, 255 Shimo-Okubo, Sakura-ku, Saitama City, Saitama 338-8570, Japan
takumikato@mail.saitama-u.ac.jp

Abstract. The need for e-learning continues to increase due to technological innovation and social changes. As service providers, it is extremely important to understand the factors that increase user satisfaction and continue to improve services. It is recognized that it is important to achieve targets to increase the satisfaction of education. However, the negative effect of not having a target has not been sufficiently discussed. Therefore, in this study, based on the results of an online survey of Japanese business persons who take the e-learning Test of English for International Communication (TOEIC), the positive effects of achieving targets and the negative effects of the non-existence of targets were evaluated. The target was to set the TOEIC score and make a judgment in relation to the achievement of the actual test. As a result of an evaluation by the logistic regression model, significant effects were confirmed. Therefore, it is considered that service providers not only improve the quality of education and the environment, but they also encourage users to set concrete targets.

Keywords: Satisfaction · e-Learning · Language study · Target · Achievement

1 Introduction

Advances in technology have raised expectations for e-learning, which provides education/training online. It is attractive that users can enjoy high-quality content at a low cost without being bound by time or place. There are service platforms not only for students but also for employees who work in various industries. For example, Coursera provides academic content provided by Stanford University and Google [1], and Skillshare provides creative content such as animation and photos [2]. In the past, because of the immaturity of technology and content, there was concern about the poor quality of e-learning. However, many technical problems have been resolved, and as the number of e-learning providers has increased, the concern has diminished. A previous study reported that students who used e-learning were more satisfied than students who used the conventional face-to-face learning method [3].

As a result, of this background, the global e-learning market size is expected to exceed US$200 billion in 2019 and exceed US$375 billion by 2026 [4]. In addition, the occurrence of the coronavirus disease 2019 has forced schools, universities, and companies to conduct remote meetings and work, so the need for e-learning is

T. Antipova (Ed.): ICCS 2020, LNNS 186, pp. 276–285, 2021.
https://doi.org/10.1007/978-3-030-66093-2_27

becoming stronger [5]. In fact, Harvard University is scheduled to reopen in the fall of 2020. However, it is expected that much of the learning will be done remotely [6].

It can easily be imagined that competition in the e-learning market will become fiercer in the future. Therefore, it is extremely important for e-learning providers to understand the factors that help increase user satisfaction and continue to improve their services. Not only the quality of education and equipment but also the usability of the system peculiar to e-learning has been clarified as a factor that enhances the satisfaction of education [7, 8]. Furthermore, it has been pointed out that the degree of achievement of the target is also an important factor [9]. However, in existing studies, there were many subjective criteria, and it was unclear whether the achievement of the above was possible. Furthermore, the negative effect of not having a target has been lacking in the discussion. Therefore, in this study, not only the positive effects of achieving targets but also the negative effects of the non-existence of targets were evaluated based on the satisfaction levels of Japanese business persons who took up e-learning. The target e-learning test was the Test of English for International Communication (TOEIC). The TOEIC is an English language proficiency test developed by the Educational Testing Service for people who speak English as a second or foreign language and has been implemented in 160 countries [10]. In Japan, it is used for university entrance examinations and company personnel evaluations, and 2,456 thousand persons took the test in the 2018 fiscal year [11]. If this study confirms the negative effects of the non-existence of targets, it is important that service providers not only improve the quality of education and the environment, but they should also encourage users to set concrete targets.

2 Factors Contributing to Education Satisfaction

Educational satisfaction is a long-standing academic study theme. Higher education, which is the source of economic and cultural development, is required to meet the needs of ever-changing environmental conditions, and so many related studies have been conducted [12]. Today, the increasing number of new higher education providers is intensifying the competition to attract and retain students. Therefore, it is important to identify the factors that contribute to student satisfaction and improve the educational program and environment [13]. It has been pointed out that satisfaction increases loyalty to service providers [14], significantly contributes to positive reviews, and is more tolerant of changes in tuition [15].

A study of British universities reported that the quality of education contributed more than physical facilities [16]. In Malaysia, knowledgeable and responsive faculties, textbooks and tuition fees, student support facilities, and relationships with teaching staff have been extracted [17]. In Turkey, the technical, functional, infrastructure, interaction, and atmosphere of higher education institutions are shown [18]. Others include a positive experience at university [19], the freedom to influence school policies and procedures, and the intellectual stimulation of peers [20].

Although it is almost the same in e-learning, perceived usability [7, 8], interaction [21], and the prevention of system failure [22] are reported as characteristic factors. In e-learning at American universities, it has been pointed out that not only instructor–

student but also student–student dialogue interaction is important [23]. From another perspective, the degree of achievement of learning is considered to influence satisfaction [24]. A study of Australian universities concluded that learning achievement is extremely important [9]. It has been proven that a Korean university also increases students' satisfaction by providing a dashboard that allows them to intuitively check the degree of learning achievement [25].

We focus on the achievement of targets. In this study [9], the criteria were subjective because the degree of achievement was evaluated by self-evaluation, and so objective evaluation was emphasized. This is because there is a concern that people who tend to respond positively to the survey will highly evaluate both achievement and satisfaction. Naturally, the reverse is also considered to occur. In addition, the effect of not only achieving the target but also not having targets is insufficiently discussed. Therefore, the hypotheses of this study are as follows:

H1: The probability of satisfaction increases when one achieves the target score in e-learning

H2: The probability of satisfaction decreases when there is no specific target score in e-learning

To objectively judge whether the target score was achieved, H1 was judged based on the target score and the actual results of the TOEIC in this study. The TOEIC is mainly divided into listening and reading tests and speaking and writing tests. The former is widely spread in Japan, so it was adopted. The test has a maximum score of 990 and is not a pass/fail judgment. Note that the average scores of examinees in each country are 871 for Canada, 798 for Germany, 684 for Russia, 673 for Korea, 520 for Japan, and 478 for Thailand [26]. It is speculated that H2 lacks targets, which makes the purpose of the lecture ambiguous and makes it difficult to recognize the value of the lecture, resulting in low satisfaction.

3 Method of Study

In this study, the "Oricon Dataset" [27], provided by Oricon ME Inc. via the IDR Dataset Service of the National Institute of Informatics, was used. As a third-party organization, Oricon regularly conducts satisfaction surveys objectively for people who use the product/service of various industries. Among them, the results of the online survey conducted from May 15th to June 5th, 2017, for those who took the e-learning TOEIC and the TOEIC within the past five years from the time of the survey were analyzed. The providers of e-learning are not specific to one company, but many major Japanese companies. The total sample size was 652.

However, infrequent attributes are difficult to analyze. Hence, respondents were screened on the condition that they were in their 20s to 50s and that their occupation was a company employee. Furthermore, to equalize the number of TOEIC target scores, 60 respondents were randomly sampled in each group: target score 400, target score 600, target score 800, and without a target score. Therefore, 240 people were the subjects of this study. Table 1 shows the distribution of the survey attributes.

The survey items used in this study were (1) satisfaction with e-learning, (2) target achievement, (3) target score, (4) gender, (5) age, (6) marriage, (7) residential area, (8) annual household income, (9) emphasized points in e-learning, and (10) situation of word of mouth (WoM).

The satisfaction score had ten levels, as shown in Table 2. In total, the mean was 6.871, and the median was seven. Looking at each target score, the group with no target score is the lowest, and the mean of satisfaction increases as the target score increases. In the analysis, those who chose 8–10 were defined as "satisfied." In total, 37.9% of cases are applicable.

Table 1. Attributes of survey subjects.

Item	Content	Number of Respondents	Item	Content	Number of Respondents
Total		240	Target score	400	60
Gender	Male	178		600	60
	Female	62		800	60
Age	20s	17		None	60
	30s	69	Time of Exam	Within 6 months	62
	40s	90		Within 1 year	57
	50s	64		Within 2 year	42
Marriage	Married	162		Within 3 year	34
	Unmarried	78		Within 4 year	14
Job	Company employee	240		Within 5 year	31

Table 2. Distribution of satisfaction.

Score of Satisfaction	Number of Respondents				
	Total	400	600	800	None
1	2	0	2	0	0
2	1	0	0	0	1
3	5	3	0	1	1
4	9	1	4	0	4
5	25	9	3	4	9
6	52	16	6	11	19
7	55	11	17	11	16
8	57	15	15	21	6
9	24	4	10	8	2
10	10	1	3	4	2
Total	240	60	60	60	60
Mean	6.871	6.617	7.133	7.433	6.300
Median	7	7	7	8	6
Satisfied	91 (37.9%)	20 (33.3%)	28 (46.7%)	33 (55.0%)	10 (16.7%)

The target achievement of (2) was judged by the actual TOEIC results asked in the survey. As shown in Table 3, when you look at the distribution of achievements by target score, the achievement rate is higher for people with a score of between 600 and 800 than for people with a low target score of 400. Those who did not have a target score were set as unachieved.

The reason for using (4)–(10) was to control the effects other than the achievement and not having a target. The emphasized points of e-learning (9) were asked with a multi-answer from a list of ten features in e-learning. As shown in Nos. 22–31 of Table 4, the features include items shown in previous studies, such as inquiry correspondence, the quality of educational materials, and appropriate tuition fees. In (10), regarding the situation of WoM concerning the e-learning that respondents took, questions related to eight patterns were asked with multiple answers, as shown in Nos. 32–39 of Table 4. WoM has a great influence on consumer behavior. Since consumers who use the products/services communicate their impressions frankly, it is easy to gain trust. Therefore, we must perform a precise evaluation.

Then, all the asked data were converted into dummy variables. The representative values of each variable are shown in Table 4. A logistic regression model was used for the evaluation. No. 1 (satisfaction) in Table 4 was the objective variable, No. 2 (target achievement), and Nos. 3–6 (target score) were explanatory variables. The attribute variables, Nos. 7–39, were input to control for other effects. The variables marked with * were not used in the analysis because they were set as the criteria for the dummy variables. Since there are many variables, the variables were selected by using the stepwise method, and the contribution was confirmed by using the odds ratio of the variables that became significant at the 5% level. In addition, 10-fold cross-validation was performed to confirm the validity of the evaluation. The statistical analysis software R was used for the analysis, and the stepAIC function of the MASS package was applied to the stepwise method.

Table 3. Target score and achievement status.

		Achievement		
		No	Yes	Total
Target Score	400	31	29	60
	600	20	40	60
	800	19	41	60
	None	60	0	60

Table 4. Variable list and representative value.

No	Dummy Variable	Item	Description (When the variable is 1)	Mean	SD
1	SatisfactionFlg	Satisfaction	Satisfied	0.379	0.486
2	Achievement	Achievement status	Achieved	0.458	0.499
3	Target_400*	Target score	400	0.250	0.434
4	Target_600		600	0.250	0.434
5	Target_800		800	0.250	0.434
6	Target_None		None	0.250	0.434
7	Female	Gender	Female	0.258	0.439
8	Age_20s*	Age	20s	0.071	0.257
9	Age_30s		30s	0.288	0.454
10	Age_40s		40s	0.375	0.485
11	Age_50s		50s	0.267	0.443
12	Marriage	Marriage	Married	0.675	0.469
13	Area_Hokkaido	Residential area	Hokkaido / Tohoku	0.017	0.128
14	Area_Kanto*		Kanto (Capital Area)	0.571	0.496
15	Area_Chubu		Chubu	0.113	0.317
16	Area_Kinki		Kinki	0.167	0.373
17	Area_Chugoku		Chugoku / Shikoku	0.058	0.235
18	Area_Kyusyu		Kyusyu	0.075	0.264
19	Income_500*	Annual household income	Less than 5 million yen	0.129	0.336
20	Income_1000		5 to ten million yen	0.513	0.501
21	Income_2000		Ten to twenty million yen	0.358	0.481
22	Feature01	Focused features	Inquiry correspondence	0.129	0.336
23	Feature02		Easy enrollment procedure	0.213	0.410
24	Feature03		Attractive enrollment campaign	0.150	0.358
25	Feature04		Preparation of trial class	0.071	0.257
26	Feature05		Variety of learning styles	0.325	0.469
27	Feature06		Abundance of curriculum	0.238	0.426
28	Feature07		Quality of educational materials	0.371	0.484
29	Feature08		Volume of educational materials	0.279	0.450
30	Feature09		Appropriate tuition fee	0.163	0.370
31	Feature10		Company credibility	0.321	0.468
32	WoM01	Status of WoM	Talk positive contents in real	0.304	0.461
33	WoM02		Talk negative contents in real	0.029	0.169
34	WoM03		Listen positive contents in real	0.133	0.341
35	WoM04		Listen negative contents in real	0.046	0.210
36	WoM05		Talk positive contents in digital	0.071	0.257
37	WoM06		Talk negative contents in digital	0.017	0.128
38	WoM07		Listen positive contents in digital	0.121	0.327
39	WoM08		Listen negative contents in digital	0.042	0.200

*Criteria for dummy variables.

4 Results and Discussion

First, as shown in Table 5, when the correlation matrix is confirmed, the achievement is 0.419, which is the highest for Satisfactionflg. As a result of the test of no correlation, p-value < 0.000, and there is a correlation between satisfaction and achievement. Conversely, Target_None had the smallest value of -0.253. As a result of the test of no correlation, p-value < 0.000, and a negative correlation was confirmed.

Table 5. Correlation matrix.

	1	2	3	4	5	6	7	8	9	10	11	12	13	14	15	16	17	18	19	20	21	22	23	24	25	26	27	28	29	30	31	32	33	34	35
1 SatisfactionFlg																																			
2 Achievement	0.42																																		
3 Target_600	0.10	0.24																																	
4 Target_800	0.20	0.26	-0.33																																
5 Target_None	-0.25	-0.53	-0.33	-0.33																															
6 Female	0.03	-0.05	-0.03	0.01	0.14																														
7 Age_30s	0.17	0.12	0.12	-0.09	-0.11	0.21																													
8 Age_40s	-0.09	0.01	-0.07	-0.03	0.07	-0.12	-0.49																												
9 Age_50s	-0.04	-0.12	-0.02	0.09	0.02	-0.21	-0.38	-0.47																											
10 Marriage	0.01	0.10	0.07	-0.07	-0.11	-0.30	0.05	0.04	0.08																										
11 Area_Hokkaido	-0.04	-0.05	-0.08	-0.08	-0.08	-0.08	0.06	0.03	-0.08	0.02																									
12 Area_Chubu	0.05	0.04	0.04	-0.02	0.04	0.09	0.04	0.00	-0.01	0.02	-0.05																								
13 Area_Kinki	-0.05	0.02	0.03	0.03	0.03	-0.01	-0.01	0.07	-0.14	-0.12	-0.06	-0.16																							
14 Area_Chugoku	0.06	0.02	-0.02	-0.10	0.06	-0.03	0.00	0.03	0.01	0.10	-0.03	-0.09	-0.11																						
15 Area_Kyusyu	0.04	0.02	-0.06	0.02	-0.02	0.09	0.06	-0.09	0.04	-0.11	-0.04	-0.10	-0.13	-0.07																					
16 Income_1000	-0.11	-0.09	-0.03	-0.09	0.02	-0.05	0.01	0.08	-0.11	-0.02	-0.07	0.00	-0.01	-0.01	0.09																				
17 Income_2000	0.08	0.13	0.07	0.11	-0.15	-0.06	-0.09	0.01	0.16	0.17	0.04	-0.02	-0.08	0.04	-0.15	-0.77																			
18 Feature01	0.24	0.17	-0.02	0.07	-0.02	0.06	0.09	-0.14	0.02	-0.05	0.05	-0.06	-0.01	0.17	0.03	-0.05	0.00																		
19 Feature02	-0.07	0.03	-0.07	-0.02	-0.16	-0.05	-0.06	0.06	-0.01	0.01	0.01	-0.09	-0.07	0.09	0.01	0.04	0.02	0.16																	
20 Feature03	0.15	0.01	0.03	0.05	0.00	0.02	0.02	-0.04	0.06	0.09	-0.06	-0.04	0.06	0.05	0.01	-0.01	0.03	0.19	0.24																
21 Feature04	0.15	0.14	-0.05	0.07	-0.05	0.06	-0.03	-0.08	0.13	0.05	-0.04	-0.05	-0.08	0.14	0.05	-0.06	0.13	0.14	0.06	0.20															
22 Feature05	0.08	-0.07	-0.01	-0.09	0.18	0.20	-0.03	0.07	0.00	-0.05	-0.09	0.03	0.02	0.02	0.01	0.00	0.00	0.00	-0.03	0.03	0.16														
23 Feature06	0.03	0.00	0.04	-0.01	0.04	0.01	0.01	-0.03	0.00	-0.01	0.08	-0.04	-0.01	-0.10	-0.01	0.02	0.01	-0.01	0.09	0.04	0.19	0.18													
24 Feature07	0.08	0.00	-0.01	0.04	0.08	0.04	0.01	-0.04	0.01	0.05	-0.03	0.00	-0.16	0.03	0.04	-0.05	0.09	0.04	-0.02	-0.03	0.12	0.33	0.32												
25 Feature08	0.03	0.02	0.03	0.03	0.05	0.06	-0.01	0.02	-0.04	0.02	-0.01	0.04	-0.18	-0.04	0.03	-0.06	0.06	0.09	0.09	-0.03	0.08	0.16	0.22	0.37											
26 Feature09	0.08	0.03	0.01	0.06	0.01	-0.03	0.05	-0.02	-0.01	-0.03	-0.06	-0.09	-0.08	0.08	0.05	-0.07	0.07	0.07	0.13	0.04	0.14	0.01	0.10	0.15	0.20										
27 Feature10	0.02	0.07	0.02	-0.05	-0.01	-0.08	-0.10	0.06	0.03	-0.10	0.12	-0.10	-0.07	0.21	0.01	-0.03	0.03	0.08	0.17	-0.04	0.05	0.15	0.10	0.21	0.25	0.08									
28 WoM01	0.29	0.21	0.04	0.10	-0.05	0.09	0.14	-0.10	-0.05	-0.10	-0.02	-0.01	0.00	0.07	0.02	-0.01	0.00	0.20	0.10	0.15	0.07	0.16	-0.01	0.06	0.11	-0.02	0.07								
29 WoM02	-0.03	-0.01	0.01	-0.04	0.13	0.07	-0.11	0.02	0.06	0.02	-0.02	-0.06	-0.01	0.06	-0.05	-0.03	0.08	0.16	0.03	0.00	0.05	0.09	-0.04	0.07	0.17	0.13	-0.01	0 .16							
30 WoM03	0.10	0.08	-0.03	0.09	0.03	-0.04	-0.01	0.05	-0.07	-0.02	0.05	-0.10	0.02	0.06	0.03	0.06	-0.06	0.21	0.10	0.08	0.04	0.12	0.04	0.08	0.08	0.09	-0.01	0.25	0.15						
31 WoM04	-0.01	-0.08	-0.04	-0.04	0.15	0.01	-0.05	0.04	-0.04	-0.06	-0.03	-0.08	-0.05	0.03	-0.06	0.01	-0.04	0.09	0.08	0.08	0.02	0.10	-0.03	0.04	0.04	0.07	0.06	0.16	0.44	0.27					
32 WoM05	0.19	0.04	0.07	0.03	0.03	0.10	0.04	0.02	-0.06	-0.02	0.09	-0.10	-0.04	-0.07	-0.02	0.01	-0.04	0.18	0.13	0.11	0.18	0.09	0.08	-0.01	0.15	0.10	0.05	0.17	0.15	0.27	0.17				
33 WoM06	0.03	0.01	0.08	0.00	-0.08	0.00	0.06	-0.03	-0.01	0.09	-0.02	-0.05	0.03	-0.03	0.09	0.00	0.04	0.05	0.09	0.13	-0.04	0.05	0.00	-0.03	-0.01	0.12	-0.02	-0.02	0.17	0.14	0.28	0.09			
34 WoM07	0.08	0.10	0.05	0.08	-0.07	0.02	-0.12	0.11	-0.08	-0.07	0.05	-0.05	-0.06	0.07	-0.01	-0.02	-0.01	0.12	0.12	-0.01	0.15	0.23	0.12	0.14	0.23	0.15	0.16	0. 09	0.09	0.27	0.16	0.35	0.05		
35 WoM08	-0.08	-0.02	0.07	-0.02	0.02	0.02	-0.09	0.10	-0.08	-0.08	-0.03	-0.01	0.02	0.13	-0.06	0.08	-0.07	0.11	0.20	0.03	0.02	0.26	0.03	0.06	0.10	0.25	0.08	0. 09	0.34	0.10	0.45	0.11	0.30	0.43	

Table 6 shows the estimation results of the logistic regression model. Among the significant variables, WoM05 had the largest odds ratio of 5.328. WoM01 also contributed 2.549, but the effect of WoM05 was greater. In other words, it can be understood that people who talk positively in the digital sphere tend to be more satisfied than people who talk in real-time. The variable showing the next highest odds ratio of WoM05 is the achievement of 4.325. Therefore, "H1: the probability of satisfaction increases when one achieves the target score in e-learning" was supported.

The variable with the lowest odds ratio is a WoM08 of 0.130.

As for the negative effects, it can be understood that viewing negative digital reviews has a greater impact than conveying negative content. The next lowest variable is 0.236, Feature02. In other words, people who attach importance to the straightforward enrollment procedure rather than the content of e-learning have little imagination after enrollment, so they are unlikely to be satisfied. The odds ratio of Target_None was 0.310, which is a negative effect. Therefore, "H2: the probability of satisfaction decreases when there is no specific target score in e-learning" was supported.

Finally, as shown in Table 7, as a result of cross-validation, the mean prediction accuracy was 81.3%, and the validity of the model was confirmed.

The limitations of this study are as follows. The first is that we judge the targets and achievements by using an online survey to ask questions only after the TOEIC has been taken. Ideally, it is desirable to ask the target score in advance and check the results, including the evidence after the test. The second is the result of the fact that only e-learning exists in relation to the TOEIC in Japan, so there is a limit on the generalization of the conclusion. In the future, it will be necessary to expand target services and countries.

Table 6. Result of the logistic regression model.

	Estimate	Odds Ratio	SE	p-value
(Intercept)	-1.952	0.142	0.386	0.000 ***
Achievement	1.464	4.325	0.387	0.000 ***
Target_None	-1.171	0.310	0.546	0.032 *
Age_30s	0.425	1.529	0.354	0.230
Feature01	1.150	3.157	0.533	0.031 *
Feature02	-1.443	0.236	0.503	0.004 **
Feature03	1.193	3.298	0.472	0.011 *
Feature05	0.732	2.079	0.377	0.052
Feature09	0.874	2.397	0.455	0.055
WoM01	0.936	2.549	0.363	0.010 *
WoM05	1.673	5.328	0.725	0.021 *
WoM08	-2.039	0.130	1.009	0.043 *
AIC		252.629		
McFadden		0.282		
Adj.McFadden		0.201		

***p<0.001; **p<0.01; *p<0.05.

Table 7. Result of cross-validation

No	Prediction accuracy
1	0.833
2	0.792
3	0.875
4	0.958
5	0.750
6	0.708
7	0.875
8	0.667
9	0.750
10	0.917
Mean	0.813

5 Conclusion

As the market size of e-learning expands, it is expected that competition will become more intense in the future. As an e-learning provider, it is important to increase user satisfaction, for the purpose of maintaining user retention rates and continuing to acquire new users by spreading positive WoM. To that end, it is extremely important to understand the factors that contribute to satisfaction and continue to improve the service. In academic studies, the improvement of educational satisfaction has been a theme that has been emphasized for a long time. So far, many factors have been shown, such as the quality of education and equipment, the user support system, and the usability of e-learning systems. Among them, it is recognized that achieving the target is very important. However, in existing studies, there were many subjective criteria, and it was unclear whether the achievement of the above was possible. Furthermore, the negative effects of not having a target were lacking in the literature. Therefore, in this study, the effects on satisfaction were evaluated for Japanese business people who took the e-learning TOEIC. As a result, significant effects were confirmed for both the positive effect of achieving the target and the negative effects of not having a target. Therefore, it is considered that service providers not only improve the quality of education and the environment, but they also encourage users to set concrete targets.

References

1. Coursera. https://www.coursera.org/. Accessed 10 July 2020
2. Skillshare. https://www.skillshare.com/. Accessed 10 July 2020
3. Liu, C.H.: The comparison of learning effectiveness between traditional face-to-face learning and e-learning among goal-oriented users. In: 6th International Conference on Digital Content, Multimedia Technology and its Applications on Proceedings, pp. 255–260. IEEE (2010)
4. Global Market Insights, eLearning market size worth over $375bn by 2026. https://www.gminsights.com/industry-analysis/elearning-market-size. Accessed 10 July 2020
5. Koksal, I.: The rise of online learning. Forbes (2020). https://www.forbes.com/sites/ilkerkoksal/2020/05/02/the-rise-of-online-learning/#7cd389fd72f3. Accessed 10 July 2020
6. Friedman, Z.: Harvard will open this fall, but it may be online. Forbes (2020). https://www.forbes.com/sites/zackfriedman/2020/04/27/harvard-colleges-covid-remote-class/#14f052b55095. Accessed 10 July 2020
7. Teo, T., Wong, S.L.: Modeling key drivers of e-learning satisfaction among student teachers. J. Educ. Comput. Res. **48**(1), 71–95 (2013)
8. Chiu, C.M., Hsu, M.H., Sun, S.Y., Lin, T.C., Sun, P.C.: Usability, quality, value and e-learning continuance decisions. Comput. Educ. **45**(4), 399–416 (2005)
9. Paechter, M., Maier, B., Macher, D.: Students' expectations of, and experiences in e-learning: Their relation to learning achievements and course satisfaction. Comput. Educ. **54** (1), 222–229 (2010)
10. The Institute for International Business Communication, About the TOEIC® Program. https://www.iibc-global.org/english/toeic/toeic_program.html. Accessed 10 July 2020

11. The Institute for International Business Communication, Data & Analysis 2019. http://www.iibc-global.org/library/default/english/toeic/official_data/pdf/DAA_english.pdf. Accessed 10 July 2020
12. de Lourdes Machado, M., Brites, R., Magalhães, A., Sá, M.J.: Satisfaction with higher education: Critical data for student development. Eur. J. Educ. **46**(3), 415–432 (2011)
13. Temizer, L., Turkyilmaz, A.: Implementation of student satisfaction index model in higher education institutions. Procedia Soc. Behav. Sci. **46**, 3802–3806 (2012)
14. Helgesen, Ø., Nesset, E.: Images, satisfaction and antecedents: drivers of student loyalty? A case study of a Norwegian university college. Corp. Reput. Rev. **10**(1), 38–59 (2007)
15. Serenko, A.: Student satisfaction with Canadian music programmes: the application of the American Customer Satisfaction Model in higher education. Assess. Eval. High. Educ. **36** (3), 281–299 (2011)
16. Douglas, J., Douglas, A., Barnes, B.: Measuring student satisfaction at a UK university. Qual. Assur. Educ. **14**(3), 251–267 (2006)
17. Yusoff, M., McLeay, F., Woodruffe-Burton, H.: Dimensions driving business student satisfaction in higher education. Qual. Assur. Educ. **23**(1), 86–104 (2015)
18. Zineldin, M., Akdag, H.C., Vasicheva, V.: Assessing quality in higher education: new criteria for evaluating students' satisfaction. Qual. High. Educ. **17**(2), 231–243 (2011)
19. DeShields, O.W., Kara, A., Kaynak, E.: Determinants of business student satisfaction and retention in higher education: applying Herzberg's two-factor theory. Int. J. Educ. Manag. **19** (2), 128–139 (2005)
20. Powers, S., Rossman, M.H.: Student satisfaction with graduate education: dimensionality and assessment in a college education. Psychology: J. Hum. Behav. **22**(2), 46–48 (1985)
21. Wu, J.H., Tennyson, R.D., Hsia, T.L.: A study of student satisfaction in a blended e-learning system environment. Comput. Educ. **55**(1), 155–164 (2010)
22. Chen, N.S., Lin, K.M., Kinshuk: Analysing users' satisfaction with e-learning using a negative critical incidents approach. Innov. Educ. Teach. Int. **45**(2), 115–126 (2008)
23. Eom, S.B., Ashill, N.: The determinants of students' perceived learning outcomes and satisfaction in university online education: an update. Decis. Sci. J. Innov. Edu. **14**(2), 185–215 (2016)
24. Owston, R., York, D., Murtha, S.: Student perceptions and achievement in a university blended learning strategic initiative. Internet High. Educ. **18**, 38–46 (2013)
25. Kim, J., Jo, I.-H., Park, Y.: Effects of learning analytics dashboard: analyzing the relations among dashboard utilization, satisfaction, and learning achievement. Asia Pacific Educ. Rev. **17**(1), 13–24 (2015). https://doi.org/10.1007/s12564-015-9403-8
26. Educational Testing Service, TOEIC L&R 2018 Report on Test Takers Worldwide. https://www.ets.org/s/toeic/pdf/2018-report-on-test-takers-worldwide.pdf. Accessed 10 July 2020
27. Oricon ME Inc.: Customer satisfaction survey data. Informatics Research Data Repository, National Institute of informatics. (dataset). https://doi.org/10.32130/idr.10.1 (2019). Accessed 10 July 2020

Transilvania University of Brasov: Brand Ambassador of Brasov

Claudiu Coman(✉) and Maria Cristina Bularca

Transilvania University of Brasov, Brasov, Romania
claudiu.coman@unitbv.ro,
maria-cristina.bularca@student.unitbv.ro

Abstract. Nowadays the process of branding has extended and it can be applied to cities and universities. In this regard, a university, along with other cultural institutions, such as museums, can contribute to building the city brand. The purpose of this paper is to promote Transilvania University of Brasov in correlation with the History Museum, integrating into the city brand this institutions, that will become in this regard, brand ambassadors of Brasov. In this way, considering theoretical aspects about cultural, city and university branding, firstly the paper presents an analysis of the current way of communication between Transilvania University of Brasov and the History Museum. Further, taking into consideration the results we propose a model of an integrated promotion campaign in which we emphasize and provide examples of tactics and strategies that the university and the museum should use, in order to send a unitary message about the city, and become brand ambassadors of the city.

Keywords: University · Branding · Brasov

1 Introduction

In this age of technology and the rapid development of society, brands, both existing and emerging, are in a position to keep up with the needs and desires of consumers, who are constantly changing and becoming more and more difficult to satisfy.

In this context, brands resort to new methods of promotion, familiarization of customers with their specificity and activity adopting strategies that help them in the positioning process. This is also true in cases where cities are the ones who want to become brands through the activities they promote, the events they organize and the institutions they collaborate with. In this regard, any city, which intends to become a brand, must have certain "brand ambassadors" whose role is to contribute to its consolidation and increase its notoriety. Thus, given the idea that Brasov, although very popular and visited, is still regarded as an industrial center rather than an educational and cultural center, the purpose of this paper is to promote, through an integrated promotion campaign, Transilvania University of Brasov, in correlation with the History Museum, by integrating into the city brand these institutions, that will become in this way, brand ambassadors of the city.

T. Antipova (Ed.): ICCS 2020, LNNS 186, pp. 286–298, 2021.
https://doi.org/10.1007/978-3-030-66093-2_28

2 Literature Review

2.1 Brand, Branding and Brand Ambassadors

The concept of brand has been defined over time in many ways, but one of the most complex definitions was provided by Philip Kotler, also considered the father of marketing. According to him, a brand represents a name, symbol, logo, design or image, or any combination of these elements, which is designed to identify the product or service [1].

From an etymological point of view, the word brand comes from the old inhabitants of Scandinavia, - *brandr* -, where it had the meaning of burning [2].

Just like the concept of brand, the process of branding has been defined over the years, in multiple ways. In this respect, branding is in essence a form of interactive communication [3] that manages to bring together people who have a common interest in a product or service. According to David Arnold, [4] the concept of branding can be seen as a process of creating a unique brand experience that the consumer can enjoy, based on each point of contact: the product, the package design, the service, advertising, promotions, the website and also reputation. Moreover, branding involves the process of making known the attributes of an organization. In order to do this, an institution must first define who and what it is, but also what are the values and principles that guide it [5].

In today's competitive market, more and more organizations, companies and institutions choose to share who they are, to promote themselves, through the help of brand ambassadors. Often, brand ambassadors are represented by famous people who advocate for a certain brand [6]. Brand ambassadors usually help increase brand awareness by sharing the brand's message, its vision and mission through various channels such as: social media, TV, events or word of mouth [7]. Even more, brand ambassadors can influence and highly impact the perception of people about the brand [8] and one of their roles is to accurately project the image of the brand and of the organization [9].

In the context of a city, brand ambassadors possess information about the city and the opportunities and experiences that it offers and they communicate this information on multiple platforms, thus aiming to attract more visitors [10] or aiming to convince people to become citizens. In this regard, place brand ambassadors, can be represented not only by people but also by the cultural and educational institutions of the city, that can act as brand ambassadors and send messages about the city, its opportunities or events.

2.2 University and City Branding

Nowadays the process of branding has expanded and it is also used by cities and higher education institutions. Thereby, cultural and place branding offers a different perspective on branding, emphasizing that it is important for brands to develop with the society and the environment they are part of [11]. Instead of focusing on the positioning process, cultural branding focuses on the identity of the brand, around which it creates a story, a myth, with which consumers can resonate [12]. Moreover, cultural branding is based on the concept of culture, and focuses on art, design, architecture, cultural heritage or entertainment [13] and takes into account the idea that the content and meanings of the city come from the minds of people [14]. In this regard, city branding can be considered the process of assigning symbols and images to a place that consists of a set of characteristics, that people have assigned to that place over time [15]. Furthermore, the purpose of city branding is to provide added value and specific meanings to a place and to create positive perceptions among people.

On the other hand, the subject of branding can also be approached in the context of higher education institutions. Today, student's preferences regarding education, study programs and universities are becoming more diverse and varied. In this respect, in order to be recognized, appreciated and to attract more students, universities have to keep up with the needs and desires of the new generations of students.

Therefore, the concept of university branding, can be defined as "a manifestation of the unique characteristics of the university, a reflection of its ability to meet the needs of students, that generates confidence in the institution's ability to provide a certain type and level of education." [16]. In the academic field, university branding can have many objectives but some of the most important objectives according to Chris Chapleo are: creating and promoting a clear image of the university, providing information about its performance and academic offer and creating a competitive advantage [17].

Thus, in order to become powerful brands, while using the process of branding, universities have to define their identity: establish a name, logo, slogan; to promote their unique elements, to identify the resources they have and establish strategies to harness those resources and last but not least, to collaborate with the city and its cultural institutions in order to provide a student experience that matches their expectations. In this context, relevant and important are the concepts of integrated communication and co-branding.

Integrated communication refers to the process of aligning the organization's policies, values, mission, vision and activities in order to send clear, concise and unitary messages about the organization [18]. Integrated marketing communication takes into account every point of contact that people may have with the services or products offered. Institutions, and brands that implement integrated communication use multiple platforms in order to communicate information to their target audience [19].

While referring to places and products, by co-branding we understand the process of associating the product with a place that has characteristics that are beneficial to the product [20]. In this regard, focusing on the city of Brasov and its institutions, such an association is made by the university, which through its name: Transilvania University of Brasov, presents itself in relation with the city.

Further, in the case of branding cities and universities, the connection between the branding strategies used by the city and by a certain university is given by their target audience, by the consumers. In other words, students, prospective students, or members of the academic community become "consumers" of the city, they become citizens whether it is for a long or a short period of time [21].

Thus, higher education institutions have an essential role in developing the city. They create professionals, future employees, who may become permanent citizens. However, in order for students to want to be part of the city and its community, the city must be promoted as a student friendly environment, it should provide support and development opportunities for students and should empower students, by giving them the possibility to express their opinion regarding certain local decisions [22].

Many universities today use integrated branding and collaborate with the city in order to promote themselves. While studying the role of the city in the choice of foreign students to study at University of Pecs, Rekettye and Pozsgai, found that among the factors that influenced their decision was the attractiveness of the city, and concluded that it essential for the university and the city to communicate, to promote each other, and to work together more efficiently, that cities who want to be seen as educational centers must meet the needs of prospective foreign students, and that such collaboration may positively affect the recruitment process [23]. Even more, the researchers also pointed that a relationship between higher education institutions and cities can be seen in the case of universities that integrate into their name, the name of the city they are part of: University of Cambridge, University of Chicago, Princeton University [23].

Another study [24] focusing on the communication and partnerships between the university and the city revealed that, for example, the city of Lüneburg from Germany and Leuphana University of Lüneburg, collaborate efficiently and develop programs in order to achieve common goals: the university involves the city in the multidisciplinary research projects developed by students. Similarly, in Milan, the city collaborates with six universities on matters regarding urban regeneration: universities are offered the chance to expose and reveal the research they conduct on this matter and other issues of the city, while city officials map and take into account their findings [25].

Together with universities, cultural institutions, such as museums, can also promote themselves in relation to the city and in relation to universities. For example, The Museum of National History in Florence, it's currently focusing on enhancing the social and cultural life of the city, it organizes events related to the issue of urban cultural development, such as conferences, offline and online, and also encourages the participation of the local community [26].

Therefore, taking into account the aspects previously mentioned, we argue that cross communication and promotion between the city and its educational and cultural institutions can help the city efficiently develop and can make its institutions act as brand ambassadors of city. Even more, from our point of view, a university or a museum can be considered a brand ambassador of the city when it sends messages that are in line with the vision of the city and when the name of the city is identified with the institutions. For example, when someone mentions the city of Oxford, we immediately associate it with University of Oxford and vice versa.

3 Short Description of the Institutions and the City

Brasov, located in the "heart of Transylvania", is a city that was founded in 1211 by Teutonic knights [27]. With a rich history and culture, Brasov offers visitors, in addition to fun, numerous tourist attractions and cultural institutions: museums, churches, theaters, but also a higher education institution, Transilvania University, and is one of the most popular and visited cities in Romania.

Transilvania University of Brasov, is the largest university in the central region [28] and offers diversity through its 18 faculties, 108 bachelor's programs and 72 masters programs. Its visual identity focuses on the concept of being in the center and urges students to study and have the courage to leave their comfort zone.

The History Museum, is one of the most important cultural institutions of Brasov [29] and its mission is to harness the historical heritage of Brasov, and to bring pride and prestige to the communities of Brasov.

4 Research Methodology

Our paper aims to analyze the online cross communication and promotion of the city of Brasov, of Transilvania University of Brasov and The History Museum in order to create a frame of reference and propose an integrated promotion campaign that should focus on the way the online and offline communication and collaboration strategies can make the university and the museum brand ambassadors of the city.

In this regard, we elaborated our paper based on the research question: How can Transilvania University of Brasov and the History Museum become brand ambassadors of the city of Brasov? Focusing our paper on this matter we argue that the integrated communication and promotion between the university and the museum can help this institutions send unitary messages about the city, they can help them better collaborate with each other as well as with the city, thus making them brand ambassadors of the city.

For the analysis of the online communication between the university and the museum, content analysis was used as method, having as an instrument an analysis grid that was previously used by two other researchers, in order to analyze the activity on Facebook of ten of the world's most famous museums in 2015 [30]. We adapted their grid so that its criteria would allow the analysis of both the university and the museum, and we created a grid, that is represented in Appendix, that contains 24 criteria which measure elements such as: *contact* and *employee information*, *information about campus*, *international mobility*, *research projects*, the use of *social media instruments*, or *real time assistance* from the institutions. We analyzed the activity of the university and the museum on their website, Facebook, Youtube and Instagram page. After establishing what type of information do the university and the museum offer online, we analyzed if and how the institutions communicate with each other online, as well as offline, if they promote each other and what types of messages do they send.

5 Results and Discussion

The paper focuses on analyzing the online communication and promotion of Transilvania University of Brasov, of The History Museum and of the city of Brasov.

The university, according to the analysis grid, on its website offers information about its history, study programs, career possibilities, contact information, but it does not offer real time assistance, information about open days or details about lifestyle and costs in Brasov, and it does not use video testimonials.

Transilvania University of Braşov, uses social media platforms such as: Facebook, Instagram and Youtube, but it does not use Twitter. On Facebook, it promotes its activity as well as the admission period, the organized events, competitions, and has more shares than comments on posts, but it does not have posts regarding its collaboration with the city or the History Museum.

The analysis of the communication on Instagram revealed that users interact less with the videos or photos posted on this platform, than with the ones posted on the Facebook page. However, on Instagram the university has diversified posts that contain photos of the city, but none of the posts mention the museum. On Youtube, the university has a reduced activity, it only has 171 subscribers and it posted the last time in 2017.

The History Museum, on its website, provides information about its history, organized events and developed projects, visiting taxes, but it does not offer real time assistance, information about internship opportunities, the meaning of the slogan and logo, and it does not use a promotional video or video testimonials.

The analysis also revealed that the museum communicates through social media platforms such as Facebook and also Youtube, but it does not have an Instagram or Twitter account.

On Facebook, the museum has posts about its events and achievements, such as awards, TV mentions, but has little public engagement on posts, most of them not having 50 likes.

On Youtube, the museum has little activity: it posted the last time in 2017, it only has 10 subscribers and it promotes only the rubric: History in 60 s.

About Braşov, the analysis revealed that it communicates online through a website, the city hall website, and through social media platforms such as: Facebook, Instagram. However, the city does not have an Instagram and Youtube account.

On its website, the city provides information about the location of some of the tourist attractions, about its history, touristic trails: medieval, architectural, cultural, about the services offered by the city hall.

On Facebook and Instagram, the city communicates through pages such as: Come to Brasov, Brasov- The city of my soul, and it does not mention neither the university nor the museum. Taking into consideration the communication between the city, the university and the History Museum, according to the analysis there is not an effective communication between them in the online environment: they do not promote each other's activity, or the events organized in collaboration, and improvements can be made in this regard.

The communication of the university with the museum takes place mainly through events realized in collaboration but also through the internship possibilities that the museum offers to the students. This programs and events take place offline but in the online environment they are very poorly promoted.

In the offline environment, the institutions communicate mainly through the events they organize, but also through the internship programs that they offer for students. An example of such a program can be considered the internship provided by the History Museum, which allows students to get into the skin of historical characters and explain to tourists the life of those characters. In this regard, its association with the city is a positive one: the museum promotes the history of the city and wants to make its citizens proud.

On the other hand, before implementing an integrated promotion campaign a research should be done, but taking into account the problem and the current communication between the institutions and the city, details about how such a campaign should be carried out, can also be provided at this stage of the project.

The preliminary research will be carried out in the county of Brasov, but it will also consider other points of interest for Transilvania University of Brașov, such as: Moldova area: Bacău, Onești. The research will be carried out through a questionnaire that will be applied to students of the 12th grade, and will contain questions about their career preferences, study domains, student life, but will also be conducted through interviews and focus groups with the citizens of Brasov and current students, to find out their opinions about the city and the university: what should be improved/changed. In this regard using the results of the online communication analysis as frame of reference, we further provide information about an integrated promotion campaign in order to improve the communication between the city and its institutions.

6 Integrated Promotion Campaign

6.1 Identifying the Problem

Regarding the problem, it is firstly emphasized that Brasov is still considered an industrial center rather than an educational center and in its promotion policies, Transilvania University of Brasov does not have the place it should have. Moreover, it is added to this problem the fact that the university does not necessarily attract students through what it has to offer: study programs, internship, international mobilities, but the students choose it because of the beauty of Brasov, and because of what the city offers them.

6.2 Target Audience

The target audience of the campaign should be structured into several segments, and it will consist of students of the 12th grade, parents of the students, citizens of Brasov, as well as from other areas of interest for the university such as: Bacău, Onești, Focșani, Galați. Romania is currently divided into seven regions: the center region, which includes Brasov, the North-East region which includes Bacau, the North-West region, West region, South West region, South region and South East region- that includes-Galati. Thus, we chose as target audience, the areas from which Transilvania University of Brasov usually gathers its students.

6.3 Purpose and Objectives

The purpose of this campaign is represented by the promotion of Transilvania University of Brasov in correlation to the History Museum Brasov, integrating into the city brand this institutions, that will become in this way, brand ambassadors of the city.

The first objective of the campaign is to improve the online communication between the city, the university and the museum. The second objective is to identify the actions that the university should take in order to become a brand ambassador of the city, and the third objective is for the city to promote in the online environment the activity and events organized by the university and the museum.

6.4 Tactics and Strategies

In order to improve the communication between the university, the city, and the History Museum and in order for these institutions to become brand ambassadors of the city, this paper provides strategies and tactics.

Firstly, there is a need for improvement in the way each institution communicates online. In this regard, it is recommended for Transilvania University of Brasov, The History Museum, and the city, to provide on their websites real time assistance in order to facilitate the communication with potential tourists and students. Another tactic would be to present on the website, video testimonials that have the role of increasing the trust of the consumers in the specificity and activity of the institutions. Moreover, the institutions should present on their websites information about the lifestyle in Brasov: costs, entertainment opportunities, cultural events, to offer students and tourists a clearer perspective on the environment in which they will live, in the short or long term.

Another strategy should take into consideration creating a Twitter account for the university, the city and the museum, in order to increase their visibility internationally, and also create an Instagram account for the museum and a Youtube account for the city. In addition, updating the content on Youtube for the university and the museum could contribute to improving their communication with the target audience, due to the fact that people today appreciate and are more interested in visual content.

In order to improve the collaboration between the institutions and the city, the university should promote on its Facebook page the events and internship opportunities offered by the museum, and in turn, the university and the city should promote each other and communicate information about the activity of the museum.

Considering the university and the city, they should develop programs in collaboration: the university should conduct research for the city, and in return Brasov should offer support in this regard. Also, to develop the cultural and educational side of the city, touristic trails, such as medieval, or architectural trails should be promoted among students. Moreover, strategies as promoting within the university's radio: Radio Campus Transilvania, the events held by the city, or inviting to the events organized in collaboration, people with online influence, can contribute to improving the communication between the city and the university, making the institutions more visible and more likely to become brand ambassadors of Brasov.

A university is strong especially because of its students. A strong university, brings to the city, in addition to economic benefits, other benefits such as: trained and qualified workforce, cultural development.

Taking into account the second objective of the campaign, a strategy for consolidating and transforming the university in a brand ambassador of Brasov would be to diversify and promote online, the study programs offered in English. In this way the university could become more visible and will attract more international students. Another strategy should consider promoting the strengths and the elements that make the university different: diversity of study programs and international mobility. Moreover, online and offline promotion through successful graduates would create among the target audience positive association with the university.

Another strategy that could transform the university in a brand ambassador of the city would be to improve the student experience outside courses: by creating recreational and socializing spaces for students such as a student cafe or by organizing themed events and parties.

6.5 Evaluation

Evaluation of the campaign is essential because the feedback can provide a clearer perspective about meeting the established objectives. The evaluation will be carried out through an evaluation questionnaire applied to the people targeted by the campaign. The questionnaire will measure if the objectives of the campaign were met, if the university communicates better with the city and the museum, if they promote each other events, if students associate the university with the city.

7 Conclusion

The aim of the paper was to analyze the communication and promotion of Transilvania University of Brasov, of The History Museum and of the city of Brasov, in order to create a frame of reference regarding the way cross communication between the institutions and the city could help the institutions become brand ambassadors of the city. In this regard through content analysis and following the criteria of the analysis grid that we created, we analyzed the online communication of the institutions and the city on their official website as well as on other platforms used by them, such as Facebook, Youtube and Instagram.

The main findings of the research reveal that the city, the university and the museum do not communicate or promote each other effectively in the online environment, the university and the museum offer online a wide variety of information, regarding contact information, taxes, research projects, news and announcements about their activity and integrate social media instruments. However, neither the museum nor the university offer online information about the city of Brasov, about the events that they hold in collaboration, or about opportunities and city lifestyle. Even more, in the offline environment the institutions communicate with each other mainly through events organized in collaboration but they do not properly promote those events so as to highlight their partnership and to emphasize the importance of the events for the development of the city. Thus, it can be inferred that currently there isn't an integrated communication between the institutions and the city, and the university even though it integrates the name of the city, it does not act as one of its ambassadors: it does not mention the city's events, it does not promote it on its social media platforms, and it's collaboration with the city is scarce. The city offers little support for the university as well as for the museum to develop research projects or organize events. Further, the results of the research provide insights on the strategies used by the city and its institutions to communicate online information about their activity, and can be used as a frame of reference in similar studies that focus on cross communication.

Therefore, with regard to this findings, and in order to improve the communication between the institutions and the city we proposed an integrated promotion campaign. The campaign focuses on providing strategies and techniques that the university and the museum can use in order to communicate effectively, to send unitary messages about their activity in relation with the city, and to act as brand ambassadors of Brasov.

7.1 Limitations and Suggestions for Future Research

While the information provided by our research is useful and can be seen as a starting point for improving communication between the university and the museum in order for them to become brand ambassadors of Brasov and promote a positive image of the city, the research also has some limitations.

One limitation is represented by the fact that our study focuses only on one university and one city. Thus by focusing only on the city of Brasov and its institutions we were not able to conduct a comparative analysis in order to find out if other universities or museums communicate with their city better or worse. Another limitation of the research is not conducting the preliminary research necessary for the implementation of the proposed campaign.

In relation to this limitations, some opportunities for future research arise. Firstly, the research should be extended to other universities and cities from Romania but also from Europe, in order to develop a comparative analysis. Secondly, after gathering information from different universities and cities, the proposed integrated promotion campaign could be improved and implemented at the level of the city of Brasov.

Appendix

Categories:	Transilvania University	The History Museum
1. Contact information	✓	✓
2.Social media instruments	✓	✓
3. Employees information	✓	✓
4. Study programs/events information	✓	✓
5. Career possibilities information	✓	✓
6. Facebook page	✓	✓
7. Twitter page	X	X
8. Instagram page	✓	X
9. Real time assistance	X	X
10. History information	✓	✓
11. Campus/ buildings information	✓	✓
12. International mobilities/projects	✓	✓
13. Information about internships	✓	X
14. Information about open days	X	✓
15. Research projects	✓	✓
16. Information about slogan	✓	X
17. Posts with video on Facebook	✓	✓
18. Posts with pictures on Facebook	✓	✓
29. Posts about competitions/contests	✓	✓
20.Video Testimonials	X	X
21. Promotional video	✓	X
22.Information about taxes	✓	✓
23.Information about city lifestyle	X	X
24. News and announcements	✓	✓

References

1. Kotler, P., Keller, K.L., Brady, M., Goodman, M., Hansen, T.: Marketing Management. Pearson Education (2009)
2. Cărămidă, C.: Brand & Branding: Visual Identity, vol. I. BrandMark Publisher (2009)
3. Ballantyne, D., Aitken, R.: Branding in B2B markets: insights from the service-dominant logic of marketing. J. Bus. Ind. Market. **22**(6), 363–371 (2007). https://doi.org/10.1108/08858620710780127
4. David, A.: The Handbook of Brand Management: The Economist Books. Edward Elgar Publishing Limited (2008)
5. Wæraas, A., Solbakk, M.N.: Defining the essence of a university: lessons from higher education branding. High. Educ. **57**(4), 449–462 (2009). https://doi.org/10.1007/s10734-008-9155-z

6. Urbaniak, M.: What Are Brand Ambassadors and Why Are They Important? (2020). https://www.meltwater.com/en/blog/what-are-brand-ambassadors-and-why-are-they-important. Accessed 02 Sept 2020
7. Dobbs, M.: Adorned with delight: Brand ambassador program guide (2018). https://www.adornedwithdelight.com/wp-content/uploads/2018/09/Brand-Ambassador-Program-Guide.pdf. Accessed 02 Sept 2020
8. Sadrabadi, A.N., Saraji, M.K., MonshiZadeh, M.: Evaluating the role of brand ambassadors in social media. J. Market. Manag. Consumer Behav. **2**(3), 54–70 (2018)
9. Alshuaibi, A.S., Mohd Shamsudin, F., Abd Aziz, N.: Developing brand ambassadors: the role of brand-centered human resource management. Int. Rev. Manag. Market. **6**(S7), 155–161 (2016)
10. Yu, E., Kim, J.: The relationship between self-city brand connection, city brand experience, and city brand ambassadors. Sustainability **12**(3), 982, 1–13 (2020). https://doi.org/10.3390/su12030982
11. Helaniemi, K.: Icons Becoming Brands: How can brands emerge as the result of culture-Case Sauna as the Iconic Brand of Finland (2016)
12. Holt, D.B., Holt, D.B.: How Brands Become Icons: The Principles of Cultural Branding. Harvard Business Press (2004)
13. Scaramanga, M.: Talking about art (s): a theoretical framework clarifying the association between culture and place branding. J. Place Manag. Dev. **5**(1), 70–80 (2012). https://doi.org/10.1108/17538331211209059
14. Sáez, L., Periáñez, I., Mediano, L.: Building brand value in major Spanish cities: an analysis through municipal websites. J. Place Manag. Dev. **6**(2), 120–143 (2013). https://doi.org/10.1108/JPMD-04-2012-0011
15. Pryor, S., Grossbart, S.: Creating meaning on main street: towards a model of place branding. Place Branding Public Diplomacy **3**(4), 291–304 (2007). https://doi.org/10.1057/palgrave.pb.6000080
16. Bennett, R., Ali-Choudhury, R.: Prospective students' perceptions of university brands: an empirical study. J. Market. High. Educ. **19**(1), 85–107 (2009). https://doi.org/10.1080/08841240902905445
17. Chapleo, C.: Exploring rationales for branding a university: should we be seeking to measure branding in UK universities? J. Brand Manag. **18**(6), 411–422 (2011). https://doi.org/10.1057/bm.2010.53
18. Christensen, L.T., Fırat, A.F., Torp, S.: The organisation of integrated communications: toward flexible integration. Eur. J. Market. **42**(3/4), 423–454 (2008). https://doi.org/10.1108/03090560810853002
19. Reid, M.: Building strong brands through the management of integrated marketing communications. Int. J. Wine Market. **14**(3), 37–52 (2002). https://doi.org/10.1108/eb008745
20. Kavaratzis, M., Ashworth, G.J.: City branding: an effective assertion of identity or a transitory marketing trick? Tijdschrift voor economische en sociale geografie **96**(5), 506–514 (2005). https://doi.org/10.1111/j.1467-9663.2005.00482.x
21. Popescu, A.I.: Branding cities as educational centres. The role of higher education institutions. Manag. Market. **7**(3), 493–512 (2012)
22. Van Den Berg, L., Russo, A.: The Student City: Strategic Planning for Student Communities in EU Cities. Routledge (2017)
23. Rekettye, G., Pozsgai, G.: University and place branding: The case of universities located in ECC (European Capital of Culture) cities. Ekonomski vjesnik: Rev. Contemp. Entrepreneurship Bus. Econ. Issues **28**(S), 13–24 (2015)

24. Withycombe Keeler, L., et al.: Transferring sustainability solutions across contexts through city–university partnerships. Sustainability **10**(9), 1–17 (2018). https://doi.org/10.3390/su10092966
25. The British Council (2019) Smart Places: How universities are shaping a new wave of smart cities. https://www.britishcouncil.org/sites/default/files/smart_places_report.pdf. Accessed 02 Sept 2020
26. Lazzeretti, L., Capone, F.: Museums as societal engines for urban renewal. The event strategy of the museum of natural history in Florence. Eur. Plan. Stud. **23**(8), 1548–1567 (2015). https://doi.org/10.1080/09654313.2013.819073
27. About Braşov (2009). http://www.simplybrasov.com/brasovguide.html. Accessed 19 Feb 2020
28. About Unitbv (2020). Transilvania University of Brasov. https://www.unitbv.ro. Accessed 19 Feb 2020
29. About museum (2020). http://www.brasovistorie.ro/despre-muzeu/prezentare. Accessed 19 Feb 2020
30. Özdemir, G., Çelebi, D.: A social media framework of cultural museum. Int. J. Akdeniz Univ. Tourism Fac. **5**(2), 101–119 (2017). https://doi.org/10.30519/ahtr.375248

Genetics Technology

Genetic Improvement for Gluten Strength in Russian Spring Durum Wheat Genotypes

V. Natoli[1,4](✉), P. Malchikov[1,2], P. De Vita[3], S. Shevchenko[2], and S. Dolaberidze[1]

[1] Skolkovo Innovation Center, Agroliga Plant Selection Center Ltd., 42 Bolshoy Boulevard, Moscow 143026, Russia
info@geneticservices.com

[2] Samara Agricultural Research Institute Volga Region, 41 Karl Marx, St., Bezenchuk pgt, Samarskaya Oblast 446254, Russia

[3] Consiglio per la Ricerca in Agricoltura e l'analisi dell'economia Agraria-Centro Cerealicoltura e Colture Industriali (CREA-CI), SS 673 km 25.2, 71122 Foggia, Italy

[4] Genetic Services Srl, Co. da Catenaccio Snc, 71026 Deliceto, FG, Italy

Abstract. Gluten strength is an important end-use quality factor in durum wheat (Triticum turgidum ssp. durum), affecting pasta manufacture and cooking quality. The aim of this work was to improve the technological performances (i.e. gluten index, GI) of Russian durum wheat varieties for pasta-making. Thirty-six spring advanced breeding lines, derived from the Agroliga Plant Selection Center and Samara Agricultural Research Institute breeding programs, were evaluated for agronomic traits at Samara in the Volga region during the 2015 growing season. Three F2BC2 populations were produced from each selected line crossed whit a high gluten quality variety (HGQ) and selected in F2, F2BC1 and F2BC2 generations for: i) genetic similarity [using 100 SSR markers] to recurrent Russian parental line; ii) presence of SDS-page HGQ HMW Glu-B1 (High Molecular Weight) band; iii) QTL located in chromosome 1A and associated to high GI value) using SSR marker Barc 148; iv) good value for SDS-micro-sedimentation test. Two durum wheat genotypes (named Taganrog and Bourbon) were selected from HGQ x 3371 Russian line F2BC2 population and multiplied in a growth chamber for increasing the quantity of seeds. The new genetic materials were tested in replicated plots at Samara, compared with local check variety (Besenchuk 209), and evaluated for gluten quality (GI) and viscoelastic properties of gluten as proportion of sodium dodecyl sulfate-unextractable polymeric proteins (%UPP). Good adaptation was observed for the new selected genotypes and a higher gluten index and %UPP values, were recorded.

Keywords: Durum wheat · Gluten strength · QTLs introgression · Russian spring environment

T. Antipova (Ed.): ICCS 2020, LNNS 186, pp. 301–312, 2021.
https://doi.org/10.1007/978-3-030-66093-2_29

1 Introduction

Durum wheat (Triticum turgidum L. ssp. durum) is an allotetraploid species with 2n = 4x = 28 (AABB genomes) and is used predominantly for food products such as pasta, couscous, and bulgur.

Approximately 30 million tons of durum wheat is produced every year in different regions of the world on an estimated cultivation area of about 16 million hectares (http://www.fas.usda.gov/pecad/highlights/2005/07/durum2005). Since durum wheat is mainly used for pasta, the varieties that meet the requirements of high-quality pasta products receive premium prices in the global market. In Russia, durum wheat is produced mainly in the conditions of the middle Volga and South Ural regions in the amount of 650,000–700,000 tons on the area of more than 0.5 million hectares. In recent years, the demand of the Russian market for high-quality pasta is estimated at 750,000–800,000 tons, which is equivalent to 1.5 million tons of durum wheat (Lozhkin et al. 2018). This new scenario could require an expansion of the current cultivated area by improving the qualitative characteristics of Russian production. Over the past six years, in fact, 62%, on average, of Russian durum wheat production fell into the poorest grain quality classes (#4 and #5), while there was practically no grain production falling in the high-quality wheat classes (#1 and #2). Since currently, the total storage capacity of the processing industries, is localized in the European part of the Russian Federation with a great potential for expansion of its cultivation, the only factor that could limit the market growth of exports could be the lack of high-quality and adapted durum wheat varieties.

The processing and end-quality of durum wheat is determined by the genotypic (G), the cultivation environment (E), the agronomic management (M) and by interaction GxExM (Mariani et al. 1995; Marchylo et al. 2001). To ensure adequate processing potential, durum wheat processor requires suppliers to meet a list of grain characteristics to high-quality standard (i.e. vitreousness, ash content, moisture content, protein content, gluten strength, semolina color). Because some qualitative parameters (i.e. gluten strength, semolina color, etc. …) are genotype-dependent, milling and pasta factories request that suppliers include only specific durum wheat varieties. Among grain quality parameters, gluten strength and tenacity, which are determined by glutenin (High Molecular Weight and Low Molecular Weight) and gliadin composition, are two of the main factors that determine pasta firmness, small starch losses during cooking, and increased tolerance to overcooking (Sissons et al. 2005; Payne 1987; Pogna et al. 1990).

In this context, one possibility to increase durum wheat production in the Volga region could be to develop new durum wheat varieties characterized by a wide adaptability and good grain quality, exploiting the Russian germplasm and introducing new genetic materials with strong gluten and excellent technological performance.

Kofa is Southwestern United States cv. released by Western Plant Breeders (Arizona) with excellent quality in terms of protein content, gluten index (GI), and SDS micro-sedimentation (SDSS) test. Protein analyses of the HMW glutenin subunits (GS) revealed the presence of Glu-B1d allele (6 + 8 subunits) in Kofa varieties

(Bubcvsky et al. 2008) on chromosome 1B identified by SSR (simple sequence repeat) markers.

The aim of this work was to improve the technological performances (i.e. gluten index, GI) of Russian durum wheat varieties for pasta-making. With this objective, a series of advanced breeding lines obtained by crossing high-gluten quality (HGQ) durum wheat varieties (Kofa) and Russian germplasm, were evaluated using biochemical and molecular markers.

2 Material and Method

Agronomic evaluation: The agronomic evaluation of Russian and Italian accessions was carried out in 2016 and 2017. In 2016, the evaluation concerned 48 accessions of which 6 were Russian varieties and 6 were Italian varieties. In boht years of evaluation the studies were carried out by the type of micro-field experiments in three-fold repetition with randomized placement of plots. The size of one plot was 10 m^2. The sowing was carried out in late April, and the rainfall during the crop cycle was scarce equal to 200 mm with average temperatures of 26 °C. Did the collection take place on August 6. A sample of each individual parcel was kept to carry out the analyzes concerning the protein content, yellow index, gluten index, SDS micro-sedimentation and one-dimensional SDS-page profile for the verification of the HMW (Hight molecular weight) glutenin bands and LMW (Low molecular weight) and the gliadin bands. In 2017 the test was carried out in the same way always in the Samara field with 18 accessions including 4 Russian tests, 6 Italian accessions and 6 best accessions selected from the 2016 test. The sowing has always been carried out in late April, but has there been more rainfall around 400 mm with an average temperature of The collection was carried out on August 10th and also in this case a sample was collected for the same qualitative analyzes as for 2016.

Population Development: After phenotyping and quality trait evaluations, in 2015 season in Samara region, three F2 populations was obtained from a cross whit Kofa, a high gluten quality variety (HGQ). Twenty plant was evaluated in F2 generation for: i) genetic similarity (using 100 SSR markers) to recurrent Russian parental line; ii) presence of SDS-page HGQ High Molecular Weight (HMW) Glu-B1 band from HCG parental line; iii) QTL located in chromosome 1A, and associated to high GI value, using SSR marker Barc 148 (HCG allele); iv) good value for SDS-micro-sedimentation test. One single line was selected from each population and crossed whit Russian parental line. The same procedure was made for F2BC2 generation production.

Selection Method: The best three resulting from the 2016 test were crossed with the Kofa variety donor of Qtls for the increase in the gluten index. The F1 generation was then self-fertilized to obtain the F2 generation. About 500 F2 plantules were obtained for each single cross (for the three crosses a total of 1500 plants) where a leaf sample was taken for the extraction of DNA at the three-leaf stage. After the evaluation for the presence of the Qtl of interest, the plants having the Kofa allelic variants, were also evaluated with 100 SSR (Single sequence Repeat) markers to select those plants more genetically similar to the Russian accession used as a female parent in the cross. The selected plants were grown in a growth chamber where through the regulation of the temperature and hours of light for increase the tillering and each tillage was in turn removed and rooted up to obtain 20 plants for each single F2 plant chosen. This is to allow to have a sufficient quantity of seed to be able to carry out an SDS micro-sedimentation and SDS-page verification test for High Molecular Weight (HMW), Low Molecular Weight, (LMW)and gliadinic gluten-free bands. 10 plants were chosen for each cross, bringing all the desired characteristics. Every single plant has been crossed with the Russian accession and on each F2Bc1 progeny only the lines with the Kofa allelic variants for the QTL of interest have been selected. As for the F2 seedlings, also here they have been multiplied so that 20 plants are available for a single accession to have enough seed for SDS micro-sedimentation and one-dimensional SDS-page evaluation. Also, in this case 10 plants were kept for re-crossing, giving the desired characteristics. The same procedure was carried out for the F2Bc2 generation and part of the seed of the chosen plants was multiplied in vitro to obtain 10,000 plants for each choice, and part is used for qualitative analyzes, this time adding the analysis % UPP (percentage of extractable protein polymers). 3 plants were chosen, one for each of the re-crosses, but only two were multiplied in vitro. The 10,000 plants for accession thus obtained have been transplanted in the field for the production of core seed which will be used for the production of commercial seed.

Genomic DNA Extraction: A standard genomic DNA extraction procedure has been implemented, applicable and repeatable on any vital cereal sample. The extraction was carried out from leaf material collected from a bulk of 20 seedlings for each accession, grown in a greenhouse on peat. The stage of collection of the leaf material coincided with the complete development of the third-fourth leaf (about 10 days from the emergency). It was decided to extract DNA from leaf material and not directly from the kernels as this ensures and simplifies the obtaining of DNA having a greater degree of purity. Immediately after collection, the material was freeze-dried and then finely ground using a mortar, pestle and sterile quartz powder.

SSR (Single Sequence Repeat) selection markers associated with the QTL (Quantitative Trait Loci) associated in the control of the gluten index character, and HMW (Hight Molecular Weight) glutenin selection band associated with the GLU-B1 locus. For this program we have transfer two of the QTLs (Quantitative Trait Loci) which is most associated with a good gluten index and SDS micro sedimentation, present in the Kofa variety and identified by Bubcvsky et al. Theor Appl Genet Sept., 2008, positioned on chromosome 1 A in the interval between the two markers SSR Barc148 and SNP (Single Nucleotide Polymorphism) BM140362 located on chromosome 1 B at the Glu-b1 locus associated with the high molecular weight glutenin band (HMW: Hight Molecular Weight). For the Qtl located on chromosama 1 B, molecular markers were not followed, but the presence of the Kofa HMW band was assessed, which was different from that carried by the selected Russian accessions (7 + 8). These two Qtls guaranteed us at least 40% of the variance for the gluten index.

SSR (Single Sequence Repeat) Selection Markers for the Genotypic Characterization of the Population Crossed Whit Russian Parental Line: About genotypic characterization, reference was made to the Ph.D. thesis of V. Natoli, 2008: from here 100 SSR (Single Sequence Repeat) Markers were, chosen based on their informativeness (PIC: Informative Information Content) and on the fact that they presented all different allelic forms between the two parents 3371 and Kofa. Furthermore, in order to have a good characterization of the accessions, the 100 pairs of primers corresponding to the same number of microsatellite markers (SSR), have also been chosen to guarantee a good coverage of the two genomes A and B of durum wheat. Generally these primers have proved highly locus-specific and, in only a few cases, the sequences used to draw the primers have been conserved at the level of the 2 homeologous genomes determining the simultaneous amplification of the microsatellite locus in both genomes A and B. The SSRs used include i) a public set available on the GrainGenes database (http://wheat.pw.usda.gov): WMC (Xwmc, Gupta et al. 2002), BARC (Xbarc, Song et al. 2002 and 2005), CFA and CFD (Xcfa and Xcfd Sourdille et al. 2001; Guyomarc'h et al. 2002) and WMS (Xgwm), (Röder et al. 1998; Somers et al. 2004; Sourdille et al. 2004). In the work of V. Natoli 2008, the level of polymorphism of each SSR locus was expressed by the Polymorphism Index, known as PIC ("Polymorphic Information Content" or "gene diversity index"), which considers both the number of alleles present at a given locus, be the relative frequency of each allele (Anderson et al. 1992; Kremer et al. 1998).

$$\mathrm{PIC} = 1 - \sum \mathrm{pj2}$$

Where pj is the frequency of the th allele in the accessions considered. The PIC coefficient represents the probability that, at a certain locus, the alleles of two random accessions, among those analyzed, are different. The observed PIC value of each locus was compared with the theoretical maximum value achievable as a function of the number of alleles present. The PIC coefficient reaches the maximum value when all the alleles are uniformly distributed in the accessions (that is, they are characterized by similar frequencies). For the selection of the SSR loci, another index complementary to the PIC (always calculated from the work of V. Natoli 2008) was taken into account: that is, the "number of effective alleles", indicated with Ae:

$$\mathrm{Ae} = 1/(1 - \mathrm{PIC}),$$

It represents the theoretical number of alleles each having a similar frequency in the material under examination at which the PIC value is equal to that actually observed at that locus. The value of Ae is always lower than the number of alleles observed. The closer the Ae value is to that observed, the higher the level of informativeness and discriminating capacity of the locus (Kremer et al. 1998).

Gluten Strength Evaluation: SDS-PAGE (Sodium Dodecyl Sulphate - PolyAcrylamide Gel Electrophoresis (electrophoresis on polyacrylamide gel in the presence of sodium dodecyl sulfate) was used for the analysis of protein extracts. The principle on which this electrophoretic technique is based is the denaturing activity of the SDS; this is able to interact with proteins in a constant ratio of 1.4 g of SDS per gram of protein. The separation takes place therefore by difference between molecular weights since the mass/charge ratio for each protein denatured with SDS remains constant. The microsedimentation index in the SDS test (Dick and Quick 1983), indicates the sediment volume obtained, under specific conditions, from a suspension of a sample of flour in a solution of lactic acid and sodium dodecyl sulfate (SDS). The method exploits the property of the reserve proteins to gelatinize in the presence of SDS; the higher the sedimentation volume, the better the rheological quality of the flour or semolina. The amount of flour used was one gram (sedimentation micro test). This index is closely associated with the gluten index (J. Ross et al. Can J. Plant 2009). This test was used to evaluate the strength of gluten in the offspring F2, F2Bc1 and FeBc2, where the quantity of grain was limited. The Gluten index (IG, AACC Method 38-12) was assessed on the Russian accessions tested in 2016 in Samara, for the evaluation of the goodness of gluten and to compare this data with the micro-sedimentation SDS index and the SDS- bands. Page, for the initial screening of materials. Size exlusion fast performance liquid chromatography (SE-FPLC), two extracts (SDS-extractable protein polymers and SDS-extractable protein polymers) were obtained from each semolina sample in accordance with Morel et al. (2000) with the modification described by Tronsmo et al. (2002). The proportion of SDS-protein polymers in extractables on the total of protein polymers (% UPP) was calculated as [F1 */(F1 * + F1)] × 100 and the percentage of protein monomers on protein polymers was calculated as (F3 + F4)/(F1 * + F1 + F2). Where F1 * = large extractable polymers, from F1 to F4 = different extractable protein polymers.

In Vitro Multiplication: In vitro multiplication was carried out starting from the seed germinated on agar containing a mixture of Murashige and Skoog basal base (medium), with the addition of gibberellic acid to favor the development of culms starting from crown cells. Germination took place at 18 °C with 8 h of light on 24. After germination the temperature was brought to 12 °C without changing the hours of light. As culms of formation were formed, these were removed and placed on another agar of the same composition and in the same growth conditions.

Analysis of Genetic Similarity Through SSR (Sigle Sequence Repeat) Markers: In the genetic similarity matrices (GSij) between the possible accession pairs, including the two crossing parents, 100 SSR (Single Sequence Repeat) markers already selected and used by the work of V. Natoli 2008 (doctoral thesis) were considered. The identifier of each allele for the single marker was converted into a number, from 1 to 4 (allelic forms found for the markers used), for analysis. The similarity between all the accession pairs were calculated using the "Simple matching" similarity coefficient: proportion of markers that have the same alleles This coefficient was calculated with the following formula:

$$\mathrm{GSij} = \mathrm{m}/\mathrm{n}$$

where m = number of allelic variants of the single stator ssr that show concordance of their state, i.e. equal variant in iej (in our case, it is equivalent to the number of markers for which our accessions have the same allele) en = total number of comparable markers. The diversity index for each of the SSR loci was calculated as:

$$GD = 1 - \sum p2j$$

where p is the frequency of the jmo allele within the population (Powell et al. 1996). The calculation of genetic similarities was performed with the SIMQUAL program of the NTSYS-pc ver package. 2.0 (Rohlf 1997. On the basis of the similarity matrices thus obtained, the cluster analysis was carried out and the corresponding dendrograms were constructed with the UPGMA method ("Unweighed Pair-Group Method Arithmetic Average") using the Jump8 program.

3 Result and Discussion

Preliminary Agronomic and Qualitative Assessment of Russian and Italian Accessions in a Russian Spring-Sown Cultivation Environment: In 2016, the evaluation concerned 48 accessions of which 6 were Russian witnesses and 6 were Italian varieties. The qualitative data inherent mainly to the gluten index showed a very low level for Russian accessions including witnesses, while Italian accessions confirmed their good quality level (Table 1). As regards the production level and adaptability to the environment, Russian accessions are clearly better than Italian accessions, confirming their genotypic adaptation (Table 1). The best lines with a fair quality level and a good grain yield were the 3363, 3379 and 3371 accesses. These three lines were identified as female parent in the crossings for the transfer of the Kofa Qtl for the increase of the gluten index and the Glu-B1 band; they were also re-evaluated in the agronomic test in Samara 2017. In this test, these accessions confirmed their good agronomic value, resulting to be the best in terms of grain yield, adaptation and resistance to diseases (Table 2). The analyzes of the SDS-Page HMW bands, related to the accessions of the test in Samara 2016, show that only three Russian accession has the same profile as the qualitative reference variety Kofa (Fig. 1), and this has confirmed that Kofa band (Glu.B1) could be easily evaluated for the transfer in the chosen Russian accessions.

Genetic Similarity Analysis of Through Single Sequence Repeat (SSR) Markers: The similarity analysis on the F2 populations developed through the crosses with the three Russian accessions chosen, showed that already in F2 there could be the possibility of identifying, among those with the Kofa allele for the Qtl of interest and the band of Kofa Glu-B1, accessions genetically more similar to the Russian accession used as female parental (Table 4a; Fig. 2a). Starting from 20 F2 individuals for each individual cross, the individuals genetically equal to the Russian line were chosen for at least 80% of the loci (except for the Qtl of Kofa and the Glu-B1 band). After the second crossbreeding, the analysis of genetic similarity shows how the goal was achieved, identifying an accession (07) genetically very similar to the Russian accession (Fig. 2b). The phenotypic evaluation confirms this similarity, in fact the chosen access is indistinguishable from the Russian accession 3371 used as a recurrent. The same was also done for the crossing with the Russian accession 3379 (data not shown).

Table 1. Yield (T/Ha) and quality evaluation during 2016 season in Samara environment. The table show Italian and Russian and Italian genotype compared with 5 Russian best local check. Very low performances of Russian genotypes were confirmed by gluten index and total dry gluten, while they showed high adaptability in terms of production.

		Harvest, Bezenchuk, 2016 year		Protein (%)	yellow index (coord. b)	Total dry gluten (%)	Dry gluten respect dry substance (%)	Gluten index (%)
		T/ha	% from check,Bezenchukskaya 210	Metodo Infratec	Metodo Minolta	Metodo UNI N. 10275 - 01/1994	Metodo UNI N. 10275 - 01/1994	Metodo UNI N. 10690 - 09/1997
Number line	Origin			± 1,00 %	± 2,00			
3.382	new Russia	3,4	130,2	14,2	28,2	9,9	11,8	63,5
3.384	new Russia	3,0	116,2	14,3	27,1	8,6	10,2	64,3
3.383	new Russia	3,0	115,1	15,0	27,6	11,7	13,9	32,1
3.380	new Russia	2,7	106,2	14,6	30,8	10,4	12,3	29,9
3.381	new Russia	2,6	102,7	14,6	26,9	10,4	12,4	65,6
3.386	new Russia	2,6	101,7	13,6	26,5	9,0	10,8	18,8
3.379	new Russia	2,4	106,1	14,6	32,1	10,7	12,7	68,5
3.385	new Russia	2,1	81,3	14,0	25,3	9,5	11,3	68,4
3.387	new Russia	2,0	93,2	14,1	25,4	7,8	9,3	66,6
3.365	new Russia	1,9	97,1	13,4	33,0	9,1	10,8	67,5
3.378	new Russia	1,8	80,3	14,5	31,8	10,5	12,5	67,5
3.364	new Russia	1,7	101,4	14,1	28,5	9,9	11,8	52,5
3.372	new Russia	1,7	117,3	13,6	27,1	9,0	10,7	68,3
3.404	new Russia	1,7	99,4	14,2	20,2	9,1	10,8	64,6
3.371	new Russia	1,6	112,2	15,1	30,0	10,5	12,5	63,5
3.375	new Russia	1,6	112,0	13,9	28,2	10,3	12,2	61,4
3.361	new Russia	1,6	114,5	13,5	31,8	9,1	10,8	61,3
3.367	new Russia	1,6	93,9	12,6	27,3	8,1	9,7	65,3
3.376	new Russia	1,6	110,3	14,6	28,1	10,2	12,2	60,3
3.395	**check**	**1,6**	**100,0**	**14,4**	**27,1**	**9,6**	**11,5**	**32,2**
3.373	new Russia	1,6	107,4	14,0	26,4	9,4	11,3	66,2
3.402	new Russia	1,6	110,3	14,6	17,2	11,0	13,1	61,3
3.366	new Russia	1,6	89,2	13,9	28,3	8,5	10,2	2,6
3.397	**check**	**1,5**	**96,3**	**14,8**	**25,6**	**8,4**	**10,0**	**68,6**
3.362	new Russia	1,5	107,5	13,3	31,7	9,4	11,2	60,5
3.363	new Russia	1,5	90,0	14,8	21,5	6,0	7,1	68,6
3.360	new Russia	1,5	105,6	14,6	25,3	7,8	9,4	60,4
3.377	new Russia	1,5	102,6	13,8	29,3	9,8	11,7	57,3
3.401	new Russia	1,5	104,8	13,1	30,4	9,6	11,5	60,5
3.370	new Russia	1,5	101,0	13,9	25,4	8,5	10,1	61,5
3.398	new Russia	1,5	85,3	13,2	29,3	9,0	10,8	49,1
3.400	new Russia	1,5	103,9	12,9	25,1	8,9	10,6	60,9
3.368	new Russia	1,5	83,7	14,1	28,3	9,5	11,4	59,5
3.374	new Russia	1,5	99,7	13,6	23,3	8,2	9,8	64,1
3.403	new Russia	1,5	102,5	15,0	19,5	11,3	13,4	67,5
3.359	new Russia	1,5	101,4	15,2	31,1	11,5	13,8	62,7
3.396	**check**	**1,4**	**90,6**	**14,9**	**29,8**	**10,8**	**12,8**	**38,4**
3.369	new Russia	1,4	82,8	14,3	22,2	9,3	11,1	67,5
3.357	**check**	**1,3**	**93,1**	**14,4**	**24,5**	**9,3**	**11,1**	**51,5**
3.394	**check**	**1,3**	**83,4**	**14,5**	**23,5**	**5,6**	**6,7**	**67,5**
3.358	new Russia	1,3	91,6	14,6	23,1	1,0	1,2	68,5
3.399	new Russia	1,0	67,3	13,6	24,9	9,6	11,4	65,6
3.392	ITALY	0,6	39,4	20,0	24,2	16,4	19,4	87,9
3.389	ITALY	0,5	34,3	20,7	22,1	15,9	18,9	88,5
3.391	ITALY	0,5	30,7	20,0	24,3	16,4	19,5	88,5
3.390	ITALY	0,5	30,4	20,0	23,6	14,8	17,6	89,6
3.393	ITALY	0,3	19,7	19,9	19,6	13,9	16,5	84,6
3.388	ITALY	0,3	18,6	19,8	19,6	12,8	15,2	90,5
Mean		**1,6**	**90,9**	**14,9**	**26,3**	**9,9**	**11,8**	**61,7**
		HCP0,05=0,21						

New Russian Genotype Evaluation: During 2017 growing season, the best genotype, identified after selection and named respectively Taganrog and Bourbon were tested in a field trial (Table 2). Anova analysis confirm the difference in yield production and flowering time (p 0,01). The new genotype obtained from recurrent selection from 3371 recurrent parental line (named Taganrog) and the second genotype obtained from recurrent selection from 3379 parental line (named Bourbon) showed significative

Table 2. Agronomic and diseases result Samara field in the 2017 season, sign. Yield, flowering time, quality trait and diseases mean values. The table show only the best genotype selected and the local check Besenciuk209. Anova analysis confirm the difference in yield production and flowering time (p 0,01). Taganrog and Bourbon showed significative increase of yield production respect to local check Besenciuck209, good level of gluten index and good resistance for principal durum wheat diseases.

Original code 2016 season	Name line/variety	Origin	Yield T/Ha	Duncan test	Heading data	Perynophora tritici-repentis %	Blumeria graminis %	Puccinia graminis %
3.379	**Bourbon**	**new Russia**	**5,2**	**k**	**03/07/2017**	**R**	**4/10.**	**4/5.**
3.371	**Taganrog**	**new Russia**	**4,9**	**i-k**	**04/07/2017**	**R**	**R**	**4/5.**
	1916D-14	new Russia	4,8	h-k	04/07/2017	R	4/10.	4/5.
	GSD-30	ITALY	4,4	g-j	04/07/2017	R	R	4/5.
3.363	1591D-21	new Russia	4,4	g-j	05/07/2017	R	R	4/5.
	GSD-399	ITALY	4,3	e-j	04/07/2017	R	R	4/5.
	SP-3/8	new Russia	4,1	d-g	06/07/2017	R	4/5.	4/5.
	1970D-2	new Russia	3,9	c-f	05/07/2017	R	4/25.	4/5.
	1874D-2	new Russia	3,9	c-f	03/07/2017	R	4/5.	4/5.
	Cannavaro	ITALY	3,8	b-f	06/07/2017	R	4/5.	4/15.
	Torrebianca	ITALY	3,3	a-c	04/07/2017	MR/MS(20)	R	4/10.
	Quadrato	ITALY	3,2	a-c	03/07/2017	S(30)	R	4/3.
	Besenciuk 209	**check**	**3,1**	**ab**	**05/07/2017**	**R**	**4/5.**	4/5.
	Pietrafitta	ITALY	2,7	a	04/07/2017	S(40)	R	4/3.
	Mean		**4,0**					
			HCP0,05=6,1					

R= Resistant; MR= Medium Resistant; MS= Medium Sensible; S= Sensible; 4/3, 4/5, 4/10, 4/15 and 4/25= level of infection

Table 3. SDS-Page results (High Molecular Weight bands detected on Russian accessions) samara season 2016. Only three Russian genotype show Kofa band.

- N, 6+8: Sample 3370/16, 3403/16, 3404/16
- N, 7+8: Sample 3381/16, 3371/16, 3402/16, 3385/16, 3397/16, 3394/16, 3363/16, 3360/16, 3358/16
- N, 7+9: Sample 3401/16
- 1, 7+8: Sample 3376/06, 3375/16, 3399/16, 3387/16
- 2*, 6+8: Sample 3372/16
- 2*, 7+8: Sample 3362/16, 3369/16, 3382/16, 3368/16, 3374/16, 3361/16, 3384/16
- 2*, 7+9: Sample 3378/16, 3367/16, 3379/16
- 2*, 6+8/7+8: Sample 3365/16

- N, 6+8 Kofa Test

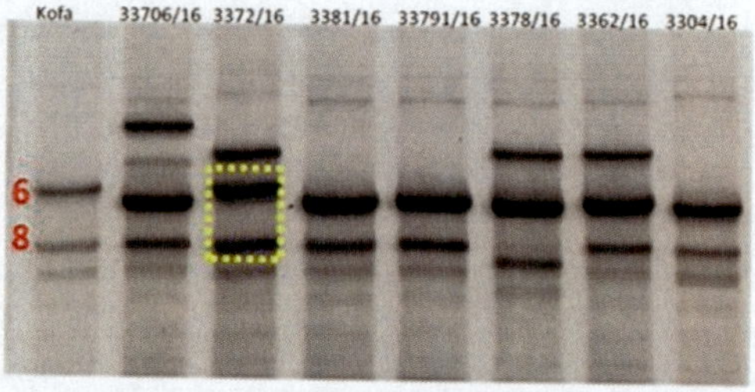

Fig. 1. Example of same SDS-Page profile in Russian accession evaluated in Samara field in 2016 season. The Kofa HMW profile is reported in the first lane in red and compared respect Russian accession.

Table 4. The table show the distance between accessions for each group of similarity: a) Distance inside population F2; b) Distance inside population F2Bc2. After second back-cross and selection (Table b) the leader line 7 of cluster 9 is associated to Russian genotype 3371 with very low distance.

a

cluster number	distance	leader line	associated line
11	2,951	4	8
10	3,302	4	10
9	3,427	2	3
8	3,564	2	6
7	**3,573**	**1**	**3371**
6	4,035	4	7
5	4,434	2	4
4	4,550	1	5
3	4,906	1	9
2	5,934	1	2
1	**6,873**	**1**	**Kofa**

b

cluster number	distance	leader line	associated line
11	0,847	6	8
10	0,847	4	10
9	**1,436**	**7**	**3371**
8	1,597	2	5
7	1,613	3	9
6	1,815	2	3
5	2,358	4	6
4	2,757	1	7
3	3,230	2	4
2	4,391	1	2
1	**12,852**	**1**	**kofa**

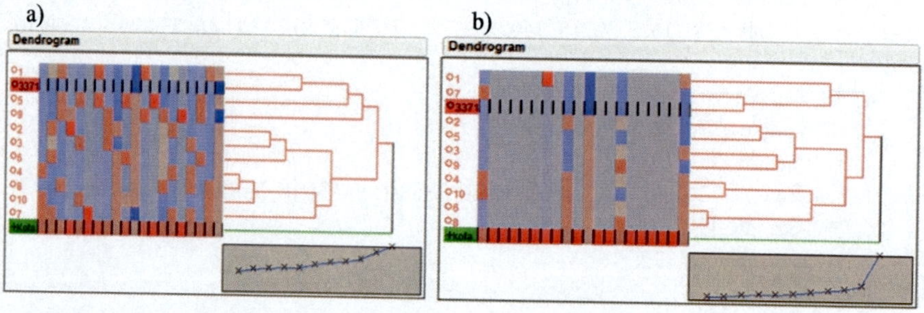

Fig. 2. Genetic similarity screening in F2 selected accessions (a) and F2BC2 selection accessions (b) generation showed three principal group of similarity, as expected, while the selection carried out for Russian genetic background, allowed recovery of about 90% recurrent lines in the new genotypes with good quality trait inherited from HGQ parental line.

increase of yield production respect to local check Besenciuck209, good level of gluten index and good resistance for principal durum wheat diseases. Currently, both genotypes are in the first year of official ministerial evaluation for registration in Russian catalog of varieties.

Qualitative Analysis SDS-Micro Sedimentation and SE-FPLC (Size Exclusion Fast Performance Liquid Chromatography): The two F2BC2 accessions selected from the two populations obtained through the crossing of Kofa with accession 3371 (named Taganrog) and accession 3379 (named Bourbon), and resulting in having the Qtl of Kofa for chromosam 1st for the gluten index, the band, also by Kofa for the Glu-B1 locus, they were analyzed through SE-FPLC (Size exclusion fast performance liquid chromatography) and SDS-Microsedimentation. The values found showed a very good quality level of gluten, even higher than Kofa (Table 5).

Table 5. Gluten quality analysis. Size exclusion fast performance liquid chromatography (SE-FPLC), showed significative (P 0,05) value increase of UPP% the for new genotypes Bourbon and Taganrog respect to local check Besenciuk209.

Sample	F1*	F1	F2	F3	F4	totale	%F1*	F1%	%F2	%F3	%F4	F1*+F1	F3+F4	F1*+F1+F2	Mon/pol	UPP%
Kofa	40,3	103	42,3	109,6	147,6	442,7	9,1	23,3	9,6	24,8	33,3	143,2	257,2	185,5	1,4	28,1
Taganrog (3371)	**22,1**	**40,2**	**33,9**	**60,2**	**117,8**	**274,1**	**8,1**	**14,7**	**12,4**	**22**	**43**	**62,2**	**178**	**96,1**	**1,9**	**35,5**
Bourbon (3379)	**36**	**40,7**	**51,1**	**87,7**	**177,2**	**392,6**	**9,2**	**10,4**	**13**	**22,3**	**45,1**	**76,7**	**264,9**	**127,8**	**2,1**	**47**

Peak areas of fractions (F) from size-exclusion in percentage of total area of the two chromatograms (sonicated (F*) and SDS-soluble fractions (F)). The UPP%, calculated as $[F1^*/(F1^* + F1)] \times 100$, and the ratio between monomeric and polymeric proteins, defined as $(F3 + F4)/(F1^* + F1 + F2)$, is also reported.

4 Conclusions

Good adaptation was observed for the new selected genotypes and a higher gluten index and %UPP values, were recorded. The marker-assisted introgression in backcross populations derived Russian varieties and HGQ, in combination with an accelerated seed multiplication program, had proved to be effective improving the qualitative performance of Russian varieties and preserving their adaptability to the local pedo-climatic conditions.

References

Bubcvsky, et al.: QTL analysis of pasta quality using a composite microsatellite and SNP map of durum wheat. Theory Appl. Genet. **117**(8), 1361–1377 (2008)

Dick and Quick: A modified screening test for rapid estimation of gluten strength in early-generation durum wheat breeding lines. Cereal Chem. **60**, 315–318 (1983)

Kremer, A., et al.: Measures of polymorphism within and among populations. In: Karp, A., Isaac, P., Ingram, D.S. (eds.) Molecular tools for Screening Biodiversity. Plants and Animals, pp. 302–311. Chapman & Hall London (1998)

Natoli, V.: Ph.D Tesis: Identificazione in Frumento duro di geni coinvolti nei meccanismi di adattamento allo stress idrico in ambienti mediterranei (2008)

Lozhkin, A.G., et al.: Spring durum wheat in the forest-steppe zone of the Chuvash Republic. Grain Econ. Russia **4**, 59 (2018)

Marchylo, D.J.E., et al.: Gluten strength, physical dough prppeties, pasta cooking quality and bread-making quality of Canadian durum wheat. Can. J. Plant Sci. **81**, 611–620 (2001)

Mariani, B.M., et al.: Durum wheat quality evaluation: influecnce of genotype and environment. Cereal Chem. **72**, 194–197 (1995)

Morel, M.H., et al.: Effects of temperature, sonication time, and power settings on size distribution and extractability of total wheat flour proteins as determined by size-exclusion high-performance liquid chromatography. Cereal Chem. **77**, 685–691 (2000)

Payne, P.I., et al.: The relationship between HMW glutenin subunit composition and the bread-making quality of British-grown wheat varieties. J. Sci. Food Agric. **40**, 51–65 (1987)

Pogna, N.E., et al.: Chromosome 1B-encoded gliadins and glutenin subunits in durum wheat: Genetics and relationship to gluten strength. J. Cereal Sci. **11**, 15–34 (1990)

Röder, M.S., et al.: A microsatellite map of wheat. Genetics **149**, 2007–2023 (1998)

Rohlf, F.J.: NTSYS-pc version 2.0.: Exter Press, Setauket (1997)

Ross, et al.: Can. J. Plant Sci. **89**, 469–476 (2009)

Sissons, M.J., et al.: New insights into the role of gluten on durum pasta quality using reconstitution method. Cereal Chem. **82**, 601–608 (2005)

Somers, D.J., et al.: A high-density microsatellite consensus map for bread wheat (Triticum aestivum L.). Theor. Appl. Genet. **109**, 1105–1114 (2004)

Sourdille, P., et al.: Microsatellite-based deletion bin system for the establishment of genetic-physical map relationships in wheat (Triticum aestivum L.) Funct. Integr. Genomics. **4**, 12–25 (2004)

Tronsmo, K.M., et al.: A study of how size distribution of gluten proteins, surface properties of gluten and dough mixing properties relate to baking propertiesof wheat flours. J. Cereal Sci. **35**, 201–214 (2002)

Innovative Applied Sciences

VLC Prototypes Developed with Off-the-Shelf Components for Wireless Indoor Data Transfer

Simona Riurean(✉)

University of Petroşani, Universitatii str., No 20, 332006 Petroşani, Romania
sriurean@yahoo.com

Abstract. Most of the surprising events, aka the black swans, brought unexpected challenges and achievements in various research areas, as humankind history reveals. As this pandemic situation shows, no effort is too big when new technologies have to be searched and developed to allow worldwide fast information exchange with the aim to maintain human health. The local and remote wireless transmission technologies based on the radio frequency spectrum are more congested every day, therefore a number of promising wireless communication technologies are based on the optical part of the electromagnetic spectrum. The key characteristics of a consistent visible light communication indoor system in a line of sight scenario are briefly discussed here. The optical channel model and its drawbacks, such as noises, signal spread delay, and signal fading, as well as two prototypes dedicated to data transfer in an optical link, are also addressed in this paper.

Keywords: Visible light communication · LED and PD key characteristics · Optical link · Prototypes

1 Introduction to VLC

The most challenging situations that we, humans have to overcome are the new and unexpected events. Most of the existing local and remote wireless communication technologies are based on the radio frequency (RF) spectrum that becomes too crowded to face a fast transfer of a high amount of Petabyte of data that turns out to be compulsory, not only necessary nowadays. On the other hand, once that the Internet of Things (IoT) shifted from concept to reality, scientists have searched for alternative technologies to keep up with the increasing data transfer needed and demanded by the sophisticated smart devices dedicated to human health and wellbeing. Some novel, promising wireless communication technologies based on the optical part of the electromagnetic spectrum are to become soon mature.

Wireless communication has gone through many paradigm changes after the advent of electromagnetic (EM) waves, remote communication such as telegraphy and the invention of radio [1, 2]. EM waves consists of radio, infrared (IR), visible, ultraviolet, etc. [2]. Among those, RF is the one that is the most used portion of EM spectrum for communication purposes.

T. Antipova (Ed.): ICCS 2020, LNNS 186, pp. 315–324, 2021.
https://doi.org/10.1007/978-3-030-66093-2_30

The RF transmission does not require a Line of Sight (LoS) topology between transmitter and receiver, can travel long distances and reach high data rates [2]. RF communication, however, faces some challenges such as interference, security issues as they can easily penetrate through solid objects, bandwidth and power limitations resulting in power inefficiency, affects to human wellbeing and safety.

Most important, the RF space has become extensively crowded lately because of the many smart devices connected simultaneously, especially indoor where the most data traffic occurs [2].

Therefore, developing alternative solutions to wireless RF transmissions especially indoor but also outdoor is imperious necessary to solve these challenges.

The Optical Wireless Communication (OWC) technology emerge as a possible powerful partner to RF technologies (Wi-Fi, Bluetooth Low Energy, Cellular Data, ZigBee, 6LoWPAN, WiMax and so on – Fig. 1) and in some specific scenarios, even the major actor.

Fig. 1. LiFi (VLC) IRversus wireless RF technologies

1.1 OWC Technologies and Applications

The optical wireless communication (OWC), with its several more or less mature applications, such as Infrared communication (IR), Visible Light Communication (VLC), Optical Camera Communication (OCC), or Light Fidelity (Li-Fi) dedicated to indoor implementations and Free Space Optics (FSO) aimed to transfer remote data outdoor, have been intensively researched and already deployed in some fields such as museums, schools, retails, wellbeing and health smart devices, Vehicle to Vehicle (V2V) communication, and many others [3].

IR wireless communication has been already searched for decades and some of its most used applications worldwide are the TV remote control, robots' control and positioning. VLC research burst once that the white LED has been widely deployed on the market after its discovery in 1990 by Shuji Nakamura from USA and in 1992 by Isamu Akasaki and Hiroshi Amano from Japan. Both teams received in 2014 the Nobel prize for physics due to the white LED discovery [4].

OCC considered in IEEE 802.15.7m standardization along with VLC, covers the transmission between light sources and cameras for low rate communication and localization [5]. LiFi, coined by Harald Haas during a Ted conference in 2011, is intended to be a networked, MIMO, full duplex (VLC download and IR upload) wireless communication technology using access points embedded into the lighting fixtures [6].

V2V wireless communication is one of the most intense researched and advanced findings in FSO area [7].

1.2 Standardization of VLC

A significant step towards the widespread commercialization of VLC networks was the advent of the IEEE 802.15.7 standard in 2011. It defines three physical layers (PHY) (from PHY I to PHY III) and the Media Access Control (MAC) layer.

PHY I layer is designed for outdoor scenarios, while PHY II and PHY III layers aims the indoor applications.

There is a total of thirty combinations of modulation and coding schemes defined for VLC. PHY I are specified uses Reed-Solomon (RS) and Convolutional Codes (CC) for Forward Error Correction (FEC), while PHY II and III rely mostly on RS codes for FEC only. Both PHY I and PHY II use On-Off Keying (OOK) and Variable Pulse Position Modulation (VPPM), while PHY III uses only color shift keying (CSK), thus PHY III is compatible only with Red Green Blue (RGB) LEDs. Depending on the modulation used, the Run-Length Limited code (RLL), the optical clock rate, the FEC code, the three PHY modes, there are different data rates: 11.67 to 266.6 Kbps, 1.25 to 96 Mbps and 12 to 96 Mbps. The physical network topologies supported by the MAC layer are star, peer-to-peer, and broadcast Three classes of devices are considered: vehicles, mobile and infrastructure. The MAC layer addresses characteristics as: support for mobility, dimming, color function and visibility, avoid flicker, support for pairing and disassociation in the VLC Personal Area Network (VPAN), generation of network beacons if the device is a coordinator, and reliability of the connection between MAC entities. Security considerations are applied at the application level. The MAC 802.15.7 layer is very close to the IEEE 802.15.4 MAC Personal Area Network (WPAN) [8].

However, many situations of using VLC technology are not considered in these standards, therefore, the IEEE Task Group 7m, first known as the IEEE 802.17.5r1, was initiated in late 2014 to revise the IEEE 802.15.7-2011 standard [9].

2 Key Characteristics of oTx and oRx in a VLC System

2.1 A General Setup of the VLC System

A VLC system (regardless the topology setup) comprises the optical transmitter (oTx) and optical receiver (oRx), two separate modules responsible for sending/receiving data piggyback by the light.

The general diagram of a generic VLC system (Fig. 2) consists of an oTx, the environment where the optical signal is affected by different noises, and the oRx.

Channel coding and modulation in oTx aim to adapt the signal to the channel and the part responsible for the conversion of the electrical signal into an optical one is the main part of the oTx, the light source.

Noises into the environment (where the optical beams of light travel) affect the signal due to the ambient light. The ambient light can be the natural light and other sources of artificial light (fluorescent, incandescent bulbs or LEDs).

The photodetector, part of the oRx, is responsible to convert the optical signal into an electrical one. The signal already converted is amplified and then demodulated and decoded to be recovered as accurately as possible.

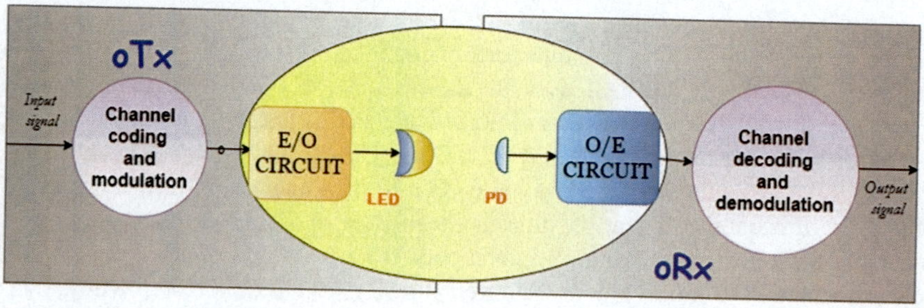

Fig. 2. A general diagram of a VLC system.

2.2 Light Sources Used in oTx Setups

VLC systems can use different types of light sources such as white LEDs (WLEDs), array or stripe of SMD LEDs, μLEDs, OLEDs, Laser Diodes (LDs), according to the requirements of the application environment. The most challenging characteristics of the LED for stable, mature VLC systems, are their nonlinearity and the limited modulation bandwidth. Numerous factors, such as quantum confined stark effect (QCSE), carrier lifetime, carrier recombination time, crystal orientation, etc. affect the modulation bandwidth of LEDs [10, 11].

The modulation bandwidth of the VLC (generally limited to MHz) is directly influenced by the frequency response of the LED, therefore, negatively affects the data rates. It depends on the lifetime (τ) of the carrier recombination on the active region and on the p-n junction capacity. The frequency response of the off-the-shelf white LEDs is fundamental constraint by the slow temporal response of the yellow phosphor layer.

There are a number of proposals in literature for addressing the LEDs' non-linearity drawback in VLC systems. The simplest method is to use binary transmission such as the On-Off Keying (OOK) or Pulse Position Modulation (PPM), when the LED is simply switched on and off at specific current values. To mitigate this effect, higher spectral efficient modulation schemes such as multi-band Carrier-less Amplitude and Phase modulation (m-CAP), Orthogonal Frequency Division Multiplexing (OFDM) or color shift keying (CSK) have been used [12]. On the other hand, using Wiener model, although demonstrated only by simulation, a valid adaptive pre-distortion technique was proposed by Qian et. al. for the VLC systems.

It consists of a linear time-invariant and memoryless nonlinear model, but with significant complexity VLC setup [13]. Since then, many other pre-distortion methods are being proposed [14, 15].

A different solution is the use of a blue filter in front of the PD, to remove the slow response of the yellow phosphorus. Both the frequency response of the LEDs and the modulation bandwidth are related to the input current level and the junction and parasitic capacity. The LED's cut off frequency is also important. It refers to the maximum frequency at which the LED's light emission decreases to half the initial light intensity. Whether the transmission system is simplex or duplex, it is important to switch the LED between ON and OFF at high speed. The requirements for the signal that determines the output color are usually lower in terms of voltage compared to their power supply.

When selecting a LED, for a robust VLC setup, the light intensity, the wavelength of light, the light scattering, the modulation speed and the power requirements must be considered.

For example, in case of using a stripe of Surface Mounted Diodes (SMD) LEDs (such as 3528s, 5050s, 2835s or 5630s) each LED can be addressed independently. Although the signal degradation occurs, the problem can be solved by connecting the LEDs to a series of synchronized controllers.

If the wavelength of LED is matched with that of the PD from the optical receiver (oRx), the noise level produced by the ambient light (natural and artificial) is decreased, therefore the correct reading of the optical signal is ensured. A high optical power requirement may also involve a cooling solution when the prototype is designed for a VLC network working in special scenarios with high ambient light noise.

2.3 Photodetectors' Characteristics for oRx Setups

Light Dependent Resistors (LDRs), imaging sensors or a photodetectors (PDs) can be used in a VLC systems use as oRx a. PDs can support higher bandwidths for higher downlink speeds than camera-based VLC.

The Positive Intrinsic Negative Photodetectors (PIN PDs), and the avalanche photodiode (APD) are the highest sensitive semiconductor PD used in VLC systems that exploits the photoelectric effect to convert light into electricity. Silicon APDs are designed to offer high conversion gain, increased sensitivity and lower noise compared to standard PIN detectors but are more sensitive to extreme temperatures than PIN PDs [16]. On the other hand, APD produce a stronger signal but add more Poisson noise.

As Si PIN PDs have low voltage operation, high tolerance to a wide range of temperature fluctuations and excessive lighting, linear response characteristics, and low costs, are most often used in VLC systems.

PDs can be configured in two modes, photovoltaic and photoconductive. In photovoltaic mode the diode is un-polarized and the relationship between voltage and the amount of light is nonlinear. Usually the photovoltaic panels operate in this way, being useful in producing electricity but limiting the sensitivity of the diode. Photovoltaic mode is preferable if we want to make accurate measurements since it is less susceptible to noise. The photoconductive mode applies a reverse bias voltage to the PD and the relationship is linear, up to 1mW optical power for a 1mm diameter diode.

The applied voltage produces a lower capacity and a higher speed of the diode to the detriment of heat production. Photoconductive circuits have applicability in bright light environments and where fast readings are needed.

2.4 Boards Suitable for VLC Development

The Arduino, Raspberry Pi, BeagleBone Black or STMicroelectronics Nucleo 32L0 are some of the development boards that proved to be suitable for making prototypes with low costs and satisfactory performance in different VLC topologies [17–21].

Arduino certainly has several options in the field of prototyping, their specialty being in robotics and small, localized systems. There are many options for compatible modules including sensors, prototyping boards, lighting, motors, and anything else that requires a short program to be controlled.

The reason the Raspberry Pi can be a proper choice for prototyping, is that it has the same behavior as a computer. It can connect to the internet, process text files, is compatible with peripheral devices (keyboard, mouse, screen), processes media files and is generally capable of anything as long as it needs enhanced processing or enhanced graphics capabilities. Last year, the newest Raspbery Pi 4B model was launched with significant improvements as 8Gb of RAM, 2 micro HDMI ports capable of 4k resolutions at 60 frames per second, Bluetooth 5.0, Gigabit Ethernet and a 1.5 GHz ARM v8 64bit processor.

The BeagleBone Black board has Sitara AM3358BZCZ100 processor, a Dual Data Rate RAM memory Micron 512 MB DDR3L or Kingston 512mB DDR3 the physical interface to the network through SMSC Ethernet PHY and a HDMI Framer that provides control for an HDMI or DVI-D display with an adapter, as well as with various connectors, LEDs and switches (DC Power, power button, 10/100 Ethernet, serial debug port, four blue LEDs, USB, miniUSB, microSD, microHDMI BOOT switch and reset button).

STM Nucleo 32L0 board is designed around the STM32 microcontrollers in a 64-pin. It provides a high power efficiency, with dynamic voltage scaling, an ultra-low-power clock oscillator and short wakeup time. The autonomous peripherals (including USART, I^2C, and touch sense controller) reduce the load of the Arm® Cortex®-M0 +core leading to fewer CPU wakeups and contribute to decreased processing time and power consumption.

However, Raspberry Pi and Arduino are the main PCBs used in VLC prototyping to keep the features high and the cost low of the final product.

3 oTx and oRx Modules for Indoor Data Transfer

The challenging journey from idea to testing a prototype has to follow few important steps [21]. The prototypes developed aim to test VLC system for indoor data transfer in a LoS topology under regular ambient light (sunlight and fluorescent light sources embedded into the lighting fixture).

In a more advanced scenario, an additive light source coming from a desk lamp with white LED placed over the VLC system aims to determine the influence of this additive noise on the system performance in terms of distance between oTx and oRx.

The prototype with the Arduino PCBs (Arduino Uno for the oTx and Arduino Mega for oRx) a text message was send to a 400 mm distance (see Fig. 3).

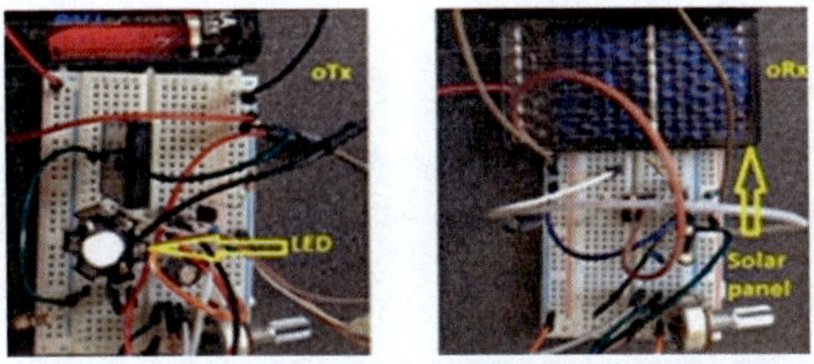

Fig. 3. oTx and oRx circuits on breadboard.

The oTx circuit was built with NPN transistors 2N3904 and 2SC5200, resistors and capacitors.

Fig. 4. VLC prototyped with Arduino PCB

The LED used is VLHW4100 with luminous intensity 90–110 LM, VF min 3.0 V max 3.6V, Max current 350 mA and CCT of 3200 - 3500K. The oRx circuit was built with the NE5532 Op Amp, resistors and capacitors. For the first scenario, I used a solar panel (Fig. 3) and then a PD type is VTB8440BH with IR filter (Fig. 4).

The prototype with the Raspberry Pi PCB (Fig. 5 and Fig. 6) has the LED type APA102C and a circuit with PNP transistors and resistors.

Fig. 5. oTx and oRx simulated with Raspberry Pi.

The APA102C is an addressable LEDs that operates on +5V power input, as well as 0–5V logic levels for clock and data. It works on 2-wire communication protocol consisting of a clock line and a data line.

It requires one more wire than standard WS2812B LEDs because the communication with the LEDs becomes timing-independent. This feature allows it to be used directly with a Raspberry Pi, for example. The oRx circuit has the buffer/line driver 74HCT125 that aims to convert the optical signal into an electrical one and ALS-PT19-315C/L177/TR8 as optical detector.

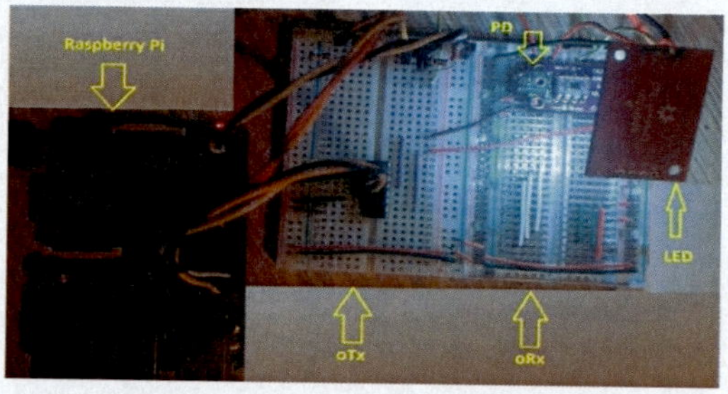

Fig. 6. VLC prototyped with Raspberry Pi PCB

The ALS-PT19-315C/L177/TR8 is a low cost 2-SMD phototransistor. It has the sensitivity wavelength range (λ_p) between 390 and 700. It has the peak sensitivity wavelength at 630 nm. The operating temperature is between −40 and +85 °C and the operating voltage 2.2 and 5 V. Its Rise time (t_r) is 0.11 ms and the fall time (t_f) 0.22 ms.

The output voltage (V_{out}) is the product of photocurrent (I_{PH}) and loading resistor (R_L) (see Eq. 1 and Eq. 2).

$$V_{out(max.)} = I_{PH(max.)} \times R_L \leqq V_{out(saturation)} = Vcc - 0.4\ V \quad (1)$$

$$R_L \times C_L \geqq 0.5 \quad (2)$$

The proper loading resistor must meet the requirement of maximum ambient light, and output saturation voltage (Figs. 7 and 8).

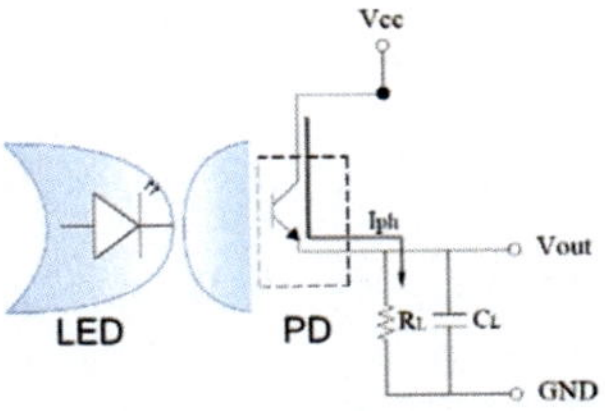

Fig. 7. Photocurrent to Voltage converter

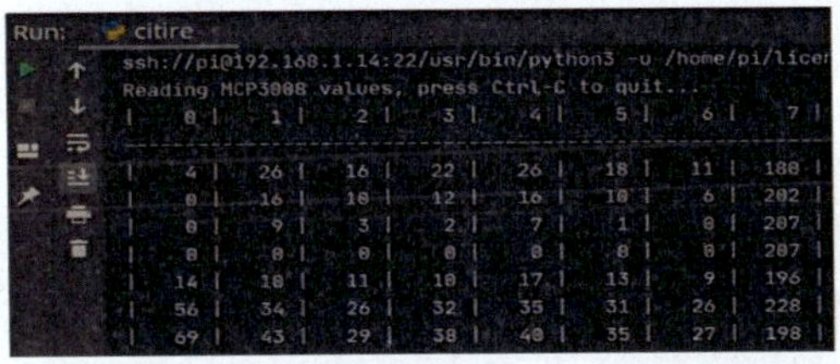

Fig. 8. Data received in the VLC system prototyped with Raspberry Pi PCB

4 Conclusions

The key characteristics of oTx and oRx for a reliable VLC data link are the most important to be considered when the future manufactured prototype is simulated. There are many drawbacks and challenges to overcome with the off-the-shelf components especially when the prototype must have a low cost with reliable data throughputs. The systems developed are tested in a LoS scenario under regular indoor ambient light with acceptable performance. The link length is not the one targeted, but, as a first solution to keep low the entire budget of the project, and increase the VLC link is to use proper optics such as biconvex lenses in front of the LED and lenses with concentrators in front of PD to increase the optical gain. Although the two prototypes developed have their particular benefits and drawbacks, with the current technological boom in this field, new, and improved spare parts are to be expected to become available on the market with low costs.

References

1. Rehman, S.U., Ullah, S., Chong, P.H., Yongchareon, S., Komosny, D.: Visible light communication: a system perspective—Overview and challenges. Sensors **19**(5), 1153 (2019)
2. Leba, M., Riurean, S., Ionica, A.: LiFi—The path to a new way of communication. In: 12th Iberian Conference on Information Systems and Technologies (CISTI), Lisbon, pp. 1–6 (2017). https://doi.org/10.23919/CISTI.2017.7975997

3. Riurean, S., Antipova, T., Rocha, Á., Leba, M., Ionica, A.: VLC, OCC, IR and LiFi reliable optical wireless technologies to be embedded in medical facilities and medical devices. J. Med. Syst. **43**, 308 (2019)
4. Nobelprize Homepage. https://www.nobelprize.org/prizes/physics/2014/press-release/. Accessed 25 Aug 2020
5. Marcu, A.E., Dobre, R.A., Vlădescu, M.: Flicker free optical camera communication for cameras capturing 30 frames per second. In: 43rd International Conference on Telecommunications and Signal Processing (TSP), Milan, Italy, pp. 166–169 (2020)
6. Ted Homepage. https://www.ted.com/talks/harald_haas_wireless_data_from_every_light_bulb. Accessed 25 Aug 2020
7. Căilean, A.M., Dimian, M., Done, A.: Enhanced design of visible light communication sensor for automotive 785 applications: experimental demonstration of a 130 meters link. In: Global LIFI Congress (GLC), Paris, pp. 1–4 (2018). 786
8. IEEE Computer Society: IEEE Standard for Local and metropolitan area networks - Part 15.7: Short-Range Wireless Optical Communication Using Visible Light, September 2011
9. IEEE 802.15 WPAN 15.7 Amendment Study Group (2015)
10. Huang, Y.M., Singh, K.J., Liu, A.C., et al.: Advances in quantum-dot-based displays. Nanomaterials (Basel) **10**(7), 1327 (2020)
11. Liao, C., et al.: Light-emitting diodes for visible light communication. In: International Wireless Communications and Mobile Computing Conference (IWCMC), Dubrovnik, pp. 665–667 (2015)
12. Burton, A., et al.: Investigation into using compensation for the nonlinear effects of the output of LEDs in visible light communication systems. In: 2nd West Asian Colloquium on Optical Wireless Communications (WACOWC), Tehran, Iran, pp. 80–84 (2019)
13. Qian, H., Yao, S.J., Cai, S.Z., Zhou, T.: Adaptive postdistortion for nonlinear LEDs in visible light communications. IEEE Photonics J. **6**, 1–8 (2014)
14. Aggarwal, P., Ahmad, R., Bohara, V.A., Srivastava, A.: Adaptive predistortion technique for nonlinear LED with dimming control in VLC system. In: 2017 IEEE International Conference on Advanced Networks and Telecommunications Systems (ANTS), Bhubaneswar (2017)
15. Ying, K., Yu, Z., Baxley, R.J., Qian, H., Chang, G.-K., Zhou, T.: Nonlinear distortion mitigation in visible light communications. IEEE Wirel. Commun. **22**(2), 36–45 (2015)
16. Lian, J., Vatansever, Z., Noshad, M., Brandt-Pearce, M.: Indoor visible light communications, networking, and applications. J. Phys.: Photonics **1**(1), 1–29 (2019)
17. Swain, K.P., Prasad, M.V.S.V., Palai, G., Sahoo, J., Mohanty, M.N.: Exploiting VLC technique for smart home automation using Arduino. In: Dash, S., Vijayakumar, K., Panigrahi, B., Das, S. (eds.) Artificial Intelligence and Evolutionary Computations in Engineering Systems. Advances in Intelligent Systems and Computing, vol. 517. Springer, Singapore (2017)
18. Wang, T., Zhao, Z.: An implementation of visible light communication based on Raspberry Pi. In: IEEE Integrated STEM Education Conference (ISEC), Princeton, NJ, pp. 218–219 (2018)
19. Riurean, S., Leba, M., Ionica, A., Nassar, Y.: Technical solution for burnout, the modern age health issue. In: IEEE 20th Mediterranean Electrotechnical Conference (MELECON), Palermo, Italy, pp. 350–353 (2020)
20. Duque, A.: Bidirectional Visible Light Communications for the Internet of Things. Networking and Internet Architecture [cs.NI]. Université de Lyon - INSA Lyon (2018)
21. Rosca, S., Riurean, S., Leba, M., Ionica, A.: An educational model of graduation project for students at automation and computer engineering. J. Digit. Sci. **1**(1) (2019). https://doi.org/10.33847/2686-8296/1.1_4

The Study of the Penzhin Bay Tidal Energy Use Possibilities for Generating Electricity

Robert Akhmetshin[1], Pavel Pichugin[2](✉), Vladimir Trushkin[3], Farida Sakhapova[1], and Olga Evgrafova[1]

[1] Kazan Federal University, Naberezhnye Chelny 423800, Russia
[2] National Research University "Moscow Power Engineering Institute", Moscow 111250, Russia
PavelPichugin999@yandex.ru
[3] Saratov State Agrarian University after N.I. Vavilov, Saratov 410012, Russia

Abstract. This article examines and deals with the experience of building tidal power plants in Russia and in the world, analyzes the possibilities of using tidal energy in the Penzhin Bay to solve the energy problems in Kamchatka region. The purpose of this work is to propose and explore a new implementation version of the Penzhin tidal power plant. The tidal power plants implemented projects analysis method revealed the advantages and disadvantages of various constructive and technological solutions, their construction conditions, which are taken into account in the proposed option were studied. The method of comparison with the Penzhin tidal power plant existing projects identified the main advantages of the proposed option. A variant which consists in the need to lay underground two conduits connecting the Penzhin Bay with the Pacific Ocean, was proposed for the construction of the Penzhin tidal power station. Near the Pacific Ocean, each conduit is divided into two siphons, in which horizontal bulb hydro-turbines are installed. It was investigated that the best place for generators installation is in the engine room of the power plant the head building, located on land. The shafts of the turbine and the generator are connected using a bevel gearing, which makes such a technical solution possible. The approximate power of turbines and generators was calculated; their brands were proposed. It was also calculated the volume of water required for the passage and during the tide. As a result, it was proved that this power plant will operate continuously, and water will always flow in one direction. The proposed option can be taken into consideration in the presence of similar geographical and technical conditions in other parts of the world. Power generation will allow for the realization of the Kamchatka region "breakthrough" plans in services, industry and agriculture.

Keywords: Tidal power plant · Energy problems · Water lines (water conduits) · Horizontal capsular turbines · Generator · Bevel gear · Nizhnekamsk hydroelectric power plant · Cheboksary hydroelectric power plant

T. Antipova (Ed.): ICCS 2020, LNNS 186, pp. 325–334, 2021.
https://doi.org/10.1007/978-3-030-66093-2_31

1 Introduction

The construction of power plants based on the marine tides use energy is one of the promising areas for the development of modern energy due to the gradual exhaustion of non-renewable resources. In Russia, one of the territories with rich renewable natural resources is Kamchatka. Here, considerable tidal energy is concentrated in the Penzhin Bay of the Okhotsk Sea, where tides reach 12.9 m [1]. Interest in unconventional renewable energy sources, including tidal energy, is dictated by the need to resolve the shortage of electrical power and a number of problems of social development in the region.

The main reasons for the difficult socio-economic situation are [2]:

a) expensive imported fuel on the peninsula, difficulties with its seasonal delivery. As an example, 84% of Kamchatka's power plants operate on fuel oil. A smaller share falls on the use of thermal energy of the subsoil.
b) extremely high tariffs for electricity and heat. The reasons are the increase in prices for fossil fuels and the poor development of transport infrastructure, as a result the transport component of the imported fuel cost from the most distant consumers can reach 80%. This situation is a deterrent in the development of the fish processing, tourism, and settlement of the region.
c) low level of consumer payments for supplied energy;
d) interruption problems in electricity and heat supply of the population, socially significant facilities and industrial enterprises. This is due to the low reliability of operation of extended power lines with a high voltage class.

Thus, "in its present form, the Kamchatka energy sector does not correspond to the modern needs of the regional economy" [3], therefore the use of renewable resources of Kamchatka is of significant importance both from an energy point of view (reduction in the use of fossil fuels) and from environmental issues (reduction in fuel and energy complex negative impact on the environment).

The Russian government drew attention to the need to take "reakthrough measures" in the development of the Far East regions, in particular, it was discussed at the international forum in Vladivostok in early September, 2018.

In this paper, we set goals to analyze the experience of tidal power plants (TPP) building in Russia and some other countries of the world, to review the existing project of the Penzhin TPP and to offer our own way to use the energy of the Penzhin Bay tides to produce electricity, featuring a new approach to the construction and location of power station hydro technical facilities.

2 Methodology

The beginning of the energy and ebbs use in Russia dates back to 1968, when the experimental TPP on the Kisloi Bay of the Barents Sea was put into operation. The construction of this power plant was facilitated by many factors: lack of ice on the direction to the bay due to the warm Gulf Stream, favorable natural relief, configuration of the shores forming a narrow passage with a steeply falling rock on the western shore and a precipice turning into a small plateau on the east shore [4]. They predetermined the convenient layout of the hydroelectric complex, and proximity to the onshore site made it possible to place open switchgear (distribution device) on it, rooms for service personnel and warehouses. The construction of the Kisloi Bay TPP was carried out by a floating method. It was the first of its kind in the world and took on the part of the world's first floating hydroelectric station with an engine room built inside the country and then towed by sea to its location and installed on a pre-built marine basement [5].

To ascend the block and occupy the project position with it, it is necessary to fill the pit (dry dock) with water, which presents difficulties with a large depth of excavation. Under the conditions of the Kisloi Bay TTP floating block, difficulties arose in the installation of hydro-power equipment due to the limited space of the block. Theoretical calculations during the design and experimental data showed that internal mechanical stresses occurred in the thin-walled structure of the TTP due to the temperature difference between the submerged and the unsubmerged part at the waterline. This circumstance required the use of thermal insulation construction.

The next experimental tidal power plant was the Kolsky TTP. The alignment of the TPP installation was chosen for it, on the Dolgaya Bay of the Kolsky Peninsula. The goal of the Kolsky TPP implementation is to test the operation of capsule units with an impeller diameter of 10 m and a power of 20 MW in various operating modes and the possibility of installing them in future, at more powerful power plants (Mezen and Tugur TPPs). Due to the small size of the tide, which averages 3.3 m, and the small area of the basin to be cut off, the construction of this TTP is not economically feasible, it is justified only by the value for the pilot tests of hydraulic unit's turbines [6].

We can clearly mention the positive experience of the launch and operation of the existing tidal power station La Rance in France. La Rance was the world's first 240 MW TPP produced using 24 capsule turbines operating in both directions of flow. It is the largest in terms of TPP production at the mouth of the Rance River. The choice of location was due to high tides at the mouth of the river (up to 13.5 m). In the 1950s–1960s, many studies were carried out, and model turbines were tested prior to the start of construction of a TPP to ensure the highest design efficiency for new capsule hydro generators [7].

The experience of putting into operation capsular generators was used in the implementation of the Kisloi Bay TTP, and the underwater part of the TTP, of this plant, became the benchmark for many other power plants. The Rance TTP is recognized as an efficient installation, the operation of which was made possible thanks to the six-stroke operation of capsular hydraulic units, the advantages of which are obvious. It is worth noting that TTP had insignificant environmental impacts and supplied the required additional power to remote regions. Methods aimed at reducing costs have led to a reduction in the price of electricity production, which is even more significant considering the TTP life cycle [8].

3 Results

Let us analyze the possibilities of constructing a tidal power plant using the energy of the Penzhin tides. There are two options for the TTP construction. The first would use the entire bay basin. It is considered the most appropriate, the construction of the TTP should be carried out in the southern section of the Okhotsk Sea Povotny – Bozhedomov capes, where the tidal range reaches 11 m. This construction assumes an installed capacity of 87 GW and an annual production of 200 TW*h of electricity. According to the second option, it is proposed to use the northern part of the basin with higher tides (13.4 m). This supposed smaller TTP can be placed at the bay section narrowing in the section line of the Sredny and Mamechensky capes (northern section).

The realization of such projects is associated with certain difficulties and limitations, including:

1) the need for expensive underwater rock excavation. For example, in the southern section with a length of 72 km on a plot of 51 km, the depth of the Penzhin Bay reaches 67 m. A multi-tiered building structure is required placing floating blocks.
2) severe climatic conditions. In winter, an active process of ice formation is observed in the north eastern part of the bay. This means that the building must be able to absorb the effects of ice.

A successful placement of the building and energy equipment of the Kisloi Bay TTP, significant hydropower potential with high tides and successful experience in using capsule hydro generators at Rance TTP gives grounds to develop the following option of the Penzhin bay tidal energy use. The location the power plant on land, which is flat on the Kamchatka Peninsula, is possible if an underground pipe in the isthmus connecting the Penzhina bay with the Bering Sea will be build. The channel for siphons should have a length of 90–100 km. Previously, such a proposal was made by N.A. Belinsky and V.S. Nazarov, but "to improve the climate" [4]. At that time, the need for a large number of excavations predetermined the economic inexpediency of this proposal. The reasons for this are in the underdeveloped technology of land large volumes excavation of and poorly studied terrain. The fact is that the isthmus is a treeless space that is not used in agricultural production, which simplifies the laying of siphons. It is possible to excavate the land by the method of a directional explosion, which has already been fully mastered, and in much smaller volumes than was required in the proposal of Belinsky and Nazarov, where there was a different goal.

The efficiency of a TTP in many respects is determined by the correct choice of a reliable power unit with high energy, operational and economic indicators. It is important to find an expedient combination of constructive and technological solutions for the turbine and the generator. From the above the Rance TTP operation experience, it can be seen that horizontal capsule hydraulic units are widely used due to their power and efficiency.

Capsule generators are called hydro generators with a horizontal shaft, enclosed in a waterproof sheath, called a capsule [9]. The capsule is mounted on the front support columns and stator columns of the turbine. In addition, the casing of the unit is fixed in the front and parts with powerful braces that create the hydraulic unit installation rigidity. The flow of water passes between the unit conical guide and the conduit wall. The rotary blades of the impeller rotate in a hemispherical working chamber [10]. The generator is also located in the capsule, the rotor of which has one support and a thrust bearing, and the turbine bearing is the second support of the unit. The wide use of such structures is explained by the fact that:

1) due to the straightening of the flow path, it is possible to minimize hydraulic losses and thereby increase the efficiency of the turbine and its capacity;
2) their use allows to reduce the length of the pressure front and the volume of construction and installation work on the building of a hydroelectric power station and to obtain the simplest outlines of the turbine unit;
3) they are capable of operating in different modes: two-sided turbine and pumping, and in two with zero power, when idling water passes [11].

Let us estimate the possible power of one turbine N, kW, calculated by the following formula [12]:

$$N = c_N Q H \tag{1}$$

where c_N – coefficient (ratio) taking into account the hydraulic losses and turbine efficiency; Q - water flow through the turbine, m^3/s; H - turbine head taking into account losses, m.

Since the expected turbine power is more than 50 MW, the coefficient c_N is taken equal to 8.8.

The water pressure in the turbine (the turbine head) H is numerically equal to the effective height of the drop in water flow, obtained by deducting from the actual height difference the equivalent height of the losses. With the actual height difference of 12.9 m, we take the effective height H = 12 m.

Water flow Q is calculated from the water flow rate in front of the turbine v and the cross-section of the siphon S:

$$Q = vS \tag{2}$$

The water flow velocity can be approximately found from the energy conservation law (g is the acceleration of free fall, g = 9.81 m/s^2):

$$v = \sqrt{2gH} = \sqrt{2 \cdot 9.81 \cdot 12} = 15.34\ \frac{\mathrm{m}}{\mathrm{s}} \tag{3}$$

Taking the diameter of the siphon to be d = 7.7 m, we find the cross-sectional area of the siphon S and the water flow Q:

$$S = \pi\frac{d^2}{4} = \pi\frac{7.7^2}{4} = 46.57\ \ \mathrm{m}^2 \tag{4}$$

$$Q = 15.34 \cdot 46.57 = 714.52 \ \frac{m^3}{s} \tag{5}$$

The power of the turbine N is equal to:

$$N = 8.8 \cdot 714.52 \cdot 12 = 75453 \ \text{kW} \tag{6}$$

In accordance with our proposed design of the Penzhin TPP it is recommended to install horizontal capsule hydraulic units with design features of the capsule turbine used at the Cheboksary Hydroelectric Power Plant (ChHPP). The flow rate of water through the turbine at the ChHPP is 725 m^3/s, which is quite close to the calculated value of the flow rate at the Penzhin TPP.

Placing the generator directly into the capsule makes it difficult to operate due to a number of factors:

1) the removal of heat from a working generator in a closed space, its repair and maintenance in cramped conditions is complicated;
2) repair of the generator requires the flow part drainage;
3) the removal of condensate formed during operation requires the use of a refrigerator and special filters that dry the air, which reduces the turnaround time and increases operating costs.

In this connection, we propose to place the generator in the upper level of the station in the machine room, and not directly in the capsule in contrast to the capsular units of the ChHPP. Since in this case the shaft of the generator is located vertically and the shaft of the turbine is horizontal, it is necessary to install a low-speed bevel gear. The Czech company Mavel has an experience in the production of such units. The model of a turbine with a bevel gear is called "KAPLAN BULB". A bevel-gear turbine capsule allows the generator to be placed on top of a turbine unit, which reduces the size of the turbine housing compared to a conventional turbine capsule when the generator is located inside [13].

Taking the average efficiency of the hydro generator $\eta = 0.965$, we can calculate the power of the hydro generator P:

$$P = \eta N = 0.965 \cdot 75453 = 72812.15 \ \text{kW} \tag{7}$$

We propose to install generators of the SV-1477/142-104 brand, used at the Nizhnekamsk Hydroelectric Power Station (NHPP) with thyristor excitation [14]. These synchronous generators have a rated power of $P = 78$ MW, which is consistent with the calculated power.

The choice of this combination of "turbine-generator" is explained by the fact that the ChHPP, NHPP and the proposed Penzhin TPP are the same in terms of the main hydraulic parameters and correlate with each other according to the marks of the upstream and downstream, as well as flat power plants.

We offer a technical solution of the Penzhin TPP in Fig. 1.

Let us analyze the principle of this TPP operation. The flow of water enters the two siphons laid in the isthmus between the Penzhin Bay and the Bering Sea of the Pacific Ocean (Fig. 1, 2) from the Penzhin Bay (point a), rotates the turbine impeller and goes

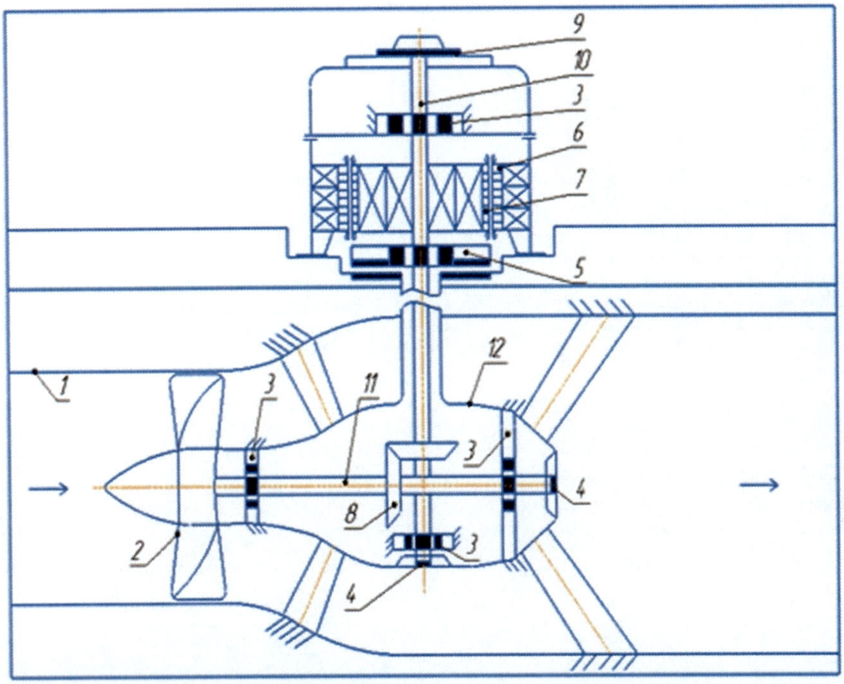

Fig. 1. The proposed hydro unit for the Penzhin TPP: 1 - water duct pipe; 2 - turbine (rotary blade); 3 - shaft bearing; 4 - thrust bearing; 5 - an umbrella generator bearing; 6 - generator stator winding; 7 - excitation winding of the generator inductor; 8 - bevel gear; 9 - brush node of voltage excitation supply; 10 - generator shaft; 11 - turbine shaft; 12 - turbine capsule.

along the path to the Bering Sea (point b). The bevel gear transmits rotation from the turbine shaft to the generator shaft, at the same time reducing the rotation speed to the required value. Thrust bearings perceive axial loads. Mechanical rotational energy is converted into electrical energy in an umbrella-type synchronous generator.

In Fig. 2 the inlet portal 1 for supplying water to the siphon 2 is indicated at point a. Each of the siphons 2 under the TPP 3 building is divided into two conduits in which capsule hydraulic units 4 are installed. The outlet (outflow) of water from the TPP is done via exit portal 5 at point b. The gates which are necessary to regulate and cut off the flow of water are installed at the inlet and outlet portals and conduits with hydraulic units.

The ratio of the tide water volume and the capacity of the conduits is such that, with a minimum water level during low tide, the height difference will be enough to continue the water flow into the siphon. By placing the siphon at an appropriate angle, you can get a unidirectional flow of water at high tides and low tides. This means that the rotation of the turbine working blades will occur only in one direction. This is an indisputable advantage, since there is no need for a complicated mechanism to rotate the blades. The slope of the siphon is determined by hydraulic calculation.

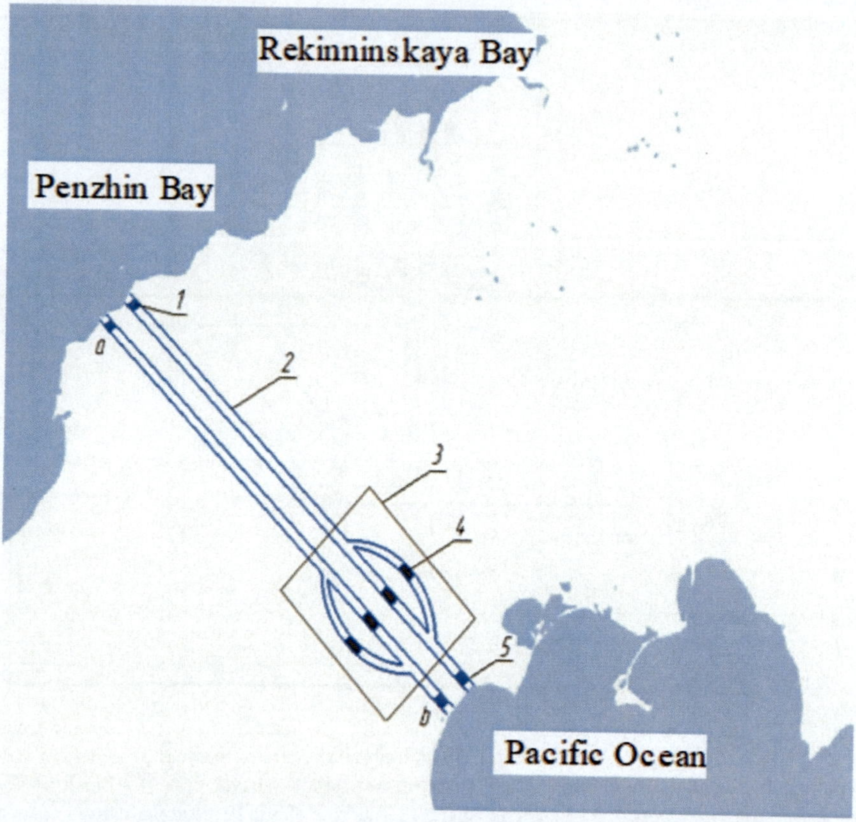

Fig. 2. The location of the siphons and the TPP building: 1 - entrance portal; 2 - siphon; 3 - TPP building; 4 - hydraulic unit; 5 - exit portal

Let's calculate the volume of water flowing during the tide into the conduits. We find that the tides in Shelikhov Bay are diurnal and, on average, 410.6 km^3 of water passes in the bay per day (one tide and one ebb). Knowing the flow rate of water Q passing through one turbine, one can find the total volume of water V_{Σ} required per day for four turbines. In days t = 86,400 s, whence it follows:

$$V_{\Sigma} = 4Qt = 4 \cdot 714.52 \cdot 86400 = 0.247 \ \text{km}^3 \tag{8}$$

From the calculation it can be seen that the required volume of water for the power plant operation is much less than the volume of water passing in the bay on average per day. This means that the flow of water into the water conduits will occur without interruption and there is no need to adjust the angle of rotation of the blades as in the case of a power plant project located in the section line of capes.

4 Conclusions

The principles of basic technical solutions for the construction of the Penzhin hydroelectric complex were considered in this article. The entrance portal from the Okhotsk Sea requires the laying of two underground conduits in an explosive open way. Four hydraulic units are operating under the building of the hydroelectric station. The generator shaft is connected to a turbine shaft by means of a bevel gear, and the generator itself is located in the building. Technological solutions and equipment of the plain ChHPP and NHPP were used at 4 energy units: 4 hydropower units with a capacity of 75 MW each. It is shown that the volume of water flowing at high tide is more than enough for the continuous operation of the power plant, taking into account the low tide (ebb) time.

It is also worth noting that the implementation of the project in the proposed technical solution will certainly not be subsidized. Since the generation of electricity does not require expensive fuel, the costs will quickly pay off. Power generation will allow for the realization of the Kamchatka region "breakthrough" plans in services, industry and agriculture.

This technical solution can be used for other potential energy facilities, where there is a similar combination of geographic and natural factors and there is the possibility of using alternative power sources to solve various energy problems.

References

1. Bystritsky, G.F.: Fundamentals of Energy. INFRA-M, Moscow (2007)
2. Collection: Development of renewable energy sources in Russia: opportunities and practices (on the example of the Kamchatka Region). OMNNO "Greenpeace Council" (2006)
3. Gajur, A.A., Nikolaeva, N.A., Osokin, Yu.V., Pokhodnya, I.L.: Energy efficiency technologies for the Kamchatka territory. "Innovations: Perspectives, Problems, Achievements". In: Collection of works of the IV International Scientific-Practical Conference, Moscow (2016)
4. Bernstein, L.B., Silakov, V.N., Gelfer, S.L., et al.: Tidal power plants. Energoatomizdat, Moscow (1987)
5. Sveinsson, N.: Profitability assessment for a tidal power plant at the mouth of Hvammsfjörður, Iceland. In: Reykjavik Energy Graduate School of Sustainable Systems Report, Reykjavik (2011)
6. Minin, V.A., Dmitriev, G.S.: Prospects for the development of innovative and renewable energy sources on the Kolsky Peninsula. In: Report of the Bellona Association, Murmansk (2007)
7. Rolandez, G., Abgottspon, A., Staubli, T.: Discharge measurements at La Rance Tidal Power Plant using current meters method. In: 10th International Conference IGHEM on Innovation in Hydraulic Efficiency Measurements, Itajuba (2014)
8. Charlier, R.H., Finkl, C.W.: Ocean Energy: Tide and Tidal Power. Springer, Heidelberg (2009)
9. Makarichev, Yu.A., Ovsyannikov, V.N.: Synchronous machines. Samara State Technical University, Samara (2010)

10. Goncharov, A.N.: Hydropower equipment of hydroelectric power plants and its installation. Energy, Moscow (1967)
11. Smirnov, I.N.: Hydraulic turbines and pumps. Higher School, Moscow (1969)
12. Giesecke, J., Heimerl, S., Mosonyi, E.: Hydroelectric power plants: design, construction and operation, 6th edn. Springer, Heidelberg (2014)
13. Kaplan Turbine. https://www.esru.strath.ac.uk/EandE/Web_sites/14-15/Tidal_barrage/turbine.html. Accessed 12 June 2020
14. Mullagaliev, I.R.: The energy of life. Nizhnekamsk HPP on the eve of the 30th anniversary. Energy of Tatarstan (3), Kazan (2009)

Mechanism for Determining the Level of Telecommunications Development in Rural Areas of the Russian Federation

Anatoly E. Shamin, Natalia Viktorovna Proskura(✉), Dmitry V. Proskura, Nadezhda V. Denisova, and Nikolay S. Volostnov

Nizhny Novgorod State University of Engineering and Economics, Knyaginino, Russia
Nvpro@mail.ru, n.d.2704ngieu@gmail.com

Abstract. Sustainable rural development is linked not only to the development of agriculture and forestry, but also to infrastructure components such as fixed telecommunications services. Universal communication services provide one of the primary needs for human life safety. The penetration of universal communication services is the basis for the existence and effective functioning of rural areas, as well as the subsequent improvement of the social and economic situation in the country. The lack of social infrastructure in rural areas, in particular, telephone communication, the Internet leads to a decrease in the level of comfort of living in rural areas, and, consequently, to an increase in the rate of urbanization on the territory of the subjects of the Russian Federation. In this regard, it is necessary to create a mechanism for determining the level of telecommunications development in rural areas, which would help solve a number of problems related to the introduction of telecommunications services in rural areas. The integrated criterion for assessing the need for telecommunications services proposed by the team of authors will help determine the degree of need for providing these services and make a step forward on the path to sustainable development of rural areas.

Keywords: Sustainable development · Rural areas · Telecommunications services · Urbanization · Universal communication services

1 Introduction

Rural development in Russia is associated or even identified with the development of agriculture, including forestry. Development of rural areas and improving the agricultural sector are the basis for ensuring food and national security of the state, promotes the preservation of historical and cultural heritage, and are necessary conditions for increasing the well-being of citizens. There is no doubt that agricultural and forest production make a great contribution to the existence of rural areas. Sustainability of development implies a complex interaction of such areas as economic, social and environmental.

T. Antipova (Ed.): ICCS 2020, LNNS 186, pp. 335–342, 2021.
https://doi.org/10.1007/978-3-030-66093-2_32

2 Materials and Methods

Issues of sustainable development of rural territories in Russia and abroad were analyzed in their works by Askarov A. A., Stovba E. V. [1] and others. Criteria for sustainable development of rural territories are considered in the Works of Tsvetnoy A. V., Shvetsov N. V. [2] and others. Problems, factors and specifics of rural development in the context of their impact on sustainable development are systematized in the works Of tuarmensky V. V., Juha V. M., Pagosyan R. R. [3], Arzumanyan M. S. [4] and others.

The theoretical basis for the research was the works of domestic and foreign economists, materials of international, all-Russian and national conferences, which studied the formation and functioning of rural territories in the digital economy. In the process of research, such methods of scientific knowledge as analysis, comparison method, logical, graphical and tabular methods were used.

3 Results

The problems of rural development are diverse. In the context of digitalization of society and the economy, much attention is paid to the search for solutions to sustainable rural development, namely from the transformation of agricultural development to integrated development. Addressing the Federal Assembly on the occasion of the 20th anniversary of the Constitution of the Russian Federation, Russian President Vladimir Putin said that rural areas of Russia should become attractive for living and working [5, 6].

In this regard, the government of the Russian Federation is implementing a number of programs aimed at comprehensive development of rural areas, the main goal of which is to maintain the share of the rural population in the total population of the Russian Federation. Thus, as of January 1, 2020, the share of the rural population in the Russian Federation as a whole is 25.3%, which is 2.03% lower than the value of this indicator calculated on average by Federal districts. This situation has a significant impact on the level of urbanization. So for the period from 1990 to 2019, there is an increase in the value of this indicator on average in the Volga Federal district by 2.9%, which is due to the desire of citizens for more comfortable living conditions that they can get in the urban environment (Table 1).

When analyzing the dynamics of the level of urbanization, it is clear that this indicator in half of the subjects of the Russian Federation is at the saturation stage. The process of urbanization in such republics as Bashkortostan, Mari El, Mordovia, and Chuvashia is still at the stage of development and growth, which is associated with the national characteristics of citizens living in these territories, and with the level of development of agriculture. There is an increase in this indicator in the Kirov and Ulyanovsk regions. The specifics of the development of urbanization in these areas lead to greater polarization, that is, to the tendency of population accumulation around large cities of 100 thousand people or more. This leads to the extinction of entire remote areas.

Table 1. Dynamics of the level of urbanization in the context of subjects of the Volga Federal district

Name of the subject that is part of the Volga Federal district	Level of urbanization, %				Deviation
	1990	2002	2010	2018	
Bashkir-Stan of the Republic	77,2	78,1	78,8	79,5	2,3
Kirov region	69,9	71,6	73,6	76,8	6,9
Republic Of Mari El	80,8	80,6	80,2	80,9	0,1
Republic Of Mordovia	56,8	59,7	59,9	60,1	3,3
Nizhny Novgorod region	66,2	67,1	67,2	67,4	1,2
Orenburg region	71,2	71,3	72,1	72,4	1,2
Penza region	63,0	64,9	66,8	68,5	5,5
Perm region	77,0	78,1	79,0	79,2	2,2
Samara region	79,8	80,2	80,5	81,1	1,3
Saratov region	77,0	75,3	74,8	77,9	0,9
Republic Of Tatarstan	73,4	73,8	75,2	76,8	3,4
ANS Of The Republic	70,1	69,7	69,3	65,8	−4,3
Ulyanovsk region	71,8	73,1	73,6	75,3	3,5
Chuvash Republic	58,3	60,5	61,3	62,1	3,8

However, even in those regions where agriculture is traditionally strong, the process of urbanization absorbs rural areas (the Republic of Tatarstan). Scientific and technological progress, as well as the dynamics of development of agricultural holdings in Russia, have directed the development of agriculture towards automation and machine production. Consequently, the process of rural migration to cities does not stop. This means that the issue of sustainable development of rural areas in Russia from the point of view of social and demographic factors remains at the level of the problem.

4 Discussion

In this regard, we can make a paradoxical conclusion that a low level of urbanization in the Russian Federation can serve as an indicator of higher, high-quality development of rural territories, and, consequently, higher-quality development of the entire subject. These territories include: Krasnodar territory, where the urbanisation rate is 54.9%, Altai territory (56.4%), Stavropol territory (58.4%), Tambov region (60.9%), Kurgan region (61.8%), Republic of Mordovia (60.1%), etc. This percentage of urbanization in these territories indicates the uniformity, and hence the sustainability of their development. Conversely, a high level of urbanisation of the territory does not necessarily indicate a high and technological development of the entire subject. The main reasons may be a large amount of abandoned land, impassability, or harsh climatic conditions.

Sustainable development of rural areas often depends on their provision with modern infrastructure, in particular social infrastructure [7]. The variety of elements of social infrastructure is a combination of numerous industries into a single whole, where special attention should be paid to the tele-communication component of rural areas.

The telecommunications component of rural areas is not only the level of telephony, but also the availability of other modern, primarily fixed telecommunications services.

The development of rural areas in the field of telecommunications was carried out with direct state participation and includes the following stages:

1. Providing rural areas with payphones for universal communication services;
2. Eliminating digital inequality and providing socially significant objects with access to the broadband Internet.

According to the Federal law "On communications" dated 07.07.2003 No. 126-FZ, article 57 "universal communication services are guaranteed in the Russian Federation". These include telephone services provided with the use of collective access facilities (payphones…); services for data transmission and access to the information and telecommunications network "Internet", provided with the use of access points.

Taksofonnaya territories provides one of the primary requirements of safety of human life. This is done by providing free access to emergency operational services. Only with state support, it became possible to install a payphone in every locality, no matter how far away it was. The importance of this service lies in the fact that it is provided to the population regardless of the level of income of the citizen, the remoteness of the locality. It is from this point of view that a universal service can serve as a starting point for the revival or sustainability of rural areas.

According to Table 2, we can conclude that it is necessary to provide telecommunications services in terms of providing payphone communication to rural areas. In urban areas, there is no need for payphones, as there is an alternative not only between fixed and mobile communications, but also competitive offers from various telecommunications service providers.

Today, a number of measures are being taken to increase the availability of communication services on the territory of the Russian Federation and in particular in rural areas. Since January 1, 2018, Rostelecom (the single operator of universal service in the Russian Federation since 2014), on its own initiative, has reset the cost of calls from payphones to landlines throughout the country. As a result, in the first three quarters of 2019 (compared to the same period of the previous year), there is a positive trend (an increase of more than 1.5 times) in the use of universal communication services.

The cost of long-distance calls to landlines was reset to zero on June 1, 2019. "Earlier - from January 1, 2018 - the fee for local telephone connections from payphones was canceled, and then - from December 1, 2018 - intra-zone calls to landlines became free."

Table 2. Comparison of the territory area and population with the availability of universal communication services (payphones) in the subjects of the Volga Federal district

Name of the subject that is part of the Volga Federal district	Population, thousand people	Territory, sq. km	Population density, people/sq. km	Number of UUS payphones, units	Number of residents per 1 payphon	Area of the territory per 1 payphone sq km
Bashkir-Stan of the Republic	4 038	142 947	28	4 562	885	31
Kirov region	1 262	120 374	10	4 451	284	27
Republic Of Mari El	679	23 375	29	1 600	425	15
Republic Of Mordovia	790	26 128	30	1 309	604	20
Nizhny Novgorod region	3 202	76 624	42	4 531	707	17
Orenburg region	1 956	123 702	16	1 763	1 110	70
Penza region	1 305	43 352	30	1 529	854	28
Perm region	2 599	160 236	16	3 676	707	44
Samara region	3 179	53 565	59	1 358	2 341	39
Saratov region	2 421	101 240	24	1 861	1 301	54
Republic Of Tatarstan	3 902	67 847	58	3 243	1 203	21
ANS Of The Republic	1 500	42 061	36	1 930	778	22
Ulyanovsk region	1 229	37 181	33	982	1 252	38
Chuvash Republic	1 217	18 343	66	1 745	698	11

These measures made it possible to change the trend of traffic from universal payphones to a positive one and increase the level of availability of telecommunications services even in the most remote localities.

These tables once again confirm the need to support the development of communication in rural areas, since in the conditions of digitalization and Informatization of the economy and society, telephone (payphone) communication and the Internet can meet the most diverse needs of the population.

The analysis of modern literature sources and regulatory documents on the problem of research [10–19] shows that in the conditions of digital transformation of the economy, it is necessary to develop a mechanism for determining the level of telecommunications development of rural territories and their needs for telecommunications services, which will help solve a number of problems related to the introduction of telecommunications services, the effectiveness of which is currently carried out on the basis of a limited number of indicators (Fig. 1).

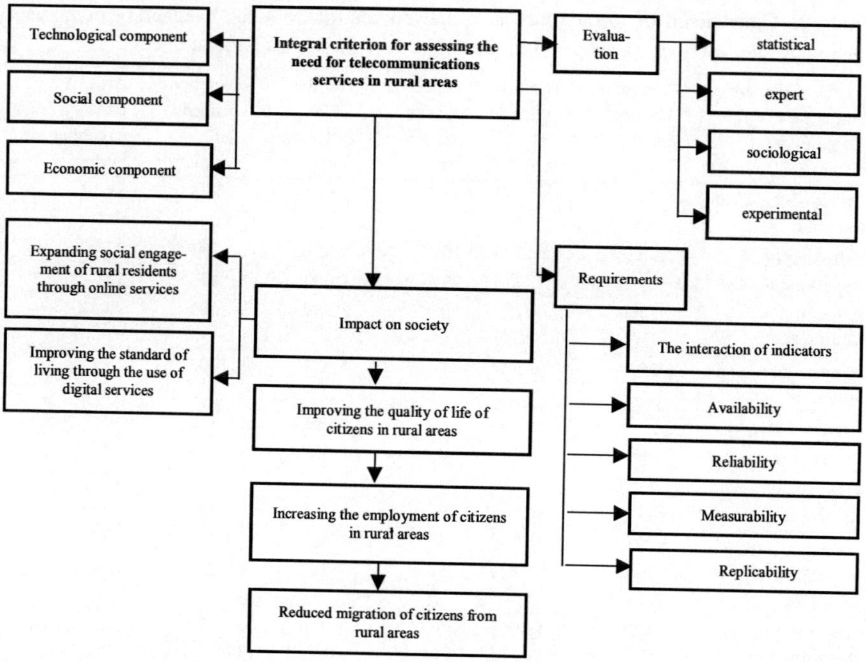

Fig.1. The mechanism for determining the level of telecommunications development in rural areas

The mechanism proposed by the team of authors in terms of increasing scientific knowledge will allow determining the level of telecommunications development in rural areas, and the proposed integrated criteria for assessing the need for telecommunications services will contribute to an objective analysis of rural development and making competent management decisions in terms of providing rural residents with social infrastructure (communications, Internet) and improving the level of comfort of their living.

Difficulties in sustainable telecommunications development of rural areas include their length, distance from urban centers, and low population density compared to other territories, which activates the processes of stratification and polarization of the village [8, 9]. On the one hand, these are successful enterprises, agricultural production cooperatives that have managed to provide themselves with material resources, show a decent financial result and work for the future. On the other hand, there are hundreds of abandoned villages with closed schools, hospitals, cultural and leisure facilities and, as a result, migration of young people and citizens who are in active working age.

5 Conclusions

Thus, the ability to use the latest services in the field of telecommunications, namely: broadband Internet access, digital interactive television, online telemetry (smart water meters and electricity meters, smoke detectors, door closures) contribute to the attachment of the population to their homes and rural areas. These services allow you to live and work comfortably in rural areas. Public services, telemedicine services, video conferencing, allow you to be in the information space of the society of the region, country, and world, while remaining on rural territories.

The polarization of rural areas in Russia continues, and the developed space is shrinking. The consolidation of settlements, which has been carried out on a large scale since January 1, 2006, has further increased the distance between localities. If earlier it was from 2 to 8 km, now the distance between localities varies from 25 to 30 km. Thus, the successful functioning of rural areas in the Russian Federation is inextricably linked to the penetration of universal communication services, that is, the telecommunications component, which, along with the infrastructure of roads and electric networks, is the basis for sustainable development of rural territories and subsequent improvement of the socio-economic situation in the country.

References

1. Askarov, A.A., Stovba, E.V.: Foreign experience of strategic planning of rural development based on foresight technologies. Vestnik NGIEI (8(99)), 77–85 (2019)
2. Tsvettsykh, A.V., Shevtsova, N.V.: Sustainable development of rural territories: essence, factors and criteria. Bull. Belgorod Univ. Coop. Econ. Law (2(81)), 280–288 (2020)
3. Dzhukha, V.M., Kuzminov, A.N., Pogosyan, R.R.: Problems and main factors of sustainable development of rural territories. Account. Stat. (3(55)), 84–91 (2019)
4. Arzumanyan, M.S.: Sustainable development of rural territories. Azimut Sci. Res.: Econ. Manag. **8**(4(29)), 57–60 (2019)
5. Message of the President of the Russian Federation V. V. Putin to the Federal Assembly. https://www.kremlin.ru/news/19825. Accessed 10 July 2020
6. Patnaik, S., Sen, S., Mahmoud, M. (eds.): Big Data for Smart Agriculture. Smart Village Technology. Modeling and Optimization in Science and Technologies, vol. 17, pp. 69–80. Springer, Cham (2020)
7. Janc, K., Czapiewski, K., Wójcik, M.: In the starting blocks for smart agriculture: the internet as a source of knowledge in transitional agriculture. Njas - Wageningen J. Life Sci. (2019). https://doi.org/10.1016/j.njas.2019.100309. 100309
8. Shamin, A., Frolova, O., Makarychev, V., Yashkova, N., Kornilova, L., Akimov, A.: Digital transformation of agricultural industry. IOP Conf. Ser.: Earth Environ. Sci. **346**, 012029 (2019). https://doi.org/10.1088/1755-1315/346/1/012029
9. Maslova, V.V., Zaruk, N.F., Fuchs, K., Avdeev, M.V.: Competitiveness of agricultural products in the Eurasian economic union. Agriculture **9**(3), 7 (2019)
10. Decree of the government of the Russian Federation of 15.07.2013 No. 598 On the Federal target program "Sustainable development of rural territories for 2014–2017 and for the period up to 2020" and amendments to it of April 29, 2014 No. 399, of January 16, 2015 No. 17, of August 2, 2017 No. 925. Accessed 08 July 2020

11. Balandin, D.A.: Theoretical aspects of rural infrastructure development. J. Econ. Theory **15** (3), 442–448 (2018)
12. Belkina, E.N., Agibalova, V.G.: Analysis of social aspects of rural development in the Krasnodar territory. Bull. Acad. Knowl. **3**(26), 50–57 (2018)
13. Big Russian encyclopedic dictionary. Repr. ed. Bolshaya Rossiyskaya EN-Cycle, Moscow (2009). 1887 p.
14. Timofeeva, N.S.: Strategic planning as the most important element of management of sustainable development of rural territories. Bull. Kemerovo State Univ. Ser.: Polit. Sociol. Econ. Sci. **4**(1), 143–148 (2019)
15. Tuarmenskiy, V.V.: Specifics of rural territories of the Russian Federation in the context of ensuring their sustainable development. In: The Age of Science, no. 19, pp. 68–73 (2019)
16. Khlevnaya, A.V., Yarotskaya, E.V.: Theoretical approaches to sustainable development of rural territories. https://www.lib.tpu.ru/fulltext/c/2015/C40/V1/028.pdf
17. Program "Digital economy of the Russian Federation". https://www.consultant.ru. Accessed 15 Mar 2020
18. Proskura, N.V.: Possibilities of application of telecommunication technologies in agriculture. In: Achievements of Science and Technology of the Agro – Industrial Complex, no. 12, pp. 66–67 (2012)
19. Strategy for sustainable development of rural territories of the Russian Federation for the period up to 2035. https://www.consultant.ru/document/cons_doc_LAW_174933

Innovative Economics

Models of Basic Operations at Enterprises in the Real Sector of the Economy

Kirill Litvinsky(✉)

Kuban State University, Stavropolskaya st., 149, 350040 Krasnodar, Russia
litvinsky@econ.kubsu.ru

Abstract. The article discusses the issues of mathematical modeling of the functioning processes of a wide variety of enterprises in the real sector of the economy based on basic operations. A basic non-alternative operation is considered as a multidimensional mathematical model of any operation with the possibility of only one successful outcome. The basic alternative operation is a multidimensional mathematical model of a logical operation, where only one of the alternatives could be selected with indication of necessary characteristics in order to assess the quality of the operation's outcome. The article provides justification of using of algorithmic networks for technological processes simulation. The basic principles of basic operations graph models creation are described. The control system is considered from the point of view of constructing a decision-making system where information sensors are created on the basis of enterprise model functional graphs.

Keywords: Basic operations · Enterprises of the real sector of the economy · Control system · Semantic and algorithmic networks · Decision-making and execution processes · Ograph · Graph model · Control of failures and errors

1 Introduction

Basic operations are those "bricks" which should ensure building of mathematical models of the functioning processes of various enterprises in the real sector of the economy (ERE). The real sector of the domestic economy is the most important component of the Russian national economy. It is an organizational and economic system that includes government bodies, industrial enterprises and organizations, as well as related industries solving important public-economic problems.

The complexity of such an organizational and economic system and the priority of its functional efficiency and safety in comparison over other production issues justify the need to use of modern scientific research methods and a special mathematical apparatus for modeling basic operations [1–3]. The modeling of ERE basic operations proceeds from alternative and non-alternative operations, based on an analysis of the experience in developing methods for describing systems and using algorithmic and semantic networks.

The structure, specifics, and development of ERE in the modern rapidly changing world are the subject of scientific discussion of both theoretical and practical directions. An objective and essential characteristic of any scientific research is the definition,

T. Antipova (Ed.): ICCS 2020, LNNS 186, pp. 345–353, 2021.
https://doi.org/10.1007/978-3-030-66093-2_33

concretization and, if possible, visualization of the main theoretical and practical concepts that determine the most general and, at the same time, essential features of the subject being studied, its content and boundaries [4–8, 12].

An analysis of used basic operations models shows that the most universal of them are the functional structural theory and the generalized structural method of prof. A.I. Gubinsky. So, in [2] models, methods, modeling technologies and consistent optimization of various indicators of functioning processes of technical and economic systems were developed.

Statement of Basic Materials

Analysis of experience in development methods of systems description shows that all available methods are focused on the description of two, usually opposed classes [2]:

- algorithmic networks which describe the sequence of operations performed typically in a rigid logical connection of sets of individual operations;
- semantic networks that well describe the nature of the relationship between their individual elements, but without a rigid fixation of the sequence of operations.

It is correct not to oppose these models classes, but to combine them into complex networks, which should combine the advantages of both semantic and algorithmic networks.

In the process of human activity, there are always elements of logical conclusions, corresponding to decision-making and modeling of future actions, as well as elements of the execution of given algorithms and programs. The first type of above mentioned actions is better described by semantic networks, and the second type is better represented in the form of algorithmic networks. In cases where these types of actions are performed in isolation, it is quite acceptable to separately use semantic networks for modeling decision-making processes and algorithmic networks for execution processes. However, when decision-making and execution processes intersect, a new hybrid modeling apparatus is required and complex networks could be one of them. Based on them ERE basic alternative and non-alternative operations are considered.

Models of Basic Non-alternative Operations at Enterprises of the Real Sector of the Economy. An ideally reliable basic non-alternative operation is a multidimensional mathematical model of any operation, when its error-free execution leads to only one possible successful outcome, provided that during the execution there were no failures and disruptions in the functioning of the elements performing this operation. The only successful outcome of such operation shall achieve the result the operation was intended, and indicate the necessary characteristics for assessing the quality of the operation [9, 11].

The mathematical model of an ideally reliable basic operation is an abstract labeled digraph consisting of two vertices, where the first (input) vertex corresponds to the event of operation beginning and the second (output) corresponds to the event of operation successful completion. They shall be connected only by one arc coming from the first vertex to the second. This arc corresponds to the complex (vector) of X characteristics, required to assess the quality of this path (see Fig. 1a).

In Fig. 1: $\boldsymbol{O}$ – event that determines the beginning of the operation; $\boldsymbol{Q}$ – an event that determines the correct end of the operation; $\boldsymbol{F}$ – event that determines the incorrect

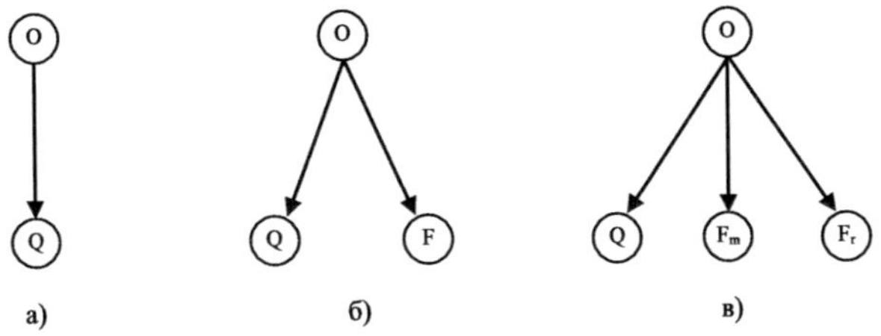

Fig. 1. Models of non-alternative operations

end of the operation; F_m – an event that determines the incorrect end of an operation due to its erroneous execution; F_r is an event that determines the abnormal end of an operation due to a failure.

However, the ideal model does not take into account possible outcomes of a working operation that arise as a result of failures and errors, where in addition to a successful outcome, there may be various options of unsuccessful outcomes. Let us introduce the concept of a real basic operation which differs from ideal by taking into account failures and errors (see Fig. 1b). For a redundant system, the following conditions are natural:

- an essential condition of performing the operation is readiness state of the system and trouble-free functioning during the entire period of operation's performance.
- neither the system as a whole, nor its individual elements have redundancy;
- any interruptions in order to eliminate the consequences of errors and failures and restore performance are not allowed during operation execution;
- control of failures and errors is absent, as a result of which failures and errors lead to unsuccessful outcomes, but remain hidden, undetected;
- the occurrence of at least one error or failure during the operation leads to its failure.

In this case, the basic mathematical model will be presented as a basic operation and will have the form shown in Fig. 1b, because due to the lack of control of failures and errors in the event of their occurrence, they will not be identified and therefore must be combined into an outcome of type F. However, due to the fact that in the event of failures and in the event of errors in the future (at the end of this operation), various actions will be required, state F can be split into two states: F_m – with a hidden error and F_r – with a hidden failure (Fig. 1c). These outcomes constitute a complete group of incompatible events. The formation logic of probabilities of these outcomes is presented in Fig. 2.

In accordance with the graph (see Fig. 2),

$$pp = R^0 \cdot M^0; \; n_m = R^0 \cdot M^1; \; n_r = R^1 \tag{1}$$

where p is the probability of operation end with a positive outcome; n_m is the probability of operation end with a negative outcome in the absence of failures; n_r is the probability of operation end with a negative outcome in the absence of errors; R^0 is the probability of system functioning without failures during the operation; R^1 is the

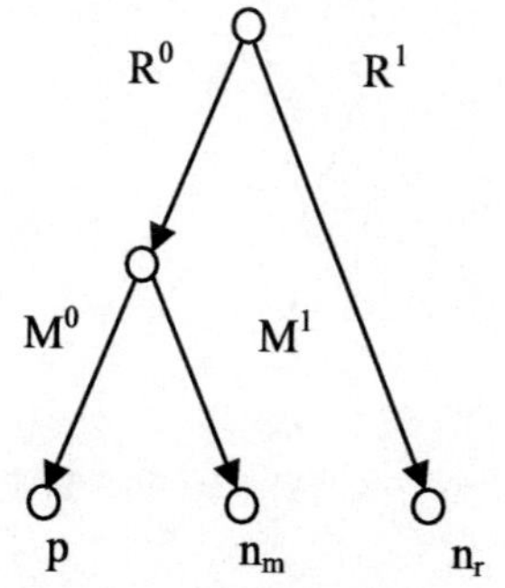

Fig. 2. The logic of the formation of outcome probabilities

probability of system functioning with at least one failure during the operation; $\boldsymbol{M^0}$ is the conditional probability of performing the operation without errors, provided that there were no failures; $\boldsymbol{M^1}$ is the conditional probability of the operation with at least one error provided that there were no failures; $(\boldsymbol{M^1} + \boldsymbol{M^0} = \boldsymbol{1})$; $(\boldsymbol{R^1} + \boldsymbol{R^0} = \boldsymbol{1})$.

Temporal characteristics should also be determined differentially for each outcome. However, in this case, due to the lack of identification of failures and errors, the operation will continue for the same time in the case of any outcome, i.e.

$$T_p = T_{nm} = T_{nr}. \tag{2}$$

The cost characteristics will be different for each of the outcomes:

$$\begin{aligned} V_R &= V_R^0 - V_R^1,\ V_{F_m} = -V_R^1 - V_{Fm}^1,\ V_{Fr} = -V_R^1 - V_{Fr}^1 \\ V_F &= n_m \cdot V_{F_m} + n_r \cdot V_{Fr} = -V_R^1 - \frac{R^0 M^1 V_{F_m}^1 + R^1 V_{F_r}^1}{1 - R^0 M^0}, \end{aligned} \tag{3}$$

where V_R^0 is the income received from performing this operation (in the case of non-manufacturing operations, it may be zero) without taking into account the costs of the operation; V_R^1 is expenses that have to be borne to ensure the performance of this operation; $V_{F_m}^1 \left(V_{F_r}^1\right)$ is losses that the system will incur due to a perfect error (the resulting failure).

Real (average) income received during the implementation of the operation is:

$$\begin{aligned} V_0 &= p \cdot V_R + n_m \cdot V_{F_m} + n_r \cdot V_{F_r} \\ &= R^0 \cdot M^0 \cdot V_R^0 - R^0 \cdot M^1 \cdot V_{F_m}^1 - R^1 \cdot V_{F_r}^1 - V_R^1 \end{aligned} \tag{4}$$

Based on these characteristics, it is also possible to determine the losses from non-reliability ("price of reliability").

The price of reliability, i.e. total loss from failure and errors is equal to:

$$\Delta V = V_R - V_0 = V_R^0 \cdot \left(1 - R^0 \cdot M^0\right) + V_{F_m}^1 \cdot R^0 \cdot M^1 + V_{F_r}^1 \cdot R^1. \quad (5)$$

In the expression obtained, the first term corresponds to the lost value due to unreliability, and the last terms correspond to additional losses associated with the consequences of error and failure.

The error price can be obtained as a special case of this formula with $\boldsymbol{R^0}$= 1

$$\Delta V_m = M^1 \cdot \left(V_R^0 + V_{F_m}^1\right) \quad (6)$$

Similarly, the price of failure can be obtained as a special case with $\boldsymbol{M^0}$= 1

$$\Delta V_r = R^1 \cdot \left(V_R^0 + V_{F_r}^1\right). \quad (7)$$

It should be noted that the price of unreliability is a non-additive function of the price of failures and the price of errors, i.e.

$$\Delta V \neq \Delta V_m + \Delta V_r.$$

Models of Basic Alternative Operations at Enterprises of the Real Sector of the Economy. An ideal basic alternative operation is a multidimensional mathematical model of a logical operation as a result of which, and if there are $\boldsymbol{n}$ alternatives: the first - "carry out only the first operation", the second - "carry out only the second operation", etc. - only one of them is selected, i.e. provided that during the execution of the alternative operation there were no failures and forced interruptions in the functioning of the system performing this operation, only one successful outcome is possible and corresponds to the achievement of the result (selection of the correct outcomes), indicating the necessary characteristics for assessing the quality of operation performance [5, 10, 13].

The mathematical model of an ideal basic alternative operation is an abstract labeled ograph in the form of a tree with one central vertex corresponding to the event of operation beginning and with $\boldsymbol{n}$ output vertices corresponding to the successful completion of the operation, i.e. the choice of one of $\boldsymbol{n}$ alternatives listed above, connected to the central vertex, each with only one arc exiting from the central vertex, with each arc corresponding to a vector of characteristics necessary for assessing the quality of this path. In Fig. 3a shows a graph corresponding to this description.

However, the ideal model does not take into account the possible paths and outcomes that appear in the event of errors and failures. To do this, we introduce the concept of a real basic alternative operation, which differs from the ideal by considering all paths and outcomes, both successful and unsuccessful due to errors and failures.

A general view of the mathematical model of reliability of a real basic alternative operation is represented by an abstract labeled digraph in the form of a tree with one input vertex, which is the central vertex of the tree corresponding to the event of the operation beginning, and with $\boldsymbol{n}$ output vertices ($\boldsymbol{n = n_1 + n_2 \geq 3}$), corresponding to various options of successful outcomes ($\boldsymbol{n_1 \geq 2}$) and various options for the

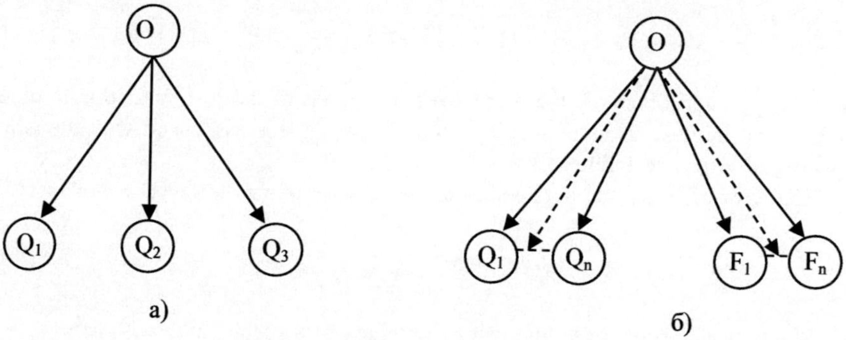

Fig. 3. Models of alternative operations

unsuccessful operation (**$n_2 \geq 1$**), while the input vertex is connected to each output with only one arc, each of which is associated with a vector of characteristics. In Fig. 3b shows a graph corresponding to this description.

Further we will use the mathematical model of reliability of the real basic alternative operation, presented in Fig. 4, which we will call the main model.

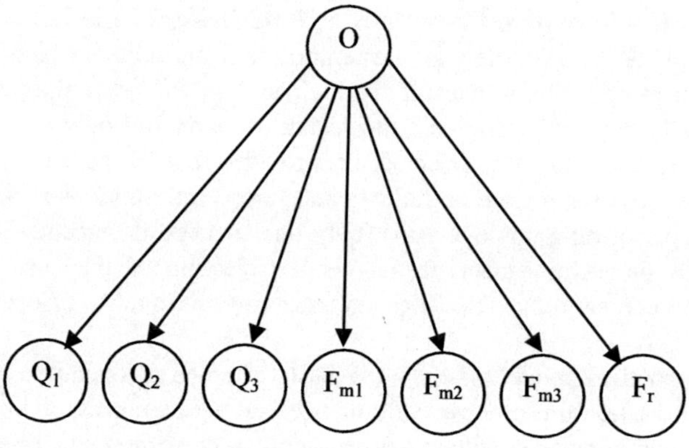

Fig. 4. Reliability model of a real basic alternative operation

The notation used below has the following meaning:

- p_{ii} is the probability of the correct choice of the i-th alternative;
- n_{ii} is the probability of the erroneous choice of the i-th alternative in the case when another alternative should have been chosen;
- n_r is the probability that the erroneous choice of an alternative is associated with element failures;
- b is the probability that errors were detected during the operation and this led to the termination of its execution;
- d is the probability that the operation was terminated due to a failure;
- T_i, V_i, W_i are time, income and resource corresponding to the i-th outcome.

Let us show how these characteristics can be determined by the example of performing an alternative operation without a redundant non-recovery system (workshop equipment of the technological chain).

In this case, the basic mathematical model of reliability of the basic alternative operation will have the form shown in Fig. 5.

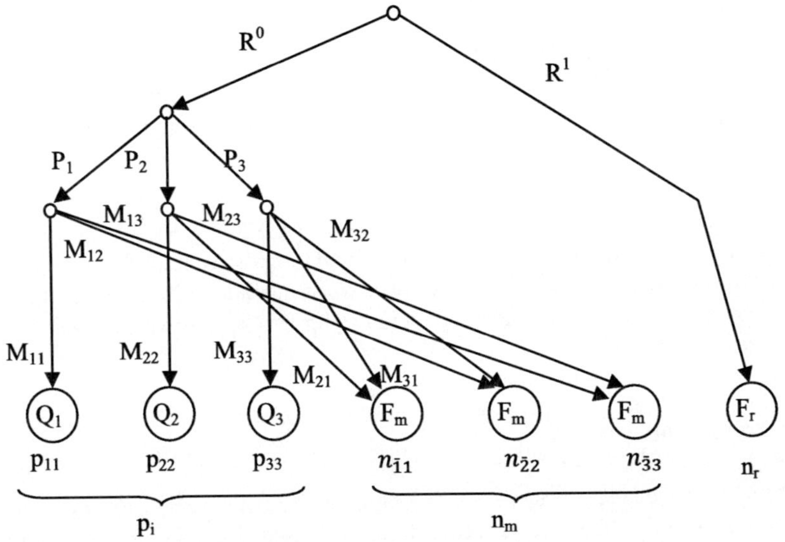

Fig. 5. Reliability model of the basic alternative operation of ERE

Since, due to the lack of control of failures and errors, if they occur, they will not be detected and therefore should be combined into outcomes $\boldsymbol{n_m}$ and $\boldsymbol{n_r}$. Moreover, due to the fact that any other outcome can be selected for errors, it is necessary to differentiate the outcome $\boldsymbol{n_m}$ into three ($\boldsymbol{n_{\bar{i}i}}$ – is the erroneous choice of the ith alternative; $\boldsymbol{i}$ = 1, 2, 3).

The formation logic of these outcomes probabilities is shown in Fig. 5, in accordance with which we obtain formulas for the probabilities of all outcomes:

$$\begin{aligned} p_{11} &= R^0 \cdot P_1 \cdot M_{11};\ p_{22} = R^0 \cdot P_2 \cdot M_{22};\ p_{33} = R^0 \cdot P_3 \cdot M_{33};\ n_r = R^0 \\ n_{\bar{1}1} &= R^0 \cdot (P_2 \cdot M_{21} + P_3 \cdot M_{31});\ n_{\bar{2}2} = R^0 \cdot (P_1 \cdot M_{12} + P_3 \cdot M_{32}); \\ n_{\bar{3}3} &= R^0 \cdot (P_1 \cdot M_{13} + P_2 \cdot M_{23}), \end{aligned} \tag{8}$$

where $\boldsymbol{P_i}$ – is the objective probability of the need to choose the $\boldsymbol{i}$-th alternative; $\boldsymbol{M_{ij}}$ is the probability that, if it is necessary to choose the $\boldsymbol{i}$-th alternative, the $\boldsymbol{j}$-th alternative is actually selected.

In those cases when the i-th alternative is uniquely defined ($P_i = 1$), due to possible errors in the implementation of the selection operation, the above formulas will take the form:

- if $P_1 = 1$

$$p_{11} = R^0 \cdot M_{11}, p_{22} = p_{33} = 0, n_r = R^1, n_{\bar{1}1} = 0, n_{\bar{2}2} = R^0 \cdot M_{12}, n_{\bar{3}3} = R^0 \cdot M_{13};$$

- if $P_2 = 1$

$$p_{22} = R^0 \cdot M_{22}, p_{11} = p_{33} = 0, n_r = R^1, n_{\bar{2}2} = 0, n_{\bar{1}1} = R^0 \cdot M_{21}, n_{\bar{3}3} = R^0 \cdot M_{23};$$

- if $P_3 = 1$

$$p_{33} = R^0 \cdot M_{33}, p_{22} = p_{11} = 0, n_r = R^1, n_{\bar{3}3} = 0, n_{\bar{2}2} = R^0 \cdot M_{32}, n_{\bar{1}1} = R^0 \cdot M_{31};$$

Time, cost and resource characteristics should be determined for each outcome, similar to how it was done in the case of a work operation. Similarly to the above, formulas are derived for alternative operations with various types of realized functions of the algebra of logic.

2 Conclusion

The article presents the approaches to modeling of basic alternative and non-alternative operations at ERE. These approaches have increased accuracy and computational efficiency compared to existing models.

General graphs of the basic operations of ERE process chain can be built using the operators and the models building principles of alternative and non-alternative operations. Complex operations could be simulated by combining several basic operations. For modeling of controlled operations it is necessary to introduce control functions. As a result of ERE basic operations modeling a complex ograph is obtained.

Building of such a graph allows to present the entire process of ERE functioning and to identify the stages where the low reliability of human or technical means functioning has the greatest impact on the failure-free functioning of the enterprise.

References

1. Litvinsky, K.: Methodology of construction management models of actors of nature. In: Antipova, T., Rocha, A. (eds.) DSIC18 2018. AISC, vol. 850, pp. 94–102. Springer, Cham (2019). https://doi.org/10.1007/978-3-030-02351-5_13
2. Gubinsky A.I. Reliability and quality of functioning of ergatic systems. 269 p, Nauka, Lenungrad (1982)
3. Litvinsky, K., Aretova, E.: The formation of data bases at the technogenic risk management system. In: Antipova, T., Rocha, Á. (eds.) DSIC 2019. AISC, vol. 1114, pp. 94–103. Springer, Cham (2020). https://doi.org/10.1007/978-3-030-37737-3_9

4. Christofedis, N.: Graph theory: an algorithmic approach: Per. from English. In: Gavrilova, G.P. (ed.) 432 p. Mir, Moskva (1978)
5. Kalaidin, E.N., Litvinsky, K.O.: The methodology of constructing models for managing natural resource subjects. Reg. Econ.: Theory Pract. **25**(352), 40–47 (2014)
6. Litvinsky, K.O.: A model of the preference structure of enterprises-nature users of the fuel and energy complex. Econ.: Theory Pract. **3**(35), 43–47 (2014)
7. Litvinsky, K.O., Nikitenko, Y.: The model of anthropogenic risk management system at industrial enterprises. Sustain. Dev. Econ. **4**(20), 139–144 (2014)
8. Malyshev, V.A., Lukin, O.V.: Models of basic operations in military-technical systems. In: The XII International Scientific-Practical Conference «Cybernetics and high technologies of the XXI century» vol. 1, 406 p. (2011)
9. Malyshev, V.A., Nikitenko Yu, V., Lukin, O.V.: Design methods and models for intelligent complexes of automated development of military-technical systems. Voronezh (2011)
10. Litvinsky, K., Aretova, E.: Matrix planning for the development of enterprises of the fuel and energy complex. In: Antipova, T. (ed.) ICIS 2019. LNNS, vol. 78, pp. 251–262. Springer, Cham (2020). https://doi.org/10.1007/978-3-030-22493-6_23
11. Burkov, V.N., Gubko, M.V., Novikov, D.A., Korgin, N.A.: Introduction to theory of control in organizations. 346 p. CRC Press, Boca Raton (2015)
12. Novikov, D.A.: Cybernetics: From Past to Future. 107 p. Springer, Heidelberg (2016). https://doi.org/10.1007/978-3-319-27397-6_1
13. Mashunin, Y.: Mathematical modeling and selection of the optimal parameters of technological process. Math. Methods Eng. Technol. **2**, 39–51 (2020)

Mathematical Modeling of Optimal Financing of Investment Modernization Projects for Fuel and Energy Complex Enterprises

Kirill Litvinsky and Elena Aretova(✉)

Kuban State University, Stavropol'Skaya St., 149, Krasnodar 350040, Russia
litvinsky@econ.kubsu.ru

Abstract. The article discusses the economic and mathematical model of optimal financing of investment modernization projects for enterprises of the fuel and energy complex (FEC) based on fixed assets balance. An analysis of the model has been carried out, which included optimal modes, optimal sequence of modes and emerging conditions for the implementation of these sequences, and, based on that, describes the choice of the optimal distribution scheme for financial flows for the stages of financing modernization projects of fuel and energy enterprises. Economic mechanisms with various schemes for financing the project stages are proposed based on the dynamic programming method.

Keywords: Fuel and energy complex · Investment project · Project financing · Optimal distribution scheme of financial flows by stages of project financing

1 Introduction

Currently the economy of FEC enterprises is in a transitional period. Falling oil prices and the national currency rate, economic sanctions from foreign countries, inflation, decreased population's demand for products and other goods force managers of these enterprises to develop and implement innovative modernization (development) projects for their production. The most reasonable and relevant issue here is finding the optimal solution for financing of such projects.

2 Review of Prior Literature

Mathematical modeling of financing of investment projects of FEC enterprises has its own specifics due to the increased share of costs for the design and development of innovations, which quickly become obsolete. The works [1, 2, 14, 15] are devoted to the problems of finding the optimal financing for such projects, as well as their modeling.

An analysis of the nature and objectives of enterprise investment management is reflected in the works of domestic scientists i.e. V.N. Livshits, I.V. Shevchenko, L.A. Voronina and others, as well as foreign scientists, such as I. Ansoff, J. Brigham, R. Owen, B. Twiss and others.

T. Antipova (Ed.): ICCS 2020, LNNS 186, pp. 354–361, 2021.
https://doi.org/10.1007/978-3-030-66093-2_34

Economic-mathematical models of optimal choice are considered in the works of I. A. Blanc, V.A. Verba, P.L. Vilensky, S.N. Ratner. Despite the presence of a large number of studies on the topic, the need for further development of theoretical and methodological principles as well as generalization of practical experience in this field is not ruled out. However, an analysis of the works published on the researched topic indicates that certain aspects of economic and mathematical modeling of optimal financing of investment projects for the modernization FEC enterprises remain insufficiently covered [3].

3 Statement of Basic Materials

The economic-mathematical model of optimal financing of investment projects for energy companies modernization includes the following processes of FEC enterprise: retirement fixed assets F_1, designed to ensure own production, to a specially created fund of fixed assets F_2 for their modernization (development). To F_2 enterprises invest their funds, allocations from private investments extrabudgetary sources, own funds (depreciation and profit) and borrowed funds in the form of certain commercial banks credit resources [4, 5].

FEC companies are ready for modernization based on the principles of project financing. Today there are self-sustaining and highly profitable projects among them. A the same time most of these projects require significant time to reach planned profitability; hence, there are the investment risks. Therefore, there is understandable desire of commercial banks to share these risks with enterprises owners or main shareholders as well as with the state [6].

Since the activity of the fuel and energy complex takes place in a market economy, its goal is to maximize profits. We will also assume that the only source of development for energy companies is their modernization.

Therefore, the model takes the following equations form of fixed assets F_1 and F_2 balance:

$$F_1(t) = -\alpha_1 F_1 \phi;\ F_1(0) = F_{10},\ \phi \geq 0, \quad (1)$$

$$F_2(t) = -\alpha_2 FF_2 - \alpha\phi;\ F_2(0) = F_{20}, \quad (2)$$

where α_i is the fixed asset retirement rate in F_i (i = 1,2), ϕ is the financial flow directed from the F_1 fund to the F_2 for modernization of energy companies, α is the coefficient of modernization efficiency of F_1 funds to F_2 fund.

Limitations on financial flow ϕ in the planned period [0, T]:

$$\int_0^T \phi dt \leq \Delta\Phi_{10}, \quad (3)$$

where Δ is the share of initial fund F_{10}, allocated to finance the modernization project of FEC enterprises.

$$dF_1 \geq p(t) + L_1(t), \tag{4}$$

Limitations on the production of FEC enterprises:
where d is the capital productivity ratio of the main fund F_1, $p(t)$ is the planned consumption of FEC enterprise products, $L(t)$ is the planned increase in output during project implementation.

The balance of capital costs during financing the modernization project of the FEC enterprise:

$$b\varphi(t) = I_1(t), \; I_1(t) \geq 0, \tag{5}$$

where b is the coefficient of capital intensity of the project fund unit F_2 ($1/b$ is the share of the costs of the project), $I_1(t)$ is the investment allocated for modernization of FEC enterprise.

FEC enterprise modernization project is invested at the expense of non-repayable basis $I_b(t)$ and repayable basis $I_v(t)$:

$$I_1(t) = I_b(t) + I_v(t); \; 0 \leq I_b(t) \leq I_{bmax}(t); \; 0 \leq I_v(t) \leq I_{vmax}(t). \tag{6}$$

In this case if funds of the enterprise $I_n(t)$, investors $I_u(t)$ and commercial banks $I_{cb}(t)$ are used:

$$\begin{gathered} I_1(t) = I_n(t) + I_u(t) + I_{cb}(t); \quad 0 \leq I_n(t) \leq I_{nmax}(t); \\ 0 \leq I_u(t) \leq I_{umax}(t); \quad 0 \leq I_{cb}(t) \leq I_{cbmax}(t). \end{gathered} \tag{7}$$

As an optimality criterion let us consider the maximum discounted profit flow minus loans payments for the planned interval [0, T] of the FEC enterprises modernization project:

$$\begin{gathered} \int = \int_0^T exp(-\gamma t)[\Pi_1 P_1(t) + \Pi_2 l(t) - (1+\rho_u)I_u(t-\theta_u) - \\ -(1+\rho_б)I_{бк}(t-\theta_{бк})]dt \rightarrow max \end{gathered} \tag{8}$$

where γ is the discount parameter, Π_1 is the final profit (net of taxes) in rubles per unit of output of the FEC enterprise, ρ_u and θ_u are percentage and term of crediting the project by investors, ρ_{cb} and θ_{cb} are percentage and term of lending the project by commercial banks (ρ_{cb}, ρ_u, θ_u, $\theta_{cb} > 0$).

The task of project financing optimization (1–8) is the linear task of optimal control with mixed restrictions on the phase coordinates (F_1, $F_2 \int_0^T \gamma(t)dt$), and the control (γ, L, I_1, I_u, I_{cb}).

In an analytical study of such problem, optimal modes, optimal sequences of modes are usually determined and conditions are formed to implement such sequences [7, 8]. However, at the same time, the issues of financial flows optimization under various financing mechanisms for the stages of FEC companies modernization projects were not considered in the above-mentioned works.

There is more general task is proposed in this work, which is the choice of the optimal distribution scheme for financial flows by stages of financing projects for the modernization of FEC enterprises.

Let us denote:

m is the number of stages of financing the modernization project for energy companies;

k_i is the coefficient of increasing the production efficiency of FEC companies at the i-th stage of the modernization project ($1 \leq i \leq m$);

Π is the planned increase in the production efficiency of the FEC enterprise from the implementation of the modernization project;

N_i is the planned increase in the production efficiency of the FEC enterprise by implementing the i-th stage of modernization.

Let us have some possibility of varying the coefficient of increasing production efficiency, i.e.

$$\mathrm{K_{imin}} \leq \mathrm{K_i} \leq \mathrm{K_{imax}},\ 1 \leq \mathrm{i} \leq \mathrm{m}, \tag{9}$$

and since

$$K_i = \frac{N_{i+1}}{N_i}, N_{m+1} = \Pi,$$

it is clear that the following conditions are also satisfied

$$\mathrm{N_{imin}} \leq \mathrm{N_i} \leq \mathrm{N_{imax}} \tag{10}$$

Then we can set the task of choosing the optimal distribution scheme for financial flows according to the stages of financing projects for the modernization of energy companies in the sense of a certain criterion. As a selection criterion, the unit cost of enterprise output is proposed. It's clear that:

$$C = f(N_i, \Pi) \tag{11}$$

where C is the cost per unit of output.

Moreover, the cost function is additive at index i, i.e.

$$C = \sum\nolimits_i C_i(N_i, \Pi) \tag{12}$$

Let us formulate the problem as follows: it is required to find such N_i values that ensure the minimum of function (12) under conditions (10).

If function (12) is linear, then the problem is a linear programming problem and it can be solved by the simplex method [9, 10, 12].

This work considers the possibility of using the dynamic programming method to solve the problem. Dynamic programming involves reducing the problem of determining the extremum of functions of many variables and a sequence of extremal problems of lower dimension.

Let us consider the following possible patterns of financial flows distribution by stages of financing modernization projects for FEC enterprises.

1. Serial Connection

This connection considers the output from the previous stage with the input of the next one (Fig. 1). Here, for convenience, the financing stages are numbered in the reverse order. We determine the objective function for the *i*-th stage as

$$C_i + C_i(N_i,\ N_{i-1}) \rightarrow \min. \tag{13}$$

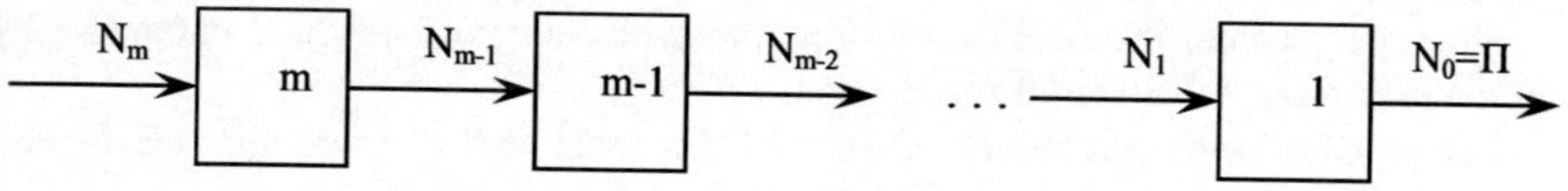

Fig. 1. Serial financing of project stages

Then the minimum of function (12) can be determined by the recurrent Bellmann formula [9, 11, 13]:

$$f_i(N_i, \Pi) = \min_{N_{i-1}}[c_i(N_i, N_{i-1}) + f_{i-1}(N_{i-1}, \Pi)] \tag{14}$$

$$N_i = N_i^{onm}(N_{i-1}, \Pi) \tag{15}$$

where $f_i(N_i, \Pi)$ is the minimum of objective function for *i* series-connected stages;

N_i^{onm} is the function that determines the optimal value of the input from the 1st stage, depending on its output.

The optimal value of N_i is obtained by the consistent application of formula (15), starting from the first to the last stage of project financing:

$$\begin{aligned} N_1^* &= N_1^{onm}(\Pi) \\ N_2^* &= N_2^{onm}(N_1^*, \Pi) \\ N_3^* &= N_3^{onm}(N_2^*, \Pi) \\ &\ldots \\ N_{m-1}^* &= N_{m-1}^{onm}(N_{m-2}^*, \Pi) \\ N_m^* &= N_m^{onm}(N_m^*, \Pi) \end{aligned} \tag{16}$$

2. Parallel Connection

Figure 2 shows the system of *m*-parallel stages of financing projects with a common output N_O. In this case the optimal output values are determined as

$$N_{onm}^* = \lambda_j N_0^{onm},\ 1 \le j \le m \tag{17}$$

where λ_i is a certain constant (determined by the results of the experiment), which can take some positive values.

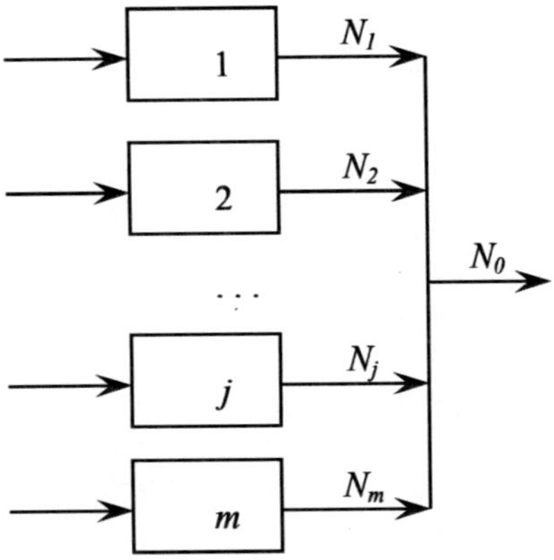

Fig. 2. Parallel financing of project stages with a common output

To apply the dynamic programming scheme, it is necessary to break the process of finding a solution into m stages for each N_j and carry out optimization for each of them.

Figure 3 shows a system of m parallel-connected stages of financing projects with a common input N_0.

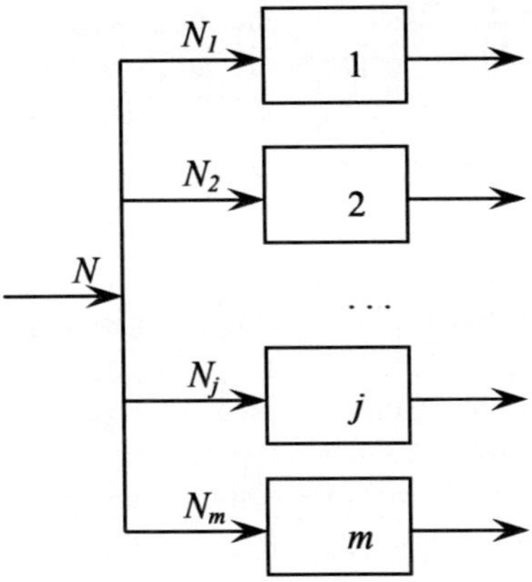

Fig. 3. Parallel financing of project phases with a common input

In this case, the optimal input value N_0 is defined as

$$N_0^{onm} = \sum\nolimits_{j=1}^{m} N_j^{onm} \quad (18)$$

i.e. the search process is grouped into one of m search steps for each N_i.

3. Parallel-Serial Connection.

In the general case, the combination of project financing stages can be represented in the form shown in Fig. 4.

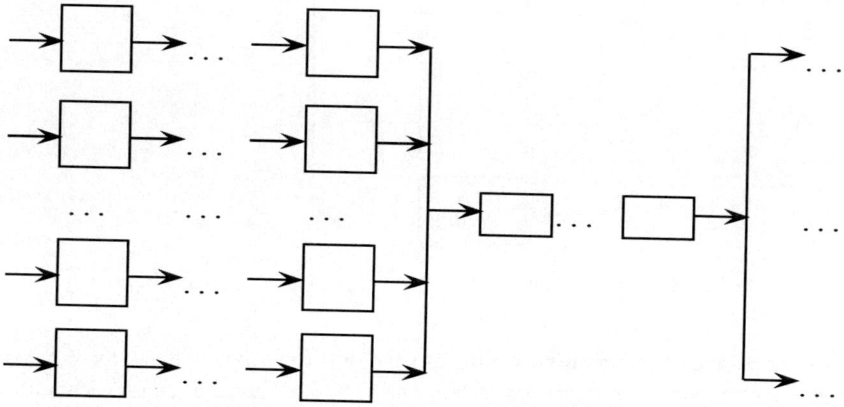

Fig. 4. Parallel-sequential financing of the stages of the project

Here the process of finding a solution starts from the last stage and gradually moves to the beginning of financing. It is clear that in this case it is necessary to take into account what has been said in paragraphs 1 and 2.

4 Conclusion

Thus, having financial flows fully optimized through the stages of financing projects for the modernization (development) of FEC companies, we will get an optimal increase in production efficiency.

This problem arises both in the design of new production (building an optimal production structure with limited financial resources) and in the management (development) of the existing one. It is clear that solving such problems is possible only with the help of computers.

References

1. Voronina, L.A., Trofimenko, E.V., Ratner, S.V.: Formalized mechanisms for evaluating and selecting innovations in the provision of project financing. Econ. Anal. Theory Pract. **5**, 8–21 (2008)
2. Litvinsky, K., Aretova, E.: Matrix planning for the development of enterprises of the fuel and energy complex. In: International Conference on Integrated Science, vol. 78, pp. 251–262. Springer, Cham (2019). https://doi.org/10.1007/978-3-030-22493-6_23

3. Litvinsky, K.O., Loginov, A.E.: Ways to increase the competitiveness of the fuel and energy complex. Econ. Theory Pract. **3**(12), 67–71 (2006)
4. Litvinsky, K.O.: The model of the preference structure of enterprises-nature users of the fuel and energy complex. Econ. Theory Pract. **3**(38), 43–47 (2014)
5. Litvinsky, K.O., Shevchenko, I.V.: Fundamentals of Production and Consumption of Goods: Theory and Practice. Ecoinvest Publishing House, Krasnodar (2010)
6. Loginov, E.L., Shevchenko, I.V., Derkach, N.L.: The problems of operation and development of the energy grid complex as a point of bifurcation of energy efficiency of the Russian economy. Natl. Interests: Priorities Secur. **28**, 2–7 (2011)
7. Kolpachev, V.N.: Models and Methods in Project Management. VGASU, Voronezh (2005)
8. Barkalov, S.A., Burkov, V.N., Kolpachev, V.N.: Heuristic algorithms of scheduling in project management. In: Modern Complex Control Systems: Sat. Scientific Tr Int. Conf, vol. 1, pp. 215–220. Voronezh (2003)
9. Taha, H.: Introduction to the Study of Operations, Vol. 1. Trans. from English. Mir, Moskva (1985)
10. Litvinsky, K.: Methodology of construction management models of actors of nature. In: The 2018 International Conference on Digital Science, vol. 850, pp. 94–102. Springer, Cham (2018). https://doi.org/10.1007/978-3-030-02351-5_13
11. Litvinsky, K., Aretova, E.: The formation of data bases at the technogenic risk management system. In: The 2018 International Conference on Digital Science, vol. 1114, pp. 94–103. Springer, Cham (2019). https://doi.org/10.1007/978-3-030-37737-3_9
12. Gajsek, B., Marolt, J., Rupnik, B., Lerher, T., Sternad, M.: Using maturity model and discrete-event simulation for industry 4.0 implementation. Int. J. Simul. Modell. **18**(3), 488–499 (2019). https://doi.org/10.2507/IJSIMM18(3)489
13. Lee, D., Gu, J.-W., Jung, H.-W.P.: Process maturity models: classification by application sectors and validities studies. J. Softw. Evol. Process. **31**(4), e2161 (2019). https://doi.org/10.1002/SMR.2161
14. Chen, S.-H.: Varieties of agents in agent-based computational economics: a historical and an interdisciplinary perspective. J. Econ. Dyn. Control. **36**, 1–25 (2012). https://doi.org/10.1016/J.JEDC.2011.09.003
15. Tikhomirov, D.V.: Financial model as management decisions tool in project finance. In: Proceedings of the Saint Petersburg state University of Economics, vol. 2, no 110 (2018)

The Protection of Industrial Innovation in Digital Age

Zhanna Mingaleva[1,2(✉)], Irina Mirskikh[3], and Vladimir Kuranov[4]

[1] Perm National Research Polytechnic University, Perm 614990, Russian Federation
[2] Perm State Agro-Technological University Named after Academician D N Pryanishnikov, Perm 614990, Russian Federation
mingall@pstu.ru
[3] Perm State National Research University, Perm 614990, Russian Federation
[4] E.A. Vagner Perm State Medical University, Perm 614990, Russian Federation

Abstract. The transition to Industry 4.0 and 5.0 in production and business is followed both by an increase in opportunities and a number of new problems, including legal regulation of new phenomena and relationship. The paper contains the analysis of problems and contradictions in legal regulation arisen under the influence of digitalization and networking of production processes. The study considers various ways of the influence of digitalization on the organization and the innovation processes in corporations. The legal aspects of such activities cover the norms of Corporate law, Civil law (in the field of intellectual property protection), information and data protection. Current legislation in the field of intellectual property rights protection does not correspond to the requirements of the digital age and needs some improvement. Having used the methods of Civil law, regulatory and comparative analysis, it is concluded that such a new object of legal regulation as "Digital Twins" needs special means of protection. In these conditions, information technology security becomes a key aspect of the corporate legal security system.

Keywords: Digitization of industries · Industry 4.0 · Confidential information · Legal (judicial) facts · Digital technologies · Digital twin · Legal protection

1 Introduction

Modern researches prove that in existing conditions of widespread digitalization of society and economy as apart of the transition to Industry 4.0, the role of information technologies (IT) and digital technologies has changed significantly if to compare their functions and applications with those that were discussed earlier a couple of decades ago. In recent years, this digital exchange has become the main form of business communication, especially in the sphere of sales (sales and supply), product and technological research, innovation, and industrial design. Modern electronic media

T. Antipova (Ed.): ICCS 2020, LNNS 186, pp. 362–370, 2021.
https://doi.org/10.1007/978-3-030-66093-2_35

makes exchange of information and access to information much easier, even for those who do not have any right for that.

That caused the change of role of IT security, as well as the legal regulation of objects connected with the digital technologies.

"Today, the fate of the enterprise depends on a functioning and secure third party access to IT, from sales to financial control, to purchasing or manufacturing; hence, it is clear that IT security has become a "top management issue" [1, p. 132].

The digitization of production processes causes many problems for the companies the most serious of which is the necessity to create an effective protection of IT infrastructure.

In particular, German law provides the following liability for unauthorized access to the contents and terms of commercial contracts between the parties resulting in any damage to one of the Contracting parties ("any damage ») – "As a result of neglect of it security obligations (responsibility) should be attributed to the negligent partner" [1, p. 132].

In the context of increasing digitalization, various aspects of corporate activities are subjected to more and more active and diverse criminal attacks by third parties. These attacks can cause serious problems for the corporations themselves, as well as their partners, contractors, and consumers (clients).

2 Theory

These studies are to lay the foundation for such currently actively developing issues as big data processing in industrial processes, the development of cognitive decision-making algorithms and ensuring the stability of production processes in cyber-physical intelligent production systems [2], and the problems of implementing innovations based on big data in production based on cyber-physical systems [3].

The problems of recognizing the rights on the results of intellectual activity of various subjects, if the discovery is made with the help of artificial intelligence, are becoming more important. In particular, in the works of Japanese scientists from the Department of Master of Intellectual Property, Tokyo University of Science Graduate School of Innovation Studies, it is examined, «who is the most appropriate inventor or patent owner of inventions interacted by Artificial Intelligence (AI)» [4, p. 772].

In the works of A. Kartskhiya and D. Makarenko "legal points of artificial intelligence technologies usage risks seen from the perspective of a new technological environment formation, digital rights and digital assets of new entity and legal regulation" are analized. The authors "proceed from the substantiate the need to create a legal framework for the practical usage and application of artificial intelligence technologies, propose the construction of a comprehensive model of legal regulation, including the formation of universal standards and rules for AI technologies utilization in legal turnover and digital virtual technological environment" [5, p. 20]

Modern researches are also aimed to understand the numerous and diverse changes that occur in the process of implementing Industry 4.0 in all spheres of life and activities [6–8].

«We are witness to big changes in all industry branches, with new business methods, product system transformation, consumption, delivery and transportation emerging, owing to the implementation of new technological discoveries that include: robotics & automation, Internet of Things (IoT), 3D printers, smart sensors, Radio Frequency Identification (RFID,) etc.» [9, p. 3].

However, it is noted that «on the other hand, there are many challenges in the coming period, such as issues of changing business paradigms, legal issues, resource planning, security issues, standardization issues, and many other» [9, p. 3].

The second important direction of research and working out practical recommendations in the sphere of science-education-production that relate to the innovative development of the economy and society is the question of ways and tools to activate innovation in industrial enterprises, create legal instruments and ways to stimulate innovation [10, 11]. In the research literature on the implementation of Industry 4.0, it is noted that «innovation is a major strategic issue for many organizations» [12, p. 326].

One of the ways to stimulate industrial innovation is to create a system of free transfer of information and intellectual activity results between production, education and scientific organizations, supported by a reliable system of patent law [13]. "To stimulate innovative development and expand the implementation of intellectual property results in real production, it is necessary to develop and implement a number of measures that provide increased participation in the management of intellectual property objects by private enterprises, as well as the active development of patent services at enterprises and universities » [14, p. 67].

However, this is not enough.

Today, more and more authors point down that it is necessary to create a conceptual understanding of the ongoing changes and to work out a new paradigm of innovation based on it. The researchers note that «The fourth industrial revolution or Industry 4.0 is already present around us, but the concept itself is not widespread» [15, p. 3]. "One question, in particular, has often been raised: is cloud-based design and manufacturing actually a new paradigm, or is it just "old wine in new bottles"? [16, p. 1]. At the same time «the overall conclusion emerging from the detailed analysis is that a focus on IT-enabled business innovation represents a new paradigm in terms of the mindset and approach involved. It amounts to a third IT paradigm, different from previous paradigms of technology implementation and a planned approach to benefits realization. Further work is required to explore this paradigm and how organizations can make the 'paradigm shift' required» [12, p. 326].

A new object that recently appeared and requires effective legal protection is Digital Twin [17, 18].

Though the official definition of Digital Twin does not exist it is possible to define it as a kind of a digital copy of any object (machine, system, process) that helps to analyze any possible problems, caused by its use and find the best solutions. The research team at the US Army Research Laboratory believes that digital twins and digital manufacturing arose from «the need to optimize designs for manufacturing, to decrease the time required to go from design to first production, and to reduce expensive downstream changes» [19, p. 3]. It is also believed that the digital twin can replace innovation. It also provides variability and increase of competitiveness of an enterprise by giving the opportunity to minimize the risk of insufficient innovations implementation [20].

3 Results of the Study

The changing of data exchange procedure between different parties in the value chain causes serious problems in the sphere of industry, innovation, and legal protection of intellectual property in digital age. The researchers consider that «developments in digital technology and manufacturing processes have expanded the horizon of designer innovation in creating products.... These collaborations are extending beyond enterprise and national boundaries, contributing to a growing concern among designers regarding the security of their sensitive information such as intellectual property (IP) and trade secrets..., this fear can inhibit participation in a collaboration even though it might result in better products or services» [21, p. 1].

The analysis of the main directions and sources of confidential information leakage related to the use of digital technologies makes it possible to reveal the two main directions of such leakage:

1. leakage of confidential information can be triggered by external attacks on the enterprise's IT network: "The increasing digitalization exposes corporations to criminal attacks by third parties which can harm corporations severely" [1, p. 132]. These can be targeted hacker attacks or accidental "releases" of commercial information to the external information environment as a result of a system failure [22];
2. confidential information may be leaked internally through the fault of a dissatisfied employee, in the case of industrial espionage or through the negligence of employees who have official access to confidential information.

Two different types and directions of confidential information leakage correspondingly define different tools and methods for its protection [23].

To protect commercial confidential information the following methods can be applied depending on the type of access to information.

External threats are traditionally protected by creating individual systems and corporate infrastructure for protecting information and data, including innovative developments. These aspects are actively studied and covered in modern literature. For example, this is "various security practices in digital domains, such as encryption and secret sharing, as well as manufacturing domains, such as physically inaccessible function (PUF) and physical part watermarking for anticounterfeiting and tamper evidence purposes" [21, p. 1].

Internal threats are prevented both by creating an IT security system, and by implementing legal regulation tools within the framework of corporate law.

IT security experts note the need to apply different methods to ensure digital protection of data and information for different stages of product development and different stages of innovation [21, p. 1].

4 Digital Twin

The concept of Digital Twin has become an essential part of the 4th Industrial Revolution [23].

Digital Twin helps to reveal the problems in production and give reliable business forecast. The development of digital technologies made it possible to combine information technologies with operational progresses. Timothy Westa and Mark Blackburnb provided a good description of the nature of the digital twin as "[a]n extensible, configurable and component enterprise-level analytical framework that seamlessly expedites the controlled interplay of authoritative technical data, software, information, and knowledge in the enterprise data information-knowledge systems, based on the Digital System Model [DSM] template, to inform decision makers throughout a system's life cycle by providing the capability to access, integrate and transform disparate data into actionable information" [24, p. 50].

Digital Twin is a computer model, a digital copy of invention, an utility model, industrial design.

Digital Twins are not patented (registered). There is no any special form for protecting such objects. That is why they are generally protected as commercial secrets or know how.

The company has to create a special strategy for protecting intellectual property including commercial secrets. Such commercial confidential information has a high potential value and can provide competitive advantages (see Fig. 1).

In order to protect Digital Twins as commercial secrets companies have to work out a number of local documents, local acts which are aimed to regulate the use of commercial confidential information. In fact, commercial secrets are not patented or registered. In order to provide confidentiality of such information the company has to take some technical and legal measures.

Source: compiled by the authors

Fig. 1. The features commercial confidential information

Technical measures include providing information technology security and the actions of special services to reveal and neutralize possible internal and external threats. Applying only technical measures can not guaranty the safety of information and will be regarded by the court as insufficient.

Legal measures are aimed to create a legal basis for possible access and use of commercial secrets.

The right holder has to:

1) Create a list of kinds of information which refer to commercial secrets.
2) Work out a special instruction how to use commercial secrets.
3) Include in the labour contract special terms concerning access and use of commercial confidential information and liability for unfair use of such kinds of information. All the employees have to sign all the necessary documents and labour contracts containing the rules of use of such commercial secrets.

All the documents, containing commercial secrets must be marked «CS» and all the employees have to be warned not to read such documents without a special permission.

Some companies even create special rules, prohibiting employees to meet and communicate with any employees from the competing enterprises in order to avoid any accidental disclosure of confidential information.

Specialists, engineers, who have access to the commercial secrets, are not allowed to take part in different conferences and make public presentations of the results of their research work.

The Civil Code of the Russian Federation (part 4) regulates the relations in the sphere of use and protection of intellectual property. Chapter 75 of the Civil Code is devoted to commercial secrets protection. Commercial confidential information is also protected by the Law "On commercial secrets" № 98-FL.

In these laws the term "trade secrets" is used as a synonym of know-how. Such legal approach contradicts the international practice where the term know-how is used to define technical secrets and unpatented inventions and utility models.

Commercial secret is regarded as a special regime of limited access to know-how and trade secrets as well as any commercial confidential information set (established) by the right holder.

The regime described is aimed to protect valuable commercial information and know-how and prevent unfair use of the results of research work conducted by the legal entities and natural persons.

The Civil Code of the Russian Federation and the Law «On Commercial secrets» formulated the basic principles of commercial confidential information protection.

In order to protect commercial confidential information any enterprise has to establish a special secret regime, limit the access to the research results, create special local acts, regulating the use and transfer of such information, take some technical protection measures.

At the same time any enterprise can take some measures stimulating the employees loyalty to protect trade secrets and know-how. The company can take some measures to create good psychological climate within the enterprise, provide some extra payments (rewards) to loyal employees, install special social guarantees for the employees (see Fig. 2).

At the same time it is recommended the terms about special contractual liability of the employees in case of misuse or disclosure of commercial confidential information

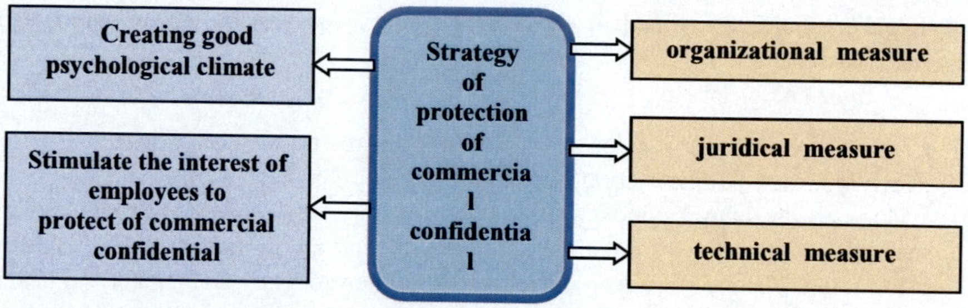

Source: compiled by the authors

Fig. 2. Strategy of protection of commercial confidential information

to be included in the employment contracts. That is why it is important to create a special system of intellectual property protection within the framework of the company.

It is important to say that by signing special agreements or implementing special rules concerning access and use of commercial confidential information a company creates new juridical facts that can change legal relations. These legal facts make foundations of intellectual property rights protection within the company.

5 Conclusions

Digital exchange has become the main form of business communication, especially in the sphere of sales (sales and supply), product and technological research, innovation, and industrial design.

The digitalization has caused many problems for the enterprises, including the necessity of creating an effective protection of IT infrastructure, protection of the commercial confidential information from external and internal threats.

A new object appeared (Digital Twin) that deserves an effective protection because it can provide variability and increase of competitiveness of an enterprise by giving the opportunity to minimize risk of insufficient implementation of innovations. Digital Twins are protected by means of intellectual property law as commercial confidential information. The protection of such information requires creating a special system of protection within the framework of the company. This system includes technical, legal and psychological measures, stimulating the loyal employees and working out a special program of intellectual property rights protection.

Acknowledgment. The work is a part of base portion of the state task of the Ministry of Education and Science of the Russian Federation to Perm National Research Polytechnic University (topic № FSNM-2020-0026).

References

1. Spindler, G.: Digitalization and corporate law – a view from Germany. Eur. Company Finan. Law Rev. **16**(1–2), 106–148 (2019)
2. Watson, A.: Industrial big data, cognitive decision-making algorithms, and sustainable production processes in cyber-physical smart manufacturing systems. J. Self-Governance Manag. Econ. **8**(2), 14–20 (2020)
3. Connolly-Barker, M., Gregova, E., Dengov, V.V., Podhorska, I.: Internet of things sensing networks, deep learning-enabled smart process planning, and big data-driven innovation in cyber-physical system-based manufacturing. Economics, Management, and Financial Markets **15**(2), 23–29 (2020)
4. Saito, A., Kobayashi, K., Yuasa, R., Hiratsuka, M.: Entity of inventions interacted by artificial intelligence and the protection by patent law. IEEJ Trans. Electron. Inf. Syst. **137** (6), 772–778 (2017)
5. Kartskhiya, A., Makarenko, D.: Status and risks of artificial intelligence: legal aspects. In: CEUR Workshop Proceedings, 10th Anniversary International Scientific and Technical Conference on Secure Information Technologies, BIT 2019, paper# 160627 (2019)
6. Haskins, P., Luţan, A.G., Tănase, A.: Sustainable value creation networks, digitized mass production, and networked information-driven technologies in industry 4.0-based manufacturing systems. Econ. Manag. Finan. Mark. **15**(2), 37–43 (2020)
7. Mingaleva, Z., Deputatova, L., Starkov, Y.: Management of organizational knowledge as a basis for the competitiveness of enterprises in the digital economy. In: Antipova, T. (ed.) ICIS 2019. LNNS, vol. 78, pp. 203–212. Springer, Cham (2020). https://doi.org/10.1007/978-3-030-22493-6_18
8. Melnik, M., Antipova, T.: Organizational aspects of digital economics management. In: Antipova, T. (ed.) ICIS 2019. LNNS, vol. 78, pp. 148–162. Springer, Cham (2020). https://doi.org/10.1007/978-3-030-22493-6_14
9. Liese, H., Stjepandic, J., Rulhoff, S.: Securing product know-how by embedding IP-protection into the organization. In: 2010 IEEE International Technology Management Conference, ICE 2010, paper# 121900 (2010)
10. Mingaleva, Z., Mirskikh, I.: Small innovative enterprise: the problems of protection of commercial confidential information and know-how, Middle East Journal of Scientific Research 13 (SPLISSUE), pp. 97–101 (2013)
11. Mingaleva, Zh., Danilina, H.: Significance of technological innovations for an increase of competitiveness of industrial companies. Life Sc. J. **11**(SPEC. ISSUE 8), 211–215 (2014)
12. Ashurst, C., Freer, A., Ekdahl, J., Gibbons, C.: Exploring IT-enabled innovation: a new paradigm? Int. J. Inf. Manag. **32**(4), 326–336 (2012)
13. Mirskikh, IYu., Mingaleva, Z.: Legal regulation and protection of intellectual property rights in russia and abroad. Vestnik Permskogo Universiteta-Naucnyj Zurnal **3**, 62–70 (2015)
14. Mingaleva, Z., Mirskikh, I.: The problems of legal regulation and protection of intellectual property. Procedia Soc. Behav. Sci. **81**, 329–333 (2013). https://doi.org/10.1016/j.sbspro.2013.06.437
15. Karabegović, I., Karabegović, E., Mahmić, M., Husak, E.: Implementation of industry 4.0 and industrial robots in the manufacturing processes. In: Karabegović, I. (ed.) NT 2019. LNNS, vol. 76, pp. 3–14. Springer, Cham (2020). https://doi.org/10.1007/978-3-030-18072-0_1
16. Wu, D., Rosen, D.W., Wang, L., Schaefer, D.: Cloud-based design and manufacturing: a new paradigm in digital manufacturing and design innovation. Comput. Aid. Des. **59**, 1–14 (2015)

17. Wright, L., Davidson, S.: How to tell the difference between a model and a digital twin. Adv. Model. Simul. Eng. Sci. **7**(1), paper# 13 (2020)
18. Alexopoulos, K., Nikolakis, N., Chryssolouris, G.: Digital twin-driven supervised machine learning for the development of artificial intelligence applications in manufacturing. Int. J. Comput. Integr. Manuf. **33**(5), 429–439 (2020)
19. Lubell, J., Frechette, S.P., Lipman, R.R., Proctor, F.M., Horst, J.A., Carlisle, M., Huang, P. J.: Model-Based Enterprise Summit Report.US Army Research Laboratory, Aberdeen Proving Ground, MD, ARL-RP- 466, NIST Technical Note 1820, 151 p. (2014)
20. Kurcz, B., Paizis, A.: Company law connecting factors and the digital age - a new Outlook. Eur. Company Finan. Law Rev. **16**(4), 434–456 (2019)
21. Chaduvula, S.C., Dachowicz, A., Atallah, M.J., Panchal, J.H.: Security in cyber-enabled design and manufacturing: a survey. J. Comput. Inf. Sci. Eng. **18**(4), paper# 040802 (2018)
22. Mingaleva, A.: Reforming Russian legislation for crimes in the digital economy. J. Dig. Sci. **1**(1), 43–50 (2019). https://doi.org/10.33847/2686-8296.1.1_5
23. Maungwa, T., Fourie, I.: Competitive intelligence failures: An information behaviour lens to key intelligence and information needs. Aslib J. Inf. Manag. **70**(4), 367–389 (2018)
24. Westa, T.D., Blackburnb, M.: Is digital thread/digital twin affordable? A systemic assessment of the cost of DoD's latest manhattan project. Procedia Computer Science **114**, 47–56 (2017)

Management Technologies and Systems

Unraveling the Secret to Success: Investigating the Link Between a Business Strategy Model and Organizational Performance

Sunet Eybers(✉)

University of Pretoria, Private Bag X20, Hatfield 0028, South Africa
sunet.eybers@up.ac.za

Abstract. Despite economic challenging times, a small insurance company (when compared to current industry leaders in developing countries) has stunned competitors by achieving extraordinary organizational success in a relative short time span. The organization is rapidly expanding their current market share and are winning awards for it. This achievement is quite noteworthy considering that prospective competitors in the insurance industry are often challenged by numerous market barriers in developing countries. These barriers include an existing saturated market with many big competitors offering insurance products at cut-throat prices. Adding to the challenge is the market's strict compliance regulation prescriptions. This sparked the interest to investigate how this organization has achieved their success.

A case study approach was used to investigate the extent to which the eight factors of a business strategy development approach – the Eight S model as proposed by Higgins [1] – was used and subsequently contributed to the overall success of the organization.

After numerous observations, the analysis of secondary documentation and open-ended interviews across different levels of the organization, the main finding was that the seven contextual factors of the Eight S model proposed by Higgins [1] namely – strategy and purpose, structure, systems and processes, management style, staff, shared values contributing to strategic performance – were to some extent implemented in the organization. However, the main contributor towards achieving their success was the re-Sources (resources) factor. The strong focus on the re-Sources factor highlighted well established IT systems which included a strong focus on data analytics, in particular a management dashboard for monitoring organizational performance.

Keywords: Organizational strategy · Data analytics · Higgins Eight S model

1 Introduction

The prospects of the current status of the global economy does not portray a rosy picture. Although an improvement in world-wide economic conditions are expected, local economic conditions in a developing country such as South Africa is expected to deteriorate even further given the current global COVID-19 pandemic. The recent

T. Antipova (Ed.): ICCS 2020, LNNS 186, pp. 373–380, 2021.
https://doi.org/10.1007/978-3-030-66093-2_36

downgrade of markets to 'junk-status' are scaring off investors and the country is officially experiencing an economic recession.

Despite economic challenging times, new organizations do emerge even though in the minority, and some of these organizations enter very competitive sectors such as the insurance sector. Strict regulatory requirements and a saturated market (to name a few) contributes to the competitiveness of this sector.

One of the very few examples of a recent entrant into the local South African short-term insurance industry who achieved success is organization X. What is even more remarkable about the success of organization X, is the market share obtained in a relatively short period of time and the continuous growth and footprint in the sector. Upon closer inspection it became apparent that organization X not only has unique organizational characteristics but also implemented elements of seven of the eight factors of the Higgins Eight S model namely strategy and purpose, structure, systems and processes, style, staff and shared values contribute to strategic performance. The eight factor, the re-Sources factor was identified as playing a vital role in their success.

The objective of this paper is to qualitatively investigate *the extent to which the eight factors of Higgins Eight S model was followed and therefore contributed to the overall organizational success*. An in-depth case study approach was adopted and included interviews across different functional levels, on-site observations and the evaluation of secondary documentation.

The paper is structured as follow: section two start with a brief introduction into business strategy models and then describes and compares the Seven S model and the Eight S model by Higgins [1]. Literature is presented on similar applications of the Eight S followed by an in-depth case study used to discuss the findings of the study. Finally, the conclusion section provides a summary of the findings and offer suggestions for future research opportunities.

2 Business Strategy Models

Business Strategy Models are popular tools used for many purposes such as developing organizational strategy; in instances where organizations wants to investigate, with the objective of improving organizational performance; to re-align the organization with new strategies or opportunities; or to examine current internal factors and interactive components of successful organizations [1, 2]. Some of these models have been around for many years, including popular models and frameworks such as the Porter Five Forces model [3], the Balanced Scorecard [4], and the McKinsey Seven S model (to name a few) [5–7]. In fact, 81 different models have been identified over the past decades until the 2010's [8, 9].

Although one can argue distinct strengths and weaknesses of various of the business strategic models the authors of this paper acknowledge this but argue that no model is perceived as being more superior to the other. In fact, as indicated by Ward and Rivani [2], Taylor [10] and Ikenedum-Dike [11], the importance of any business strategy model is in the execution or implementation of the strategy and not on which business strategy model is used. Also, for the purpose of this study, the Higgins Eight S model was selected as a logical option given the current organizational setup. The

McKinsey 7-S model, on which the Higgins Eight S model is based, is a popular model adopted extensively globally. Furthermore, the model focus on both human resources related issues (also referred to as the 'soft' issues such as skills, shared values, style and staff) and the interrelationship with the hard issues (such as strategy, structure and systems). Finally, although the organization under investigation might be hugely successful now, future adjustments might be required to sustain the anticipated future growth rate. The result of this investigation can therefore contribute to the organization's future expansion objective.

2.1 McKinsey Seven S Model and Higgins Eight S Model

The Eight s model was introduced by Higgins in 2005 [12], based on the well-known Seven S model introduced by management consulting firm McKinsey in the 1980's [5–7]. The original McKinsey Seven S model contained seven factors, each starting with the letter 'S'. The model postulated that organizational strategy is not the only determining factor to achieve organizational success. Rather, other factors, such as organizational structure, systems, management style, the skills of staff members as well as staff and their shared values play a key role in increasing organizational performance. In 2005 Higgins [12] proposed a revised version of the model and introduced eight factors, similar to the seven postulated by the McKinsey Seven S model, namely strategy and purpose, structure, systems and processes, style, staff, re-Sources (resources) and shared values contributing to strategic performance [12].

The skills factor contained in the McKinsey Seven S model was replaced with the re-Sources factor in the Higgins Eight S model. Whilst skills referred to the know-how of staff to perform job functions, resources refer to the ability of an organization to implement the strategy, focusing on the skills of staff, the financial ability or resources and technology, including management information systems [1, 12]. This factor seemed more relevant for current organizations and one of the main reasons for the adoption of the Eight S model over the Seven S model.

The addition of the Strategy Performance factor by Higgins [12] (Eight S model) was included to ensure that the organization focus on the goal of achieving it's objectives, measured in the actual results or organizational output. The remainder or the factors are similar to the original McKinsey model, and are described below:

- The Strategy factor considers the influence of the external environment on the organization and its planned direction and refers to the execution plan for the organization to achieve its objectives.
- The Organizational Structure refers to the organizational hierarchy specifying the business divisions and departments inside the organization.
- Systems and Processes (only referred to as Systems in the Seven S model) refers to formal organizational processes and procedures used by the organization prescribing how an organization should function.
- Shared Values refer to the mutual values and beliefs amongst employees which makes the organization unique and governs staff behavior.

- Staff refers to the workforce required to reach the organization's mission and vision. The factor focuses on the number of employees required, the training required and the background required for specific positions.
- Management Style refers to how managers lead and guide staff to achieve goals.

The Structure, Systems, Shared values, Staff and Style factors are referred to contextual or internal factors that influence how the organization is internally structured to achieve its strategy goal of sharing values. The contextual factors are further subdivided into organizational 'hard' factors namely Strategy, Structure, Systems; and 'soft' factors focusing on 'people characteristics' namely Skills, Staff, Style and Shared Values.

3 Case Study

The organization under investigation (hereafter referred to as Organization X) is a new, privately owned insurance company in South Africa. They entered a very competitive short-term insurance industry eight years ago employing only twenty employees. Today the organization offers employment to more than seven hundred employees based in their head office in Gauteng, the economic capital of the country.

The client base of Organization X has grown from 5,000 policy holders within three months of its establishment in 2012. In 2014 they reached another milestone of 100,000 policies sold. To date, more than 102,000 clients make them one of the fastest growing short-term insurance companies in the country.

The successful entrance to the market of Organization X has been recognized by the local South African Service Awards where they were awarded the "Best Short-Term Insurer" as well as "Best New Comer" (across all industries). In fact, they have received an astonishing 10 awards in 2019. In an interview with the General Manager of Data Analytics, the question was obvious: How does this organization achieve success despite challenging economic circumstances? The answer, according to him, was easy: a strong strategic drive within the organization supported by the introduction of metrics to track various aspects of the strategy. Other interviews conducted with key resources on middle management as well as multiple observations were further conducted to complete the study. Although the Eight S factors of Higgins was not explicitly mentioned during the interviews, the focus was on these factors as main theoretical lens for the purpose of this study.

Strategy and Purpose: The organizational strategy and purpose was not explicitly communicated during the interviews conducted. There was no organizational mission or vision statements anywhere in the organization visible. But despite this the all the employees new exactly what they wanted to achieve – to become one of the largest insurance providers in the country.

Structure: Due to the relatively small size of the organization (less than 80 employees with only a few (less than 5 per region) employees in other main provinces of the country the organization adopted a relatively centralized structure. The organizational hierarchy was relatively 'flat' with a top management structure defined, few middle managers followed by the rest of the employees. This was evident in the physical seating of the various managers, i.e. the top managers' (CIO and Departmental Managers for example) desks were part of the open plan seating area (with no special office space). This contributed to the unique culture of the organization described in more detail later.

Systems and Processes: Strong operational systems were implemented to support their relatively simplistic business processes. This include a sophisticated Risk Rating System that allowed agents to create a very accurate individual or business risk profile for clients. As a result, competitive insurance premiums could be presented to potential customers. The system also allowed for a continuous adjustment in premiums according to certain conditions. Daily targets for agents are well communicated and displayed. When agents reach their insurance sales targets, mechanisms are used to communicate it to the team. Such performances are therefore acknowledged and celebrated.

Re-Sources: Another key differentiator, according to the interviewees, was the organization's success in their adoption and standardization of technology products (such as Microsoft products) that is readily available. Being a young organization, no legacy systems exist and where possible, no customized systems were used. The technological model is therefore simplistic which enables the organization to quickly adapt to environmental changes.

Shared Values, Staff and Management Style: The organization has a strong focus on adopting a fun, relaxed culture marked by *laissez-faire* leadership. For example, employees are allowed to take breaks whenever needed, enjoy informal games in a play area and enjoy free popcorn during breaks. One of the managers indicated during the interview that this is one of the most important values of the organization when new recruits are considered. Irrespective of the skills level of potential employees, they should first fit into the culture. The culture is one of the most important aspects to the business as this is a vital part of their corporate identity. Recruitment drives are therefore focused not only on the correct people-technology fit, but also a strong cultural fit.

Satisfied, happy employees are so important to the organization that they have employed 'happiness' meters at various strategic places in office spaces. The logic behind these meters are that happy employees will transfer their sense of happiness to current and prospective customers. Employees may, at any time of the day, indicate their level of happiness (or satisfaction with the organization) by simply selecting three indicators: a happy face (☺), a neutral face (😐) or a sad face (☹) on physical 'happiness meters'. The level of happiness/unhappiness is recorded by the length of time the button is pressed. For example, ecstatic employees will hold down the button for a couple of seconds to indicate that they are extremely happy, whilst relatively happy employees will press the happy face button briefly. These responses are captured and available immediately to managers to take action. Weekly accumulation of these results are also displayed and analyzed on flat screen monitors mounted at strategic places in the organization. One of the main contributions to the successful implementation of these meters are the simplistic business process and technological architecture behind the devices based on readily available Microsoft technology.

According to one of the interviewees happy employees contributes directly to meeting, and in this case, achieving strategic performance objectives.

3.1 Key Findings

There are three key findings to the study:-

- Firstly, evidence of all eight elements of the Higgins Eight S model were evidence in the organization. However, some elements, such as the focus on the skills of staff, re-Sources in support of organizational process and a distinct management style had a huge influence on organizational culture which positively affect the organization.
- Secondly, each department has adopted a set of well-defined metrics. These metrics constitutes of individual metrics on the lowest level which feed into a higher level departmental set of metrics. For example, individual targets are set for call center agents. Agents should sell at least three insurance policies per day. Once a policy has been sold by an agent, a LED bar above each call center agent's chair light up to indicate progress made towards achieving the target of three policies. Once all three LED bars are lit the call center agent has achieve its target. These LED bars are visible to fellow colleagues. It therefore motivates fellow colleagues to achieve the same result. The individual metrics contribute to the overall departmental metrics. Progress towards achieving departmental metrics are visible using dashboards displayed on screens feeding from real-time data. Therefore the organization has invested heavily in reporting and data analytics systems to communicate the progress made in achieving targets.
- Thirdly the focus on the morale of employee was enormous. The organization was characterized by positive, friendly and happy employees. One of the interviewees stated that this is one of the most important aspects of their organization as this is the image of their organization they portray to current as well as prospective clients. They are a young organization with limited funds available for marketing – hence they invest in their employees to promote their products through their interaction with clients. It was therefore important to the organization to identify and address unhappy employees as a matter of urgency. The utilization of physical employee satisfaction measurement devices were key towards the continuous monitoring of employee morale. This is an important aspect linked to organizational performance as also highlighted by a study conducted focusing on a developing country such as Kenya, where the authors found the strategic leadership and human resources to be statistically significant linked to organizational performance [13]. Interesting, although organizational culture and organizational structure were identified as contributing factors to organizational performance, these were not statistically significant.

A study conducted by Msabeni [14] found that management competency, the skills and education of staff and subsequent performance management as well as the importance of organizational cultured played a vital role in the performance of insurance organizations in Kenya[1]. Noteworthy is that all these factors were identified in the case study and prevailed in the organization under investigation.

4 Conclusion

The eight elements of the Higgins Eight S model proved to be a good indicator of how Organization X achieved success in a relative short time period. It also indicates that all factors should be present in successful organizations, although early studies indicate that some factors might contribute more towards achieving organizational success.

The study has highlighted the importance of data analytics in organizations that can, if implemented correctly, contribute to a rich organizational culture and the achievement of organizational success. It will be interesting to compare the result of the study with similar cases in the insurance study of organizations that are also successful or even less successful. A more in-depth study into the data analytics adoption in the organization can also be of value.

The final recommendation for organizations who wish to be successful in a highly competitive environment is to adopt a balanced approach to their business. Although one aspect of the business was highlighted – namely resources – the other elements contributed to the success of the organization. It can therefore not be neglected.

References

1. Bhatti, O.K.: Strategy implementation: an alternative choice of 8S'S. Ann. Manag. Res. **1**, 52–59 (2011)
2. Ward, D., Rivani, E.: An overview of strategy development models and the Ward-Rivani Model (2005)
3. Porter, M.E.: How competitive forces shape strategy. Harv. Bus. Rev. **57**, 137–145 (1979)
4. Kaplan, R.S., Norton, D.P.: The balanced scorecard - measures that drive performance. Harv. Bus. Rev. **83**, 172 (1992)
5. Waterman Jr., R., Peters, T., Phillips, J.R.: Structure is not organization. Bus. Horiz. **23**, 14–26 (1980)
6. Pascale, R., Athos, A.: The Art of Japanese Management. Penguin Books, London (1981)
7. Peters, T.J., Waterman, H.: In Search of Excellence. Harper and Row, New York (1982)
8. Ghemawat, P.: Competition and Business Strategy in Historical Perspective. Bus. Hist. Rev. **76** (2002)
9. Freedman, L.: Strategy: A History. University Press, Oxford (2015)
10. Taylor, A.: Business Strategy Models - Which one for your business? https://www.smestrategy.net/blog/business-strategy-model-which-one-for-you. Accessed 23 June 2020
11. Ikenedum-Dike, O.M.: Strategy as configuration: strategy structure, micro foundations of capability configurations and their effects on execution gaps under volatility (2017)
12. Higgins, J.M.: The Eight 'S's of Successful Strategy Execution. J. Change Manag. **5**, 3–13 (2005)

[1] The study did not explicitly claim to adopt the Higgins Eight S model, but did mentioned it briefly.

13. Kilile, F., Munga, J., Were, E.: The influence of strategy implementation on organizational performance of public sector organisations in the Ministry of Tourism in Kenya. Int. J. Bus. Manag. Finance. **2** (2018)
14. Msabeni, C.M.: Analysis of growth strategy choice and the performance insurance firms: a study of Nairobi Central Business District (2019)

Logistic Performance in Mexican SMEs

Abel García-Villagrán(✉), Patricia Cano-Olivos, José Luis Martínez-Flores, and Diana Sánchez-Partida

Universidad Popular Autónoma Del Estado de Puebla, 72410 Puebla, Mexico
abel.garcia@upaep.edu.mx, {patricia.cano, joseluis.martinez01, diana.sanchez}@upaep.mx

Abstract. Logistics plays a crucial role in all Small and Medium-sized Enterprises as a fundamental factor of competitiveness. Many manufacturing SMEs are experiencing new ways to manage their supply chains as a result of changes that are being experienced globally and that directly impact their operations. The purpose of this paper is to understand the logistics structure of manufacturing SMEs in México, to design better tools that can be applied and to increase their competitiveness. This study explores with a survey a crucial data collected from logistics managerial staff in Manufacturing SMEs of the metropolitan area of Puebla, Mexico. A sample of SMEs was selected using simple random sampling, seeking to reflect the characteristics of the companies under study accurately. Mexico has the full potential to become one of the leading players, but, in the logistics field, its performance has not been among the best. The lack of resilience plans in SMEs in the face of any disruption to the supply chain represents a critical aspect. SMEs due to their size they do not allow the incorporation of exclusive personnel in each area. It becomes difficult to understand how they work to influence and improve processes because they operate without logistic indicators. This paper examines logistics and supply chain management in SMEs seeking to present the reality of these companies and find areas in which they can be supported to improve their performance.

Keywords: SMEs · Logistic · Supply chain

1 Introduction

Small and medium enterprises (SMEs) are essential for the economic growth, innovation and development of countries around the world. For Latin America, SMEs are a fundamental component of existing companies, their importance is manifested both in the participation by the number of companies and the creation of jobs, they have become the central part of increasing business growth, these represent the 99% of the total economic units in the region and generate 67% of jobs. However, the contribution to GDP and its overall productivity are low (OCDE 2012). According to (CEPAL 2012), SMEs only reach between 16 and 36% of the productivity of large companies. For the World Bank (2019), these companies play an essential role in most economies, particularly in developing countries, contributing 60% of total employment and 40% of Gross Domestic Product (GDP) in emerging economies. They support and improve the quality of life of the world population. The main characteristic that limits the growth of

T. Antipova (Ed.): ICCS 2020, LNNS 186, pp. 381–390, 2021.
https://doi.org/10.1007/978-3-030-66093-2_37

most SMEs is that they are family businesses and the administration is carried out by the owners, although they do not have the appropriate knowledge for decision making, and the profits are used to meet the needs of the family; usually, no qualified personnel is hired not to pay competitive salaries, the quality of the products has deficiencies because the controls are minimal or nonexistent.

SMEs are commercial, industrial or other companies that have a small number of workers and with moderate-income (ProMéxico 2014), directed by the owners in a personalized and autonomous way (Cardozo et al. 2012), and that with organization and coordination of the factors (capital and labour), they seek to obtain a benefit (Andersen et al. 1999). Globalization does not exclude these companies to participate in international markets. Currently, the competition is between supply chains and SMEs have become an indispensable part, so they must offer products at lower prices without neglecting quality, this will increase the competitiveness of the supply chain (Mishra 2016). Globalization has forced SMEs to change business models by adopting new innovative capabilities. It is necessary for SMEs to be continuously updated, periodically modifying their strategies in the areas of production, promotion, marketing, distribution, financing and can satisfy their customers by preventing them from leaving with their competitors (Martínez and Alvarado 2013).

2 SMEs in Mexico

For Mexico, SMEs are the backbone of the country's economy, since they contribute 52% of GDP and also contribute to the generation of employment with 78.5%, this information being relevant if we consider that they represent 99.8% of the total economic units (García 2014). According to the National Survey on Productivity and Competitiveness of Micro, Small and Medium Enterprises, the number of SMEs in Mexico is shown in (Fig. 1):

Size	Company		Employment	Size by number of workers and economic sector		
	Number of SMEs	(%)	Participation (%)	Industry	Comerce	Services
Micro	3,952,422	97.6	75.4	1 – 10	1 – 10	1 – 10
Small	79,367	2.0	13.5	11 – 50	11 – 30	11 – 50
Medium	16,754	0.4	11.1	51 – 250	31 – 100	51 – 100
Total	**4,048,543**	**100**	**100**			

Fig. 1. Employment generation and SMEs classification in Mexico (INEGI 2016a, b).

This information allows governments to design strategies that boost their growth and plan the necessary public policies and generate financing programs to improve these companies. The strengths of these economic units must be used to make a competitive advantage for the nation to attract new investments and strengthen products both inside and outside the country (Salazar and Mancera 2017). In Mexico, SMEs have had several obstacles, including poor organization, financial restrictions that prevent them from having profitable projects and diminish the possibilities of growth. Besides, the lack of leadership of the people who are directing these companies makes that do not have competitiveness (Heredia 2014). In the OCDE report (2012), it is mentioned that Mexican SMEs represent 72.3% of the country's employment sources. However, their productivity is low (OCDE 2013), since their production only constitutes 18% of the Mexican output, in contrast to 40% of employment.

It was in the eighties of the last century when companies were organized into four primary segments: micro, small, medium and large, where the first three were classified into a single category that was declared as the priority "sector" of the government action (Palacios and José 2013). Between 2001 and 2005, the Law for the Development of Competitiveness for micro, small and medium enterprises emerged (Ferraro and Stumpo 2010). For 2007, Mexico promoted entrepreneurship through nuclear programs: New National Entrepreneurs Program, National Microenterprise Program, National SME Program (SME Fund), National Gazelle Business Program. In 2009, the Mexico Entrepreneur program was designed to promote viable projects that encouraged job creation (Saavedra and Tapia 2012). In 2013, the National Entrepreneur Institute (DOF 2013) and the Innovative Development Sector Program 2013–2018 was created to support SMEs in financial advice, exports, technological innovation, training and consulting, logistics processes, among others (García et al. 2017). However, most of these programs and strategies have failed to land because public officials see these programs like the opportunity to do their own business without solving the real problem (Roldan 2004). Today, as part of the National Development Plan 2019–2024, the Mexican government will promote the creation of jobs by facilitating access to credit to small and medium enterprises through the reduction and simplification of requirements.

It should be noted that only 1.5% of all SMEs in the country have had access to financing sources (Saavedra and Tapia 2012). This situation makes these economic units, in which the majority of the population lives, more critical each year. That is why the high mortality of these companies is another indicator of the poor results (Palacios and José 2013). It is essential to mention that of every 100 SMEs that are formed, 90 of them do not reach two years of life, this is a very high index compared to industrialized countries since these only reach 40% of business mortality (Molina et al. 2014). Mexican SMEs share recurring characteristics such as mismanagement, lack of financial management, lack of strategic planning, lack of experience, unethical behaviours, financing problems, lack of business culture, poor access to the foreign market, little trained staff and deficiencies in production processes (Acosta et al. 2015). It is also important to note that the productivity of SMEs in the industrial sector is low because they use manual and artisanal processes in their production process, while large companies are highly technology (García 2014). SMEs are vulnerable because of their limited resources, which is why they must partner or build networks that allow them to obtain the necessary information at a lower cost. Most of the time these networks are informal (friends, social contacts, professionals) and become more structured and formalized as the company develops and structures its surveillance (Molina et al. 2014). That is why logistics plays a crucial role in all Small and Medium-sized Enterprises in Mexico as a fundamental factor of competitiveness. Every day, many manufacturing SMEs are experiencing new ways to manage their supply chains as a result of changes that are being experienced globally and that directly impact their operations.

3 Logistic Performance in Mexican SMEs

Logistical performance in international trade as well as in the domestic market is critical to economic growth. The competitiveness of countries with efficient logistics connects people and businesses with markets and opportunities, achieving higher levels of productivity and well-being. Unfortunately, the gap between poor and rich countries in the issue of logistical performance continues to exist (World Bank 2016). Although, Mexico has the full potential to become one of the leading players in international logistics for its geographical position, available talent, outstanding capabilities for product design, engineering and manufacturing that have led it to stand out from the other countries. In the logistics field, its performance has not been among the best. Progress has been made but at a lower rate than other nations. Logistics in the economy accounts for between 8% and 15% of countries' gross domestic product, so it is vital for productivity and the development of local markets (ProMéxico 2018). The logistical efficiency of a country or region is directly related to the development of infrastructure, the regulatory framework, as well as the quality and accessibility of logistics services (Kirby and Brosa 2011). The World Bank assesses every two years the logistical performance of 168 countries, through 6 indicators (Customs, Infrastructure, International shipments, Logistics services quality, Tracking & tracing and Timeliness) on a scale of 1 to 5 (where 1 is the least efficient level, and 5 is the most suitable level). Mexico ranks 51st with an average rating of 3.05 points (World Bank 2019). Mexico needs a logistical reform that allows creating a better tax policy, reduction in bureaucracy in customs, investment in infrastructure, boost by the government of companies specialized in logistics to have a better performance as a country. Competitiveness is an increasingly important aspect in companies, which stems from the demands of today's economic environment with the process of globalization.

It is normal for events that threaten to disrupt operations and endanger the supply chain, but the better the logistical capabilities, the greater resilience (Cotte 2018). Logistics services allow companies to perform better in the supply chain (Enríquez et al. 2015), from raw material extraction to consumption (Castorena 2017), due to the rapid delivery of the products to customers (Chase and Jacobs 2010). According to the Council of Supply Chain Management Professional (CSCMP 2019), logistics is the process of planning, execution and efficient control procedures in the transport and storage of goods. It includes services and related information from the point of origin to the point of consumption according to customer requirements. SMEs are immersed in globalization and must offer high-quality products with added value, to achieve this they need to increase their productivity, profitability and improve the quality of life of their workers (Cardozo et al. 2012). As a critical element for improving logistics in SMEs, it is vital to use logistics management indicators to diagnose and make decisions (Becerra 2015). On the other hand, collaboration, information exchange, supplier and customer engagement in decision-making will also help improve supply chain coordination (Kumar and Kumar 2017).

Manufacturing SMEs, because of the nature of their operational activities, show that managing supply chain management has a significant performance influence (Enríquez et al. 2015). SMEs need to build partnerships to support and control the supply of resources to the maximum. These companies can be part of a network of customers and suppliers linked to large companies as they understand the operation of the new dynamic of working in a "global chain of supplies" and with the "full package" model for certain products (Olivos et al. 2015). Based on current supply needs for globalization and dynamic market demand, appropriate information and communication technology systems must be implemented and used (Aguilera et al. 2015). Information systems such as Enterprise Resource Planning (ERP) are tools that organizations use to manage their data and resources. But in SMEs, it is necessary to carry out feasibility studies before considering purchasing an ERP (Rivera and Salazar 2013). The most common problem for SMEs is the control of inventories (Osorio 2013) the existence of excesses or lack of inventory, in addition to security inventories and their reordering points, are determined solely based on average demand, ignoring variability. SMEs put logistics activities in their marketing, production and management processes, due to their size they do not allow the incorporation of exclusive personnel in each area. The important thing is the integration in such a way that the coordinated management of the flows of goods and information is carried out (Ortiz and Izquierdo Ojeda 2012).

4 Research in Industrial SMEs of Puebla, México

The objective of this research is to represent a sample of the logistics structure of manufacturing SMEs in the Puebla-Tlaxcala metropolitan area. The following research methodology was used for this purpose. First, literature review. Second, selection and sample size. Third, survey design. Fourth, interpretation of results and conclusion.

4.1 Selection and Size of the Sample

With the information obtained from the literature review, research was carried out in some Manufacturing SMEs in the state of Puebla, Mexico. The results of the economic censuses (INEGI 2016a, b) showed that in the State of Puebla, 98.6% of the economic units correspond to the sectors: Manufacturing, Commerce and Non-Financial Services. These three sectors accounted for 94.1% of the staff occupied. In total gross production, manufacturing is of the utmost importance as it has 68.3% of the total. According to (García 2016) there are 950 Manufacturing SMEs in the Puebla-Tlaxcala metropolitan area. Considering the size of SMEs in the region, the population was limited to manufacturing SMEs with a recognized track record. Simple random sampling was used, this technique allows each member of the population to be selected as the subject of study, this permit to effectively determine that a sample accurately reflects the characteristics of the population studied (Rabolini 2009). To obtain information directly from companies, the information was available from the database of the University's linking department and the survey was applied to staff with some managerial position in the logistics activities of manufacturing SMEs of the Puebla-Tlaxcala metropolitan region. It is assumed that 100% of the SMEs that answered the survey have an acceptable level of the degree of performance of their activities. Questionnaires were submitted and interviews with participants were scheduled to deepen understanding of the results. After several days of the exercise with the companies, it was obtained as a result that only answer 21% of the companies, between October 1st and November 15th, 2019. The 32 manufacturing SMEs surveyed, 6% are micro-enterprises, 31% are small, and 63% are medium-sized enterprises, which have an average of 101 workers. The industrial activity of SMEs in the studio are: Automotive (8), Plastic injection (5), Furniture (2), Food industry (4), Glass (2), Textile (3), Machinery and equipment (5) and Others (3).

4.2 Survey Design

The survey is about 36 questions from the strategic areas of the supply chain. The Likert scale is used for this research, where the number 1 represents the lowest value and the highest number 5. The results of the questionnaire show aspects of the strength and weaknesses of logistics processes in the SME supply chain (Fig. 2):

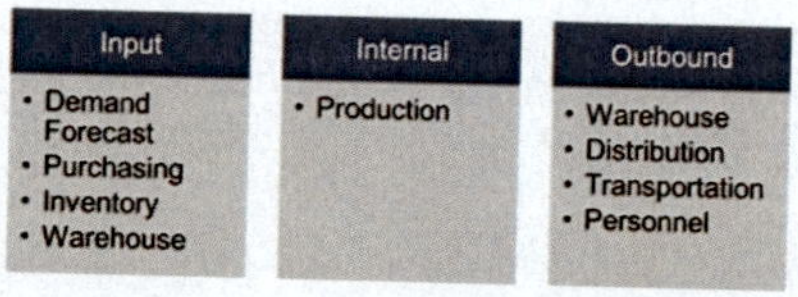

Fig. 2. Areas of logistics in SMEs survey. Source: Own Design

4.3 Interpretation of Results

SMEs are seeking to reduce costs in operations, without affecting the level of service and responding to consumer demand. Having a forecast with a degree of certainty plays an essential role in production programs to achieve the objectives set out (Narváez 2016). According to the results, 63% of SMEs make forecasts from a mathematical method to more accurately determine the demand for products, the rest of the companies makes it based on the experience of its staff.

Indicators allow measuring the performance of the different logistics processes in the supply chain, which is the basis for controlling the use of the resources used, monitoring the fulfilment of objectives and identification of areas of improvement (Zuluaga et al. 2014). According to the results of the companies surveyed, areas that lack indicators are production, purchasing, transport and distribution. It is common to find a lack of indicators that measure the performance of SMEs. The most affected area is purchasing. It is not possible to know how long it takes for a purchase order from the time it is made until the raw materials are received in the warehouse. Compliance to the procurement budget is also not evaluated so that expenses cannot be controlled correctly.

Inventory management must determine the quantity to store, how much to receive, record the required materials, and place orders. Effective inventory management ensures the quality of the products and the supply of the materials in the chain (Salazar and

Mancera 2017). In the 32 companies surveyed; 85% have a supervisor who manages and controls inventories, 72% define material locations through layout to make material handling efficiency, and 56% have inventory policies that support decision-making and reduce financial problems. Most of the organizations realize progress in the production processes to increase the reliability and in SMEs is the area that better controls its processes. This fact allows for improving the performance of the operations of production. The highest qualifications registered in the areas of production planning and scheduling, traceability and quality control. It should be noted that 56% of SMEs know the costs of each stage of the production process. Although the production area is strength in SMEs, only half of these companies have plans to build resilience to any cause or source of risk that compromises delivery times with customers.

Another important area is Distribution and Transport, in which 56% of SMEs do have loading and transport procedures. On the other hand, 54% of SMEs are interested in knowing the level of service they offer their customers, but only 65% of them make continuous improvement in processes. 65% of companies take out insurance that covers damages to finished product loads and operators. Finally, human capital is an essential factor for competitiveness in the field of knowledge. However, it is one of the least considered for the development of SMEs (Heredia 2014). The results show that employees in logistical roles are assigned according to the requirements of the position profile. Still, once they start working, only 57% of companies provide training to employees.

5 Conclusion

SME´s play a significant role in the economy of countries. Global processes force them to seek new strategies for their survival and development. These companies are deficient in several areas because of their lack of experience and financial resources. They have limitations for the implementation of strategies in the logistics area; it is necessary the support of different academic and governmental entities, as well as business pacts to help these companies to make a better management of their supply chain, increase their productivity and competitiveness (Becerra 2015). Logistics is one of the main affected areas in SMEs because it is present in all the company's operations. Without indicators, it becomes difficult to understand how they work to influence and improve processes. It is easy to realize that there is no sharing culture between SMEs and of these with large companies.

The lack of resilience plans in SMEs in the face of any disruption to the supply chain represents a critical aspect that needs to be expanded and studied in future research to support them and provide them with tools to help them survive in the face of unexpected events. These investigations should include components such as the company's adaptive capacity, vulnerability, probability of event occurrence, time and level of recovery in addition to the company's competitive position (Sanchis and Poler 2011). Knowing the logistical obstacles of manufacturing SMEs in Mexico helps to design better tools that can be applied and that support to increase their competitiveness. The logistics areas with opportunities identified in the research are the following (Fig. 3):

Logistics area	Opportunities in SMEs
Demand Forecast	Demand forecasts are needed using quantitative methods and not just rely on workers' experience.
Purchasing	90% of SMEs use information technologies to carry out purchase orders. They do not measure the time it takes for these purchase orders to complete. Only half of them perform regular evaluations on their suppliers.
Inventory	Half of the companies have inventory policies; this lack makes decision-making complicated and can bring future problems by not knowing exactly how much to buy, or when to reorder goods.
Indicators	The lack of indicators that measure performance in logistical areas is common in SMEs.
Production	It is the area best evaluated, but in case there is any disruption due to reasons outside the company, there are no resilience plans. Future research is needed to learn how to support them and provide tools to help prevent internal and external risks.
Warehouse	SMEs warehouses have Layout. There is little information about logistical costs, and in most cases they are unknown.
Procedures	The lack of procedures in all Logistics areas is recurrent. The processes are not documented and the few that are documented are not properly complied with.
Transport and Distribution	Only 65% of companies have insurance for their transport units and warehouses. SMEs are not prepared for theft, accidents or natural disasters, and this can lead to poor service to customers.
Customer Service	Half of SMEs conduct surveys to measure customer service; unfortunately, only 65% use that information to make improvements to their logistics and production processes.
Personnel	SMEs provide little or no training to their workers. They have a lot of staff turnover, and new workers are not correctly inducing their jobs. With this constant change of workers, new procedures are redesigned, resulting in loss of valuable information.

Fig. 3. Opportunities in Mexican SMEs. Source: Own Design

SMEs are fundamental to the Mexican economy, many of them can compete with companies from other countries in the world. This strength must be used to have a nation that can attract investment and increase the presence of Mexican products abroad. But it cannot be achieved if the logistical opportunities found are not corrected. It is necessary that each link in the supply chains of SMEs be improved, from demand planning to delivery to the customer, avoiding any disruption. These companies must understand the potential that logistics represents, any improvement in their processes will have a direct impact on their performance.

References

Acosta, C.R., Maquitico, L.B., Lugo, N.A.O.: Agenda para estudio de las problemáticas de las MiPyMES Mexicanas. FACCEA **5**(2), 107–117 (2015)

Aguilera, L., Hernández, O., Pérez, O.: Relationship supply chain management competitiveness of manufacturing smes in aguascalientes. RIA&F **8**(1), 79–91 (2015)

Andersen, A.: Diccionario de Economía y Negocios. ESPASA, España (1999)
Becerra, Y.L.: Propuesta metodológica para la definición de estrategias de mejoramiento en logística de pymes. Ingeniería Investigación y Desarrollo **15**(1), 48–57 (2015)
Cardozo, E., Velásquez de Naime, Y., Monroy, R.: Definition of SMEs in Latin America. In: 6th International Conference on Industrial Engineering and Management, pp. 1345–1352 (2012)
Castorena, O.H.: Proveedores y modelos de gestión en la cadena de suministro: Pymes manufactureras de Aguascalientes (México). FACCEA **7**(1), 21–28 (2017)
CEPAL: Perspectivas económicas de América Latina 2013. Políticas de Pymes para el cambio estructural. París, Francia. No. 189 (2012)
Chase, R.B., Jacobs, F.R.: Administración de operaciones: producción y cadena de suministros. McGraw Hill, México (2010)
Cotte Urrutia, A.M.: Cadenas de suministro resilientes (2018). www.repository.unimilitar.edu.co
CSCMP: Council of supply chain management profesionals. CSCMP's Definition of Logis-tics Management. CSCMP (2019). https://cscmp.org
DOF: Diario Oficial de la Federación. Instituto Nacional del Emprendedor (2013) http://dof.gob.mx/nota_detalle.php?codigo=5295903&fecha=15/04/2013
Enríquez, L.A., Castorena, O.H., Adame, M.G.: La gestión de la Cadena de Suministro, vol. 10, no. 1, p. 516. Institute for Business & Finance Research (2015)
Ferraro, C., Stumpo, G.: Políticas de apoyo a las pymes en América Latina. CEPAL & Cooperazione Italiana, Santiago de Chile (2010)
García Caudillo, P.M.: Las Pymes en la zona Metropolitana de Puebla-Tlaxcala. Un análisis de los encadenamientos productivos de la industria de alimentos. BUAP, México (2016)
García, M.L.S.: La Pyme como generadora de empleo en México. Clío América **8**(16), 153–172 (2014)
García, J.I.U., Mendoza, J.J.C., Sánchez, P.R.: Innovación y Desarrollo Regional en México. Revista Global de Negocios **5**(7), 85–95 (2017)
Heredia, E.Á.: Las Pymes en México: desarrollo y competitividad. Observatorio de la Economía Latinoamericana, 201 (2014)
INEGI: Gobierno de México. Sitio web (2016a). https://www.gob.mx/se/prensa/se-difunden-estadisticas-detalladas-sobre-las-micro-pequenas-y-medianas-empresas-del-pais-46847
INEGI: Censos Económicos 2014. 3 dic 2019, INEGI Sitio web (2016b). https://www.inegi.org.mx/app/biblioteca/ficha.html?upc=702825082666
Kumar, R., Kumar Singh, R.: Coordination and responsiveness issues in SME supply chains: a review. Benchmark.: Int. J. **24**(3), 635–650 (2017)
Kirby, C., Brosa, N.: La logística como factor de competitividad de las Pymes en las Américas. Inter-American Development Bank (2011)
Martínez, M.M.A., Alvarado, K.I.M.: Las PYMES ante el proceso de la globalización. Observatorio de la Economía Latinoamericana **185**, 4–9 (2013)
Mishra, K.P.: Supply chain insights for small and medium enterprises (SMEs) in Malaysia, Doctoral dissertation (2016)
Molina Sánchez, R., López Salazar, A., Contreras Soto, R.: El emprendimiento y crecimiento de las Pymes. Acta Universitaria **24**(1), 59–72 (2014)
Narváez, A.J.V.: Logística de Aprovisionamiento para Micro, Pequeñas y Medianas Empresas (MIPYMES). Ingeniería, Innovación y Desarrollo Sostenible **1**(1), 7–22 (2016)
Olivos, P.C., Carrasco, F.O., Flores, J.L.M., Moreno, Y.M., Nava, G.L.: Modelo de gestión logística para Pymes en México. Contaduría y Admon **60**(1), 181–203 (2015)
OCDE: Políticas Pymes (2012). https://www.oecd.org/newsroom/nuevaspoliticasparapymesserequierenparaimpulsarelcrecimientoenamericalatinasegunlaocdeylacepal.htm
OCDE: Organización para la cooperación y el Desarrollo Económico. Temas y políticas clave sobre PYMEs en México (2013). https://doi.org/10.1787/9789264204591-es

Ortiz, A., Izquierdo Ojeda, H.E., Rodríguez Monroy, C.: Modelo de gestión logística para pymes industriales (2012)

Osorio, C.: Modelos para el control de Inventarios en las Pymes. Resultados de Investigación de la Facultad de Ingeniería y Ciencias Básicas **2**(6), 4–10 (2013)

Palacios, L., & José, J. (2013). Más allá del discurso económico liberal. Por una política industrial para el siglo XXI en México. Espiral (Guadalajara), 20(56), 69–105

ProMéxico: Pymes, eslabón fundamental (2014). http://www.promexico.gob.mx/negocios-internacionales/pymes-eslabon-fundamental-para-elcrecimiento-en-mexico.html

ProMéxico: Mapa de Ruta Nacional de Logística (2018). https://www.gob.mx/promexico/articulos/promexico-presenta-el-mapa-de-ruta-nacional-de-logistica?idiom=es

Rabolini, N.M.: Técnicas de muestreo y determinación del tamaño de la muestra en investigación cuantitativa. Revista argentina de humanidades y ciencias sociales, **2** (2009)

Rivera, I., Salazar, M.R.P.: Guía de selección de ERP en las pequeñas y medianas empresas mexicanas. Arbor **189**(760), 025 (2013)

Roldan, J.: Desafiemos al Dragón Chino. Ediciones Medialuna, México (2004)

Saavedra, M.L., Tapia, B.: El Entorno Sociocultural y la Competitividad de la PYME en México. Panorama socioeconómico **30**(44), 4–24 (2012)

Salazar, M.D.C.T., Mancera, P.G.: Administración de inventarios, un desafío para las Pymes. Inventio, la génesis de la cultura universitaria en Morelos **13**(29), 31–38 (2017)

Sanchis, R., Poler, R.: Medición de la resiliencia empresarial ante eventos disruptivos. In: 5th International Conference on Industrial Engineering and Management, pp. 104–113 (2011)

World Bank (2016). http://www.bancomundial.org/es/news/press-release/2016/06/28/germany-tops-2016-logistics-performance-index

World Bank: International LPI. World Bank (2019). https://lpi.worldbank.org/international/global

World Bank: Small and Medium Enterprises (SMEs) Finance. World Bank (2019). https://www.worldbank.org/en/topic/smefinance

Zuluaga, M.A., Gómez, M.R., Fernández, H.S.: Indicadores logísticos en la cadena de suministro como apoyo al modelo Scor. Clío América **8**(15), 90–110 (2014)

CPA-LR Human Resource Management Model for Ensuring the Innovation Process in the Enterprise

Zhanna Mingaleva[1,2(✉)], Yuliya Zinnurova[1], and Elena Shironina[1]

[1] Perm National Research Polytechnic University, Perm 614990, Russian Federation
mingall@pstu.ru
[2] Perm State Agro-Technological University Named after Academician D N Pryanishnikov, Perm 614990, Russian Federation

Abstract. The aim of the research is to test the CPA-LR model as a model of personnel management for innovatively oriented industrial enterprises. The model was tested using a field study. The processing of the survey results was carried out using general logical research methods: induction, deduction, analysis, synthesis, scientific abstraction, comparison and juxtaposition. The model is formed on the basis of a point assessment of two classification signs: the level of responsibility and the level of complexity of the performed labor functions. A different combination of classification features provides a classification of personnel and allows you to build a matrix of complexity of labor functions to determine the willingness of personnel to innovate. This allows you to develop and scientifically substantiate individual adaptation programs to innovation. The study was conducted on the basis of the supply department, whose employees perform functions that are most easily digitalized and networked. At the same time, the quality of the procurement department has a significant impact on the production of a wide range of products, including composite materials. The study revealed that the innovation process involves a certain sequence of personnel movement. The article shows that the transition from lower to upper groups is carried out only after optimization of the functional and number of the first two groups. The model is tested on the example of a large Russian industrial enterprise introducing the production of products from composite materials.

Keywords: Innovation processes · Organizational change · CPA-LR model · Complexity of professional activity · Level of responsibility · Change management

1 Introduction

The importance of organizational change as the basis of innovation processes has long been recognized, but remains relevant today [1–3]. "The key to success in manufacturing today is the ability to adapt. Constant change has become the norm. Manufacturers must implement new solutions that allow them to respond quickly to customer

T. Antipova (Ed.): ICCS 2020, LNNS 186, pp. 391–399, 2021.
https://doi.org/10.1007/978-3-030-66093-2_38

demands and competition" [4, p. 3]. This is confirmed by Russian researchers, noting that the current lag of the domestic machine building industry "in terms of production of new structural materials from leading foreign countries is due to the weak economic interest of the manufacturer and the consumer, the lack of a scientifically grounded program for the development, production and use of new materials and products based on them" [5, p. 169]. "This means that the processes of strategic planning of the activities of mechanical engineering enterprises must … be able to flexibly adapt to the emerging innovative values" [6, p. 179]. The high uncertainty of the process of determining the material property in the structure of the structure plays a significant role in this process, and "the influence of technological factors, such as characteristics of raw materials and material components, parameters of technological processes, is difficult to take into account in real production" [7, p. 27]. A special role in the implementation of innovations is played by the personnel of enterprises and organizations. At the same time, digitization processes make significant additions to both personnel management in general and to the composition of its labor functions and qualification characteristics.

Given the existing need to take into account the increasing complexity and uncertainty of the processes taking place at different hierarchical levels of management, it is necessary to develop a methodology and model for studying the distribution of personnel, its functionality, innovation potential, and opportunities for more efficient use for the enterprise to gain competitive advantages [8]. In our previous research, we have highlighted the importance of HR management for innovative enterprises [9–12].

This article presents the results of a theoretical justification and construction of a model for assessing the qualitative characteristics of the personnel of an industrial enterprise, taking into account the stimulation of innovative processes.

2 Theoretical and Research Methodology

The Theoretical Foundations

Although there is no common understanding of change management, there is recognition that changes occur in all types of activities, shapes and sizes of organizations and enterprises [13–15]. A dynamic production system is a fully-integrated, collaborative manufacturing system that responds in real time to changing demands and conditions in the factory and in the supply network. The primary characteristic of a dynamic production system is the ability to easily and rapidly reconfigure factory production and supply networks to optimize system performance. Such systems deal effectively with uncertainty and abnormal events, and learn from past experience to enable continuous improvement [4, p. 3]. Any enterprise needs innovative changes to develop its competitiveness [16–18].

Many authors have proven that only ever-changing organizations can hope for survival [19–21]. Change is an essential feature of organizational life, both at the operational and strategic levels [13, 22–26]. For any organization, the ability to determine the desired future and manage the changes necessary to achieve it is important. However, By indicates that change for the sake of change is harmful [27]. A change is only required if it is related to the goals of the organization and consistent with organizational values [24, 25].

Research Data
A total of 24 employees of the supply department of a large enterprise were examined. Procurement operations are the most convenient and the fastest to be digitized and networked [28].

The questionnaire survey was conducted in January–February 2020.

The processing of the survey results was done in March–April 2020.

Further, recommendations were developed for the management of the enterprise in the field of personnel policy, personnel motivation, training and changes in professional status.

Analysis Method
The design of the model is based on the theory of social dynamics in the organization.

The field study included interviews, observation, a photograph of the working day, and analysis of job descriptions.

Processing the results of the field study was carried out using general logical research methods: induction, deduction, analysis, synthesis, scientific abstraction, comparison and juxtaposition.

The model was developed according to two key assessment parameters: the complexity of professional activity and the degree of responsibility. The complexity of professional activity is determined by the components: type of activity, variety of work, controllability of work, scale and complexity of management. The degree of responsibility is determined by the components: independence, initiative, commitment to the result, significance of the results of work. The model is built on the basis of a score depending on the level of complexity of the tasks and labor functions and the level of responsibility. The result of modeling is the selection of 4 groups of specialists: "Implementers", "Specialists", "Experts" and "Leader" (Fig. 1).

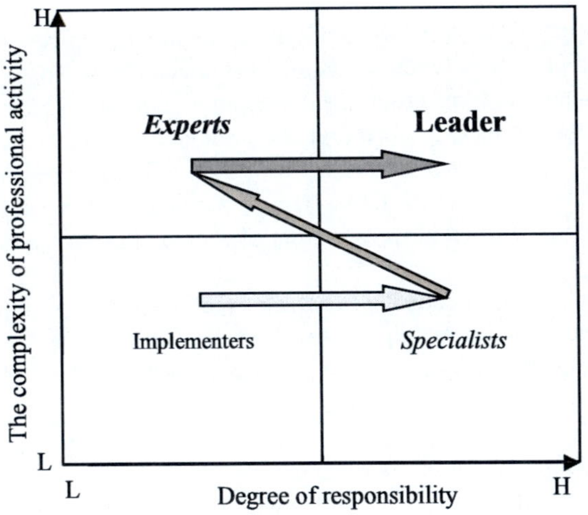

Fig. 1. CPA-LR model (theoretical description). Source: refined and adapted by authors from [11, p. 14]

A high and a low level are distinguished for each assessment parameter (the complexity of professional activity and degree of responsibility).

For each component, four characteristics are determined, each of which is assigned a score: from 1 - the minimum level, to 6 - a high level. For the amount of 4–10 points, a low level of responsibility/complexity is defined. For the amount of 11–24 points, a high level of responsibility/complexity is established.

3 Findings and Discussion

Four groups of personnel were formed depending on the combination of the results of the assessment.

The first group "The implementers" is characterized by a low level of responsibility and low complexity of the work performed. Employees from this group can be easily replaced without noticeable changes in the quality and stability of the processes implemented in the organization. The main functionality of representatives of this category is routine, repetitive processes that can be optimized or eliminated [24, p. 15].

The second group of "The specialists" is characterized by low level of complexity at a high responsibility. The staff of this group do not have exclusive knowledge, and the basic skills they need to do the job can be formed in the course of training or internship directly at the workplace (during the probationary period). Simple staff (ordinary employees) from this group, unlike representatives of the next two groups, are quite easy to replace and quickly prepare new ones. Evaluation of the quality of work of ordinary specialists is carried out using quantitative methods, timing of the working day, determining the number of operations performed per unit of time [24, p. 15]. The procedure for optimizing the function of representatives of this group is easily carried out on the basis of digital technologies and a network structure of activity.

The third group "Experts" is characterized by a low level of responsibility with a fairly high complexity of work. Representatives of this group perform technically or intellectually complex labor functions that require special knowledge. The results of the work of representatives of this group are important in general for planning the activities of the organization and its individual divisions, but errors in their work will not have a significant impact on the result of the work of the entire organization, since the final decisions are not made by them, but their conclusions are double-checked [24, p. 14]. It should be noted that such employees are not always present in the organization or its divisions.

The fourth group of "Leader" is characterized by both high complexity of performed labor functions and the highest level of responsibility. Representatives of this group have high qualifications, special professional skills, and a high level of education. The influence of the labor activity of representatives of this group on the results of the organization's work is very significant. The mistakes of the representatives of this group entail problems for the entire organization. The efficiency of the representatives of this group directly affects the result of the work of the entire organization [24, p. 14].

As already noted in the methodology section, the study assessed the complexity of professional activities and the degree of responsibility of employees of the procurement department of a large manufacturing enterprise. The results of the assessment and the distribution of employees into 4 selected groups are shown in the Fig. 2.

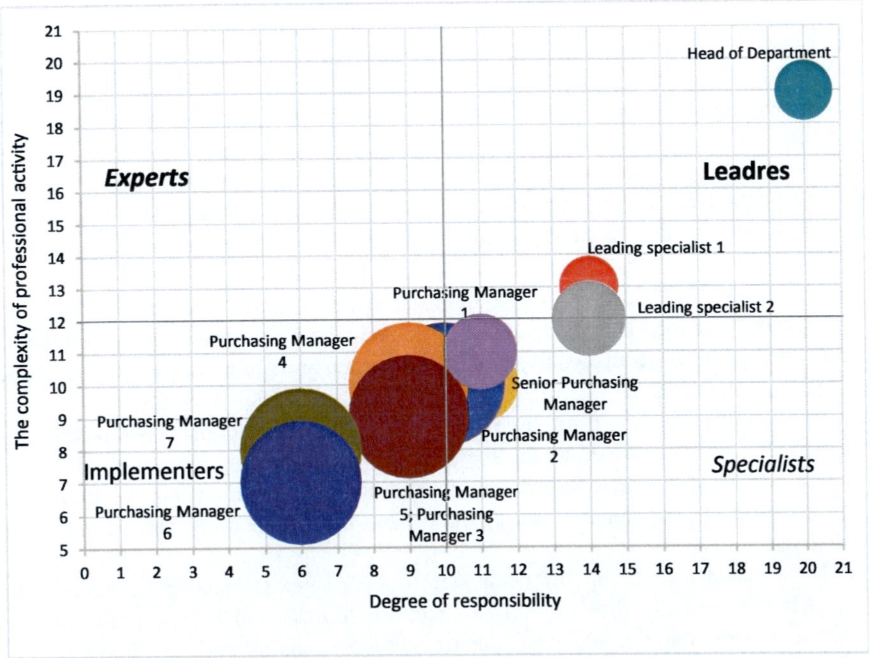

Fig. 2. CPA-LR model (practical description). Source: compiled by the authors based on research materials at the enterprise

As a result of the study, it was established that not a single person was included in the group of Experts. All employees of the procurement department were distributed between the groups "Leaders", "Specialists" and "Implementers".

The "Leaders" group (3 employees) includes the functional of the department head, since the complexity of his work is determined by the assessment and decision-making at the department level, the work related to the organization of activities for which the goals and results of the department are determined. Intellectual complex labor functions are performed, requiring analytical thinking and professional judgment, which make a great contribution to the overall work, and also a high responsibility for the work of the entire department. In addition to the position of head of department, this group also includes the position of lead specialist 1, the functionality of which includes a variety of work related to labor functions, which require a significant number of sequential actions according to a given algorithm of actions. The duties of this position include guidance in the direction of "basic materials", so the complexity of the work is determined by the requirement of in-depth and extensive knowledge of the group of materials.

The “Leaders” group unites employees who hold positions of leading specialists and the head of the department. These employees have a higher complexity of work compared to others, their work cannot be considered routine and repetitive, as it requires analytical thinking and professional judgment. For these employees, it is not so much the amount of remuneration that is important as to be involved and inspired by their work.

At the intersection of the “Specialists” and “Leaders” groups, there is the position of the leading specialist 2, who also performs various works and is the head of the “auxiliary materials” direction, but this direction refers to work with less complexity, since it does not require in-depth knowledge with technical bias. Responsibility is at the same level and can be considered quite high due to independence in relation to labor functions, also related to the organization of activities of other employees.

The positions of the procurement manager1 and the senior procurement manager also belong to the “Specialists” group (6 employees), have a rather high degree of responsibility, as they are personally responsible to the management for performing the designated labor functions within the direction group and participate in the preparation of the draft decision.

In the transition zone from “Implementers” to “Specialists” there is a procurement manager 2, this is explained by the specifics of working with a group of materials that require certain specialized knowledge, in addition, this position has the function of replacing other procurement managers during the absence of the corresponding employee.

If we consider the group “Implementers” (15 employees), then we can note that this group is conditionally divided into two parts with a difference in responsibility and complexity of work. The zone with the greatest responsibility and complexity includes procurement managers 3, 4, 5, who generalize, describe the work, prepare calculations and group indicators and work results in accordance with a given algorithm within the framework of the direction for which certain special knowledge is required to work with a group of materials. The zone with the least responsibility and complexity includes procurement managers 6 and 7, whose positions do not involve participation in decision-making of the department. They perform simple routine functions and repetitive processes within the “auxiliary materials” direction, including stationery, household equipment, cleaning and washing products, and laboratory materials that do not require special knowledge to work with a group of materials.

4 Conclusions

The proposed approach to assessing the qualifications and job responsibilities of employees of enterprises, from the point of view of the possibility of successful implementation of modern methods of managing innovation processes in the context of digitalization and networkization, includes the distribution of elements of the organizational structure (departments and individual employees) in accordance with the proposed CPA-LR model (“complexity of professional activity” – “level of responsibility”).

To build a model for each group of personnel, specific levels of responsibility were established and individual and group levels of complexity of the labor functions performed were determined. The CPA-LR matrix was built on the basis of expert data.

Based on the results of the assessment and the data from the CPA-LR matrix, decisions regarding staffing should be made and implemented. In particular, these are decisions on optimizing the number of personnel in units that do not have high complexity and responsibility, since their work does not significantly affect the achievement of organizational results. For personnel characterized by the complexity of the tasks performed and a high level of responsibility, it is necessary to develop special development programs, including various ways to motivate personnel [29].

Based on the need for digitalization of production operations, the possibility of their networkization and the need to strengthen the innovative focus of the main divisions of the enterprise for employees of all groups, the main directions for developing programs for individual and corporate development were determined.

Firstly, it is necessary to strengthen the personal orientation of employees for high-quality performance of work with an increased emphasis on independence in decision-making and search activities. This implies an increase in the level of knowledge, abilities, skills, improvement of competence and competencies, concern for intellectual and general physical development.

Secondly, it is necessary to strengthen the personal orientation towards status in the organization - taking care of one's own social and professional status, developing organizational skills and leadership qualities. This will help the development of personal career growth of employees.

Third, it is necessary to strengthen the focus on interpersonal relationships. It involves the development of communication skills, the development of emotional culture, awareness of corporate identity.

The development of development programs based on the allocation and analysis of the basic elements of labor and innovation activities of employees of various departments of enterprises is aimed at ensuring strategic coordination of actions and enhancing innovation.

Further research suggests a more detailed analysis of those groups of employees that are characterized by a low level of complexity. This will help to identify (and eliminate in the future) such organizational problems as uneven workload and managerial load, duplication of functions, lack of control of one process with excessive control of another. These problems can arise as a result of the digitalization and networkization of labor operations.

Acknowledgment. The work is a part of base portion of the state task of the Ministry of Education and Science of the Russian Federation to Perm National Research Polytechnic University (topic № FSNM-2020-0026).

References

1. Amis, J.: Understanding organization change and innovation: a conversation with Mike Tushman. J. Change Manag. **18**(1), 23–34 (2018)

2. Ha, K.M.: Digital business leadership: digital transformation, business model innovation, agile organization, change management. R&D Manag. **50**(2), 171–172 (2020)
3. Strasser, T., de Kraker, J., Kemp, R.: Three dimensions of transformative impact and capacity: a conceptual framework applied in social innovation practice. Sustainability **12**(11) (2020). Paper # 4742
4. Lubell, J., Frechette, S.P., Lipman, R.R., Proctor, F.M., Horst, J.A., Carlisle, M., Huang, P. J.: Model-Based Enterprise Summit Report.US Army Research Laboratory, Aberdeen Proving Ground, MD, ARL-RP- 466, NIST Technical Note 1820, 151 p. (2014)
5. Vaytsekhovich, S.M., Vlasov, Y.V.: A model of economic efficiency of structural materials production and consumption in machine building industry. Vestnik PNIPU. Socio-Econ. Sci. **2**, 168–194 (2020)
6. Lobova, E.S., Mamaeva, L.N., Mentemirova, T.G., Shishkina, K.A.: Prospects for the strategic development of enterprises of the machine-building complex in the framework of the priority directions of the country's development. Vestnik PNIPU. Socio-Econ. Sci. **1**, 178–192 (2020)
7. Kosheleva, N.A., Shipunov, G.S., Voronkov, A.A., Merkusheva, N.P., Tikhonova, A.A.: Experimental studies to determine the deformations of samples from a polymer composite material using fiber-optic sensors, Bulletin of the Perm National Research Polytechnic University. Aerosp. Eng. **50**, 26–35 (2017)
8. Silva, E., Lopes, P.: A study on market intelligence: the professional, the applicability of information technologies to innovate and gain competitive advantage. J. Digit. Sci. **1**(1), 51–62 (2019)
9. Danilina, E.I., Mingaleva, Z.A., Malikova, Y.I.: Strategic personnel management within innovational development of companies. J. Adv. Res. Law Econ. VII **5**(19), 1004–1013 (2016)
10. Mingaleva, Z., Deputatova, L., Starkov, Y.: Values and norms in the modern organization as the basis for innovative development. Int. J. Appl. Bus. Econ. Res. **14**(10), 124–133 (2016)
11. Zinnurova, Y.A., Shironina, E.M.: Theoretical and methodical approach to assessment of human resources potential. Vestnik universiteta I **5**, 12–17 (2020)
12. Zinnurova, Yu., Shironina, E.: Theoretical aspects of human resources capacity development. Vestnik universiteta I **12**, 17–22 (2019)
13. Burnes, B., Jackson, P.: Success and failure in organizational change: an exploration of the role of values. J. Change Manag. **11**(2), 133–162 (2011)
14. Mingaleva, Z., Mirskikh, I.: Small innovative enterprise: the problems of protection of commercial confidential information and know-how. Middle East J. Sci. Res. **13** (SPLISSUE), 97–101 (2013)
15. Luecke, R.: Managing Change and Transition. Harvard Business School Press, Boston (2003)
16. Kotter, J.: Leading Change. Harvard Business School Press, Boston (1996)
17. Mingaleva, Zh., Danilina, H.: Significance of technological innovations for an increase of competitiveness of industrial companies. Life Scie. J. **11**(Spec. Issue 8), 211–215 (2014)
18. Mingaleva, Z., Deputatova, L., Starkov, Y.: Management of organizational knowledge as a basis for the competitiveness of enterprises in the digital economy. In: Antipova, T. (eds.) Integrated Science in Digital Age, ICIS 2019. Lecture Notes in Networks and Systems, vol. 78, Springer, Cham (2020)
19. Parry, W., Kirsch, C., Carey, P., Shaw, D.: Empirical development of a model of performance drivers in organizational change projects. J. Change Manag. **14**(1), 99–125 (2014)
20. O'Reilly, C., Tushman, M.: Lead and Disrupt: How to Solve the Innovator's Dilemma. Stanford University Press, Palo Alto (2016)

21. Linnenluecke, M.K., Griffiths, A.: Corporate sustainability and organizational culture. J. World Bus. **45**(4), 357–366 (2010)
22. Burnes, B.: Managing Change: A Strategic Approach to Organisational Dynamics, 4th edn. Prentice Hall, Harlow (2004)
23. Balogun, J., Hope Hailey, V.: Exploring Strategic Change, 4th edn. Pearson, London (2004)
24. By, R.: Organizational change and leadership: out of the Quagmire. J. Change Manag. **20**(1), 1–6 (2020)
25. By, R.: Organisational change management: a critical review. J. Change Manag. **5**(4), 369–380 (2005)
26. Carnall, C.: Managing Change in Organizations, 4th edn. Prentice Hall, Harlow (2003)
27. By, R.T.: New year's resolution. J. Change Manag. **19**(1), 1–5 (2019)
28. Mingaleva, Z., Bykova, E., Plotnikova, E.: Potential of the network concept for an assessment of organizational structure. Proc.-Soc. Behav. Sci. **81**, 126–130 (2013)
29. Zakharov, A.: The impact of motivation on labor resources reproduction. In: ICIS 2019. LNNS, vol. 78, pp. 3–10 (2020)

Current Trends of Megascience Facilities Utilization

Artem Balyakin, Nurzhan Nurakhov, and Marina Nurbina(✉)

NRC Kurchatov Institute, 1, ac. Kurchatov sq., Moscow 123182, Russia
Nurbina_MV@nrcki.ru

Abstract. The paper considers the specifics of unique scientific facilities functioning at the present time. A number of Russian and European megascience facilities are considered. It is shown that currently megascience facilities utilization focuses at the maximum standardization and automation of processes, with the transfer of a significant part of the activity to the remote access format. These trends have received an additional impetus due to the quarantine regime during coronavirus pandemic. The necessary changes to business processes related to the operation of megascience facilities in distant mode are discussed. The role of e-Infrastructure (including the construction of digital twins) as an essential part of unique scientific facilities is revealed. The idea of a possible new form of international scientific research is formulated.

Keywords: Megascience · Reactor PIK · International scientific cooperation · Remote access · E-infrastructure · Digital twins

1 Introduction

Nowadays globalization process reflects itself in any aspect of life, with science being no exceptions. New scientific globalized knowledge is produced by joint research conducted by a consortium of scientists in specialized facilities (research infrastructures), the financing of the creation and operation of which goes beyond the capabilities of separate entities (with regard to countries, companies, institutions, etc.) [1].

Research infrastructures can be of various types: distributed (with its parts allocated in different organizations and/or countries); localized in one place (single-sited) or virtual (e-infrastructures). They may include scientific equipment, scientific collections, archives, databases or any other unique object that can be used for research purposes [2, 3]. Some studies also include methods and approaches of research in the definition of "research infrastructures", as the obtained results are tightly connected with the way of thinking, analyzing and handling the data [4].

In order to emphasize the uniqueness of the performed work, the term "megascience facilities" is often used as a complimentary to "unique (specialized) research infrastructure". As a rule, megascience facilities are international research complexes that have no analogues in the world. Despite their size and complexity of organisation they function as a whole, and are focused on obtaining scientific results that cannot be achieved at other facilities in the world.

T. Antipova (Ed.): ICCS 2020, LNNS 186, pp. 400–409, 2021.
https://doi.org/10.1007/978-3-030-66093-2_39

Institutionally, megascience facility can be defined as a scientific "supranational" organization with "independent representations" [5] or as organizational and managerial innovation [6]. Financing, construction and operation of such installations is implemented largely on the basis of international scientific and technical cooperation. Even in the case of the construction of such installations by one country independently (e.g., ISSI-4 in Russia, that will be the fourth generation X-ray source of synchrotron radiation in Protvino) their use goes beyond the capabilities of one country, and from the planning stage it is supposed to attract research teams from other countries.

It should be understood that the results obtained at megascience facilities actually find their practical application in various fields. Thus, the development of a special legal algorithms concerning only and exclusively above mentioned technologies proves to be impossible. As a rule, several countries participate in megascience projects. This immediately causes a number of problems associated with the existing legal restrictions on the work of foreign citizens in the countries where the megascience facility is located, as well as the need (opportunity) to optimize certain operations, carried out in the process of working within the framework of a megascience project. The last also happens due to the need to optimize business processes in the provision of services using high-tech tools.

2 Main Objectives of the Study

The use of megascience facilities involves solving many related problems: from the technical maintenance of the project, to its methodological support. One of the pressing issues is the organization of scientists' access to research infrastructures and – in the future – to the results of scientific research [6, 7].

In this paper (carried out as a part of the RFBR project No. 18-29-15015), we consider the features of the unique scientific equipment in remote access mode. This issue can be treated as a specific highly specialized service (product), and it plays crucial role for developing mass remote access technologies, not only for scientific purposes, but also for social tasks and commercial applications.

The need for remote access to research infrastructure (including unique facilities), has existed since the advent of the technical feasibility of implementing such access. It is accompanied with a number of solutions highly demanded in various fields. For example, the growing trend towards the expansion of peripheral computing is based on the idea that local distributed traffic will reduce the delay in data transfer. Another idea is to attract as many participants as possible while expecting obtaining different solutions to one problem.

This option (work on megascience facilities in remote access mode) was assumed in most installations initially, but quarantine restrictions associated with the spread of coronavirus infection in 2020 made the remote access format absolutely vital. In addition to solving the current tasks of carrying out research projects, the coronavirus pandemic has demonstrated the relevance of the task of providing remote access to research and educational infrastructure, as well as providing the possibility of remote work in principle, from a completely different perspective: this task should and can be solved in order to ensure public health safety across the globe.

Thus, in the framework of this study, the following tasks were set:

- to study the use of remote access techniques in the operation of megascience facilities.
- to consider the novelty of remote access technology (in regard to megascience projects). How new and fundamentally changing the nature of research is the "remote approach" compared to traditional research methods?
- to identify new trends in the work organization at megascience facilities, and to determine the required changes in business processes that accompany work at megascience facilities.

3 Megscience Facilities and Their Operation

Essentially, megasience projects are the program of concentration and effective implementation of intellectual property objects. Recent studies from the European Commission show that research projects in the field of ICT and energy, as well as grants from the European Council for Research and Development (ERC) for basic research, create a large amount of intellectual property, including patents, copyrights and trademarks [8].

In the Russian Federation, 7 megascience projects are planned or implemented at present time: NICA (Nuclotron-based Ion Collider Facility) complex of superconducting rings on colliding heavy ion beams; International Center for Neutron Research based on the PIK high-flux research reactor (ICNR PIK); Tokamak with a strong magnetic field (Ignitor); Accelerator complex with colliding electron-positron beams "VEPP-5 Complex" (Super Charm-Tau Factory); International Center for Extreme Light Field Research (CIES); Fourth Generation X-ray Synchrotron Radiation Source (ISSI-4); Siberian Ring Source of Photons (SKIF) [6, 9].

Our studies focused on the possibility of implementing international experience at ICNR PIK. The PIK reactor will serve as a powerful source of neutrons that slow down to the required energy, and are removed from the reactor through special channels and transported through the neutron guide system to experimental facilities for research. In its parameters and experimental capabilities, the PIK reactor surpasses all existing research reactors, including the world's only analogue - the HFR reactor at the European Center for Neutron Research - International Institute Laue Langevin (ILL, Grenoble, France). By 2025, it is expected to commission 20 major appliances.

The construction of the ICNR PIK is carried out in Gatchina, Leningrad Region, on the basis of the St. Petersburg Institute of Nuclear Physics named after B. P. Konstantinov, a unit of the Kurchatov Institute [10, 11].

Also, the NRC Kurchatov Institute, on behalf of the Russian government, provides scientific guidance in a number of foreign research projects. For example, this is the ITER (International Thermonuclear Experimental Reactor), the European Radar Free X-ray Laser (Eu-XFEL), and the Large Hadron Collider (LHC) at CERN.

Despite the high role of fundamental research in megascience facilities activities, there are also significant prospects for their practical implementation (mostly medicine and material science). Thereby, synchrotron sources facilities contribute to various sectors of the real economy [9, 12]. The use of super-bright X-ray beams in the study of matter allows us to see how extremely complex systems (e.g., proteins) are arranged, how energy is released in living cells (e.g., human brain cells), how new artificially produced material function (so called metamaterials are examined by the powerful beams of different nature), how nanoparticles move and interact [13].

The results obtained at megascience facilities are used by research teams around the world. Regarding the data gained while working on research infrastructures, 2 approaches can be noticed. As part of the first approach (performed, e.g., in CERN), an open data policy is implemented when the scientific results obtained by the collaboration are published freely and are disseminated in the public domain. At the same time open access to raw (unprocessed) data is not expected. The data itself (in processed format) is stored for a long time and is available for re-analysis.

The second approach, implemented in European XFEL, is the exemption from the policy of open scientific data for private research. All raw data, and associated metadata, as well as raw data analysis results obtained from private research, will belong exclusively to the client, who has gained access and is not subject to the European XFEL Scientific Data Policy. Whereas the results of purely scientific experiments are available after a period of embargo (3 years). The processed data and the results of the interim analysis and the associated metadata are not considered by EuXFEL as long-term storage facilities (5 years or more).

Totally, regardless of the approach adapted at the facility, the research infrastructure produces the excessive amount of data (we recall here that "big data" originated from CERN experiments). The large volume of the obtained results, the need for their comprehensive processing, storage and analysis have led to the fact that working with data has been removed from the general list of activities and has become a separate factor in the megascience facilities operation. After minimal initial processing, the information collected was being sent for analysis to the data centers of the project participants.

The process of obtaining data on megascience facilities (taking CERN as an instance) can be represented in general form as follows [14]. During the experiment, data is generated by detectors. Data from the detectors arrives into a temporary storage, where preliminary data can be pre-filtered. Temporary storage is usually a system with fast access of a limited volume, which must be freed up to receive the next portion of experimental data at first demand. Data from temporary storage is moved to permanent storage, and additional processing of data may or may not occur depending on the experiment. The location of the permanent storage is not tied to a specific installation, and can be anywhere in the world. Data transferred to persistent storage is processed and analyzed to obtain on-site results (without involving specialists from the original megascience facility). At all stages of obtaining and primary processing, access to data belongs (usually) only to members of a limited team of scientists participating in the experiment or members of the collaboration.

The organization of these works required the optimization of the relevant business processes and the consideration of the specifics of remote access in title documents. Thereafter, in the case of CERN, back in 1998, the MONARC project (Models of Networked Analysis at Regional Centers for LHC Experiments) was created, one of the results of which was the concept development of hierarchy of processing centers, modeling and data analysis. There are currently 4 levels of processing centers. At the bottom there is the zero level (Tier-0 CERN Computing Center), which is engaged in the primary reconstruction of events, calibration, permanent storage and archiving of a complete set of raw and simulated data. Next come Tier-1 (13 centers), Tier-2 (about 170 centers), and finally, Tier-3 (about 50 centers), which are university clusters, or centers that provide resources on a voluntary basis, physical data analysis. Tier-1, 2, 3 facilities are located all over the world, and connected within the special net.

A similar data policy was implemented in the case of the global neutrino network (GNN) [15]. This network can be interpreted as a distributed scientific infrastructure, the elements of which can be individual research infrastructure and collaborations (for example, IceCube collaboration - 47 organizations from 12 countries or the Dubna multi-megaton deep-sea neutrino telescope).

As infrastructure GNN is a set of experimental facilities, and a number of scientific institutions (institutes, universities, research centers, etc.) distributed throughout Europe, America and Asia that are engaged in the setting up of experiments and processing of the obtained data. Thus, in this case we are talking about the existence of a unique scientific facility (neutrino telescope), which implements a wide research program in remote access mode (sometimes in real time). Neutrino telescopes can also be considered as sources of information that is distributed within the network between its participants for further processing.

4 Trends and Prospects

Without exception, all unique research infrastructures include the following main components: physical (actually a complex of scientific equipment); digital (or informational) infrastructure (including distributed one), which provides the research and development activities of the installation.

Thus, the activity on megascience facility can be divided into 2 components. In the first aspect, this is the solution of engineering (applied) problems that arise during the creation, operation and modernization of the installation. In the second, "scientific" aspect, the facility turns to perform the solution of scientific and practical problems, setting up an experiment and analyzing the data obtained. In fact, the second definition can be attributed to the "research infrastructure" in a broad sense.

As part of the work in the first aspect, the physical implementation of the experiment and its support (installation, configuration, repair and commissioning of equipment, maintenance of activities) are carried out. Activity on the second aspect is the collection and processing of the results.

To solve the first problem, in the immediate vicinity of the installation there is a group on duty, which is responsible for ensuring the operability of the facility (see examples on CERN above). The scope of tasks of this group is limited, and mainly consists in starting and, if necessary, setting up the installation software [11]. That is, this group does not directly interfere with the installation itself.

In the work of all astronomical projects (Square Kilometer Array, GNN, SETI), the second part is more important and noticeable. Distributed analysis of data obtained from installations is carried out remotely by scientific groups within their work places using their own computing power (for example, as described above for GNN). For example, in the international Boreksino project [16], an additional, independent data collection system (implemented by NRC Kurchatov Institute) based on fast waveform digitizers operates, in order to expand the dynamic range of the spectrometric measurements of the detector in an area inaccessible to the main electronics to ~60 MeV. The complex allows, inter alia, collecting data remotely without the need for scientific teams to travel to the location of the detector.

It can be seen that remote access is a necessary element of many mega-science facilities, which has firmly entered into scientific practice. The development of digital infrastructure gave rise to the new phenomenon – e-Infrastructure. Its creation was forced by the aims to optimize the remote work, and maximize the involvement of the scientific community. Another aspect that was not considered crucial at the beginning, was the commercialization of the results, that was much facilitated by introduction of the e-Infrastructure. We note that digital infrastructure does not duplicate and does not replace the "physical" one.

Noteworthy to stress that this approach (digitalization of the scientific research) is a method of organizing modern scientific research. For example, for the European Union, the main trends in this area are related to the formation of a single scientific space based on open access and the formation of the e-Science system. The first step should be the unification of science, data collection systems and access to them. e-Infrastructure is thought to be the tool for implementing EU policies in science, when the achievements of the Internet, grid systems, cloud computing and databases are assembled in a new infrastructure. The first step in the development of digital infrastructure in the EU was the creation of the open scientific portal EOSC (European Open Science Cloud), launched in 2016, in order to increase the growth potential of the EU digital economy. Based on the EOSC work experience, the Go FAIR initiative is being prepared to put ideas and proposals related to digital science into practice [17].

To solve the applied problems of processing scientific information in the EU, a number of specialized data processing projects are planned: GEANT (management of scientific and educational network projects), EGI - Advanced computing for research (providing calculation options for CERN, EMBL projects), PRACE (providing computing power, 465 projects at the moment). In addition to the above, in order to codify and standardize digital infrastructure, within the framework of the Horizon 2020 program, the European Union launched the e-Standards project [18].

It is expected that the first consumers of e-Infrastructure will be representatives of the natural sciences, however, the greatest impact and the most significant results will be manifested in the field of humanitarian knowledge, which forces participants in the process to develop appropriate assessment methods and approaches today, simultaneously solving legal issues.

It can be seen that the digitalization of scientific research, the high role of remote access to megascience facilities and/or to data obtained from unique research infrastructures, is a long-term trend that has been under development and implementation for a long time. Its actualization in 2020 due to the coronavirus pandemic seems to be only the intensification in existing dynamics.

However, in order to take into account these changes and new requirements, it is necessary to introduce a number of modifications to some business processes. Hereafter we discuss some ideas that can be implemented in order to improve current practices. For an appropriate analysis, it is advisable to consider a project of "megascience" class as a high-tech tool designed to solve applied problems and provide research services. In this regard, the organization of remote activities, in our opinion, requires taking into account the following points:

Firstly, it is necessary to minimize the need for the customer to be present at the facility where the megascience facility is located. It includes:

- remote approval and conclusion of contracts, adoption of digital signature. Signing additional agreements or adaptation of previous ones;
- delivery of materials for research from verified (approved) suppliers (which means, inter alia, the possible rejection of their own experimental samples). Formation of relevant specialized delivery service, standardization and examination services;
- transfer of most of the research work at the facility to certified personnel (permanent staff of the facility). Development and approval of work algorithms

Secondly, it is necessary to develop an appropriate digital infrastructure. This means both the creation of new digital infrastructure objects (data centers, processing algorithms, etc.), and the integration of megascience facilities into existing e-Infrastructure elements.

Thirdly, legal and methodological support for the operation of megascience facilities in remote access should be provided. So, to ensure a remote format of the research infrastructure, the creation of an appropriate information system is required. Some aspects of creating such a system were considered by us earlier: taking into account socio-economic aspects [19], the need to create a system of information support for the circulation of intellectual property objects [20], and taking into account risks and challenges in ensuring national security [21].

Fourth, it is necessary to train engineering personnel to solve the related tasks of maintaining and ensuring the remote access mode. One of the solutions to this problem is the use of digital twins of existing projects. Such super-powerful projects could solve several important tasks at once: reconstruction and restoration of various physical processes with unverified characteristics; ensuring the participation of a wide range of scientists in long-term experiments in real time; a fundamental reduction in the risks of conducting dangerous experiments or experiments that could adversely affect the environment; use of event generators in the learning process; the ability to adjust technical tasks for the construction of real megascience facilities; the ability to reduce costs in the construction of real megascience facilities.

Currently, data obtained by computer-generated events are already used in conjunction with physical installations on some megascience facilities. They usually solve some local problems of specific experiments. For example, ATLAS (conducted in CERN) uses data obtained by computer-generated event generation (by Monte-Carlo method). Corresponding data set (EVNT) is generated on the computing resources of any type of site. It all depends on which sites the tasks were sent to generate this data. After generating EVNT data, they are copied to other sites for subsequent processing, which, in essence, is similar to processing raw data [22].

Obviously, the creation of a complete digital twin of a mega-science project will be a mega-science project by itself and will require the infusion of comparable amounts of funds and intellectual activity. It is assumed that to create and support the work of digital counterparts, it is possible to attract business participants (that would also reduce associated costs). The creation of a digital twin will also make it possible to solve the problem of providing a remote work format for a large number of specialists in the scientific and educational sphere. Such democratization of technology will provide easy access (including one for non-specialists) to knowledge in the field of technology and business without long or expensive training. This policy, dubbed "citizen access", already finds its application in the development of applications, data and analytics, design and knowledge [23].

To ensure the continuity of the educational trajectory for students (both primary and high school), it is proposed to use existing game shells, launching add-ons with the possibility of modeling the studied processes. Minecraft game shell seems to be a suitable tool for digital megascience world construction.

Fifthly, it is necessary to conduct a separate study of the risks and challenges that accompany the active transfer of work at megascience facilities to a remote format.

In our opinion, the main problem will be related to ensuring the safety of the transfer and use of information and maintaining reputation (improving the reputation of companies collecting, transmitting and processing data and research samples). These two difficulties are derived from the high rate of evolution of technological progress that creates a crisis of confidence. This trend requires focusing on the key elements of trust: honesty, openness, accountability, competence and consistency.

At present, there is an understanding of the importance of this issue, but there is a delegation of decision-making in favor of the same digital services that they serve. So, many states and business players use elements of artificial intelligence to make decisions and there is a temptation to use them in the case of scientific data. It seems that the general direction will be to take into account the social, moral and ethical components in this matter, which will inevitably lead to the adoption of a number of regulatory acts similar to the European Union General Data Protection Regulation [24].

5 Conclusions

In general, the current trends in the use of megascience facilities continue the previous scenarios, and the coronavirus pandemic acted as a catalyst for existing trends. We expect an increase in the importance of digital technologies, recognition of the role of e-infrastructure (as compared to "physical" one) and a gradual transition from solving technical issues (access, processing and transmission of information) to understanding of their socio-economic consequences [25]. This implies a detailed study of both the legal mechanisms of digital technologies and the consideration of moral and ethical risks and challenges associated with accelerated digitalization in combination with a remote format of work in the scientific field.

Our analysis shows that megascience facility in nearest future will consist of both digital and "physical" components. E-infrastructure will become an inevitable part of any scientific project, and should be treated not as an addition or ersatz to the real one, but as an integral part of the facility as a whole. In practice, megascience facilities will always have their digital twins. Ad hoc this process will be accompanied by an enhanced role for e-Infrastructure as a response to current challenges.

The preference for a remote interaction format and at the same time the requirement for uninterrupted operation of installations will lead to standardization of activities. From the point of view of scientific research, megascience facilities, converted to the remote access format, will lose their uniqueness and become a common place. This fact means that they will inevitably be replaced by a new generation of new unique scientific attitudes.

A high degree of interconnectedness and inclusion in science and beyond will lead to the formation of a new over-infrastructure (super facility). It is expected that the basis of such a structure can be elements of the scientific infrastructures of ESFRI in the EU, a global network in the Russian Federation, or GRAIN project elements in BRICS.

Now it is not possible to predict what new unique scientific settings requiring the personal presence of researchers will look like. However, it is clear that scientific and technological progress is impossible without taking into account the interaction of man, society, science and technology. As one of the options, the converged technologies approach can be considered (a.k.a. NBICS technologies) [26].

In the Russian Federation, work is already underway to integrate ICNR PIK into the digital future. Currently, the prospect of creating a unified research infrastructure of the Union State (primarily based on the megascience facilities), is being discussed with a view to consolidating resources and expanding opportunities for scientific research. Additionally, it is planned to widely involve third-party organizations (NAS of the Republic of Belarus, German scientific institutions and funds, etc.) to participate in the implementation of projects to create research infrastructure of the mega-science class (first of all, to participate in work at synchrotron-neutron research centers in the Russian Federation, including the International Center neutron research based on the PIK high-flux reactor, and others). The convergence of science and technology will bring together scientists from different countries and will facilitate the coordination of research and development aimed at overcoming global challenges, including the development of methods to prevent the spread of pandemics based on genetic research.

Acknowledgments. The authors are grateful to Taranenko S.B. for useful discussions. This work was supported by the RFBR grant No. 18-29-15015.

References

1. Fotakis, C.: FP7 Interim Evaluation, Analyses of FP7 supported Research Infrastructures initiatives in the context of the European Research Area, Final Report 12 November 2010 (2010)
2. Florio, M., Sirtori, E.: Social benefits and costs of large scale research infrastructures. Technol. Forecast. Soc. Chang. **112**, 65–78 (2016)
3. Lami, S.: Challenges and new requirements for international collaborations. Sci. Diplomacy **6**(2) (2017). https://www.sciencediplomacy.org/article/2017/mega-science-collaborations
4. European Research Infrastructures. https://ec.europa.eu/info/research-and-innovation/strategy/european-research-infrastructures_en. Accessed 20 June 2020

5. Crease, R.P., Martin, J.D., Pesic, P.: Megascience. In: Physics in Perspective, vol.18, pp. 355–356. Springer (2016)
6. Karlik, A.E., Platonov, V.V.: Conceptual foundations of the study of megascience as an organizational and management innovation. Innovation **10**(228), 11–16 (2017)
7. Zaplatina, T.S.: Professional qualifications' recognition and megascience projects. J. Phys: Conf. Ser. **1406**(2019), 012020 (2019)
8. Balyakin, A., Mun, D.: Features of development of european science: the program horizon 2020. Inf. Innov. **1–2**(2017), 75–78 (2017)
9. Soldatov, A.V.: Mega-science facilities as an important instrument for synergy of world level education and science. High. Educ. Russia **8–9**(2015), 94–98 (2015)
10. Gulyaeva, T.K., Zaplatina, T.S., Nurahov, N.N.: The aspects of a draft model of international scientific and technical cooperation of the international center for neutron research PIK. J. Phys. Conf. Ser. **1406** (2019)
11. Nurakhov, N.N., Kozlov, Y.V., Petrov, A.A., Kravchuk, V.L.: Life cycle of the project of creation and operation of the megascience facility. J. Phys. Conf. Ser. **1406** (2019)
12. Hoddeson, L., Kolb, A.W., Westfall, C.: Fermilab: Physics, the Frontier and Megascience, p. 497. The University of Chicago Press, Chicago (2008)
13. Anokhin, E.O., et al.: Silica coated hard-magnetic strontium hexaferrite nanoparticles. Adv. Powder Technol. **30**(9), 1976–1984 (2019)
14. CERN. https://home.cern/science/computing. Accessed 19 June 2020
15. Global Neutrino Network (GNN). https://www.globalneutrinonetwork.org/. Accessed 30 June 2020
16. Agostini, M., et al.: (Borexino collaboration): comprehensive geoneutrino analysis with Borexino. Phys. Rev. D **101**(1), 012009 (2020)
17. Dutch Techcentre for Life sciences. https://www.dtls.nl/fair-data/go-fair/. Accessed 10 July 2020
18. Report on the EU funded eStandards project. https://ec.europa.eu/digital-single-market/en/news/report-eu-funded-estandards-project. Accessed 14 July 2017
19. Balyakin, A.A., Malyshev, A.S., Nurbina. M.V., Titov M.A.: Security issues of scientific based big data circulation analysis. In: 8th International Conference on Data Science, Technology and Applications DATA 2019, Prague, Czech Republic, pp. 168–173 (2019). http://www.dataconference.org
20. Nurakhov, N.N.: Integrity of Innovation Management and INSO Inventory, p. 156. MST, Moscow (2010)
21. Ivanov, K.V., et al.: Big data technologies as a tool for ensuring national security. St. Petersburg State Polytechn. Univ. J. Sci. Tech. Sheets Econ. Sci. **13**(1), 7–19 (2020)
22. Ay, C., et al.: Monte Carlo generators in ATLAS software. In: Journal of Physics: Conference Series, CHEP09, Prague, Czech Republic, vol. 219 (2010). https://iopscience.iop.org/article/10.1088/1742-6596/219/3/032001
23. Gartner's Top 10 Strategic Technology Trends for 2020. https://www.pcmag.com/news/gartners-top-10-strategic-technology-trends-for-2020. Accessed 21 Feb 2020
24. The EU General Data Protection Regulation (GDPR). https://gdpr-info.eu. Accessed 18 Nov 2019, 17 Mar 2020
25. Balyakin, A.A., Malyshev, A.S., Nurbina, M.V., Titov, M.A.: Big data: nil novo sub luna. In: Antipova, T. (ed.) Integrated Science in Digital Age. ICIS 2019. Lecture Notes in Networks and Systems, vol. 78, pp. 364–373. Springer, Cham (2020)
26. Balyakin, A.A., Taranenko, S.B., Nurbina, M.V., Titov, M.A.: Social aspects of big data technology implementation. ICS J. Digit. Sci. **1**(1), 15–24 (2019)

Medicine, Public Health and Rehabilitation

Model of Individual Human Behavior in Health Care Safety Management System

Yuriy Voskanyan[1], Fedor Kidalov[2], Irina Shikina[3,4](✉), Sergey Kurdyukov[1], and Olga Andreeva[3,5]

[1] Russian Medical Academy of Continuing Professional Education of the Ministry of Health of Russia, Bld. 1, 2/1 Barrikadnaya St., Moscow 125993, Russia

[2] Moscow State Budgetary Institution "Information and Analytical Center of Healthcare", Bld. 10, Basmannaya Novaya St., Moscow 107078, Russia

[3] Central Research Institute for Organization and Informatization of Health Care at the Ministry of Health of Russia, 127254, Bld. 11, Dobrolyubova Str., Moscow 127254, Russia
shikina@mednet.ru

[4] Central State Medical Academy Office of the President of the Russian Federation, Bld.1A, 19, Marshal Tymoshenko St., 121359 Moscow, Russia

[5] National Research University Higher School of Economics, 20 Myasnitskaya ulitsa, Moscow 101000, Russia

Abstract. The work attempts to build a model of human behavior on the basis of which it becomes possible to develop effective solutions to reduce the probability, early detection, and prevention of the health professionals' mistakes consequences, who are on the "sharp edge" of interaction with the patient - active threats. It is shown that human errors are a consequence. Modeling of human behavior allows to identify the main actions, events caused by them, complex interrelations and interdependencies, without understanding which it is impossible to understand the nature of human mistakes occurrence. Taking into account understanding of human behavior mechanism the main strategies of human error control are considered.

Keywords: Medical care safety · Adverse events · Human behavior · Human and medical errors

1 Introduction

Adverse events, to which we attribute cases of unintentional mental or physical harm resulting in temporary or permanent disability, death, extended hospital treatment and most likely associated (according to experts' opinion) with medical care than with the course of the underlying disease or associated conditions [1], develop in 12.7% of hospitalized patients [2], accounting for 0.71% of deaths [3]. The direct cause of adverse events are active threats in the form of errors and violations of medical personnel [4–6]. The probability of these errors is dramatically increased in the case of so-called latent failures (hidden states or root causes), which are attributed to the roots

T. Antipova (Ed.): ICCS 2020, LNNS 186, pp. 413–423, 2021.
https://doi.org/10.1007/978-3-030-66093-2_40

causes associated with the individual behavior of health professionals; team work-related root causes; organizational root causes (organizational management; human resource management; organizational environment; non-organizational root causes (external microenvironment, medical domain and macro factors); patient behavior-related root causes [5, 7–11]. Medical errors are random, difficult to predict and, therefore, very difficult to control. In contrast, latent threats are characterized by relative constancy and can be easily detected that opens up great prospects in the sphere of medical care safety management due to the possibility to influence latent threats and through them on the probability of medical errors and their consequences severity [4–7, 12, 13]. The work attempts to analyze one of the latent threats related to the individual behavior of a person, in particular a medical worker.

The process of human behavior consists of five sub-processes: perception; information processing; decision making; planning; behavioral act or explicit action [5, 13]. The behavioral act ends with a result that is once again perceived by man as a signal. Thus, human behavior is a closed cycle, which begins with the perception of signals of the current situation and ends with the perception of signals of the result of the behavioral act in the form of feedback (Fig. 1).

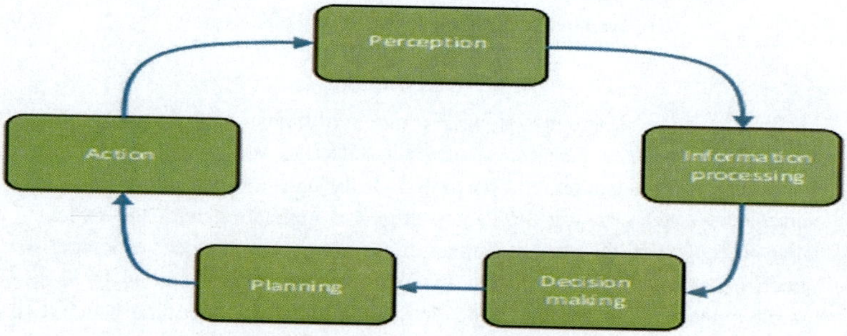

Fig. 1. Human behavioral cycle.

Thus, visible human behavior is the result of invisible autonomous and regulated psychological processes. Thus, visible human behavior is the result of invisible autonomous and regulated psychological processes. Psychological regulation of behavior does not follow strictly logical arguments but is the result of interaction of thinking process, emotions, mental attitudes, and motivation [14].

Individual human behavior (the human factor), which corresponds to the situational context, is considered normal or correct, if not, then they talk about errors and violations [5, 13, 15–17]. In order to reduce the impact of such a latent failure as individual behavior, it is necessary to identify the most vulnerable points of psychological regulation of behavior and then develop effective solutions that will reduce the impact of these vulnerabilities on the end result.

2 Materials and Methods

The work represents the literature review of individual human behavior problem as a source of errors, and also analytical research with the purpose to construct a human behavior model which in the subsequent will be accepted as a basis in decisions development directed on reduction of human errors probability and severity of their consequences. Information search was implemented by two independent researchers over the period 1980–2020 using medical databases MEDLINE, Cochrane Collaboration; EMBASE, SCOPUS, ISI Web of Science. Prospective and retrospective observation studies of high methodological quality, analytical collections and original scientific articles were used for the analysis.

The technology of simulation modeling in the form of agent-based modeling was used to build a model of human behavior [18]. In the course of modelling two levels are allocated: the first - detalization of behavior process at subprocesses level, the second level - detalization subprocesses human behavior at actions level. To describe the processes and subprocesses we used the notation EPC of ARIS Integrated Modeling Environment [19] modified by the authors with the following elements: start event (starts the behavior process) - rectangle with red filling; final event (the result of a behavioral act that completes the behavioral process) - a rectangle with a purple fill; intermediate event (an intermediate result of a behavioral act that makes sense) - rectangle with pink filling; action (work performed) - rectangle with green filling; influence factors (constant, independent of previous actions, elements of psychological regulation of behavior within the three contours described above) - a rectangle with a yellow fill; control flow arrows (indicate direction of action and connect action to intermediate or final event) → - blue arrows; information flow arrows (connect start and final event with the subsequent action) → - red arrows; c impact arrows (connect impact factors to the action they affect) → - brown arrows; waiting time for the action start (an icon inside the action that indicates the need for multiple preconditions to occur at different time periods); with two or more logical operators waiting for action); ^ - logical operator "AND"; **v** - logical operator "OR". ⊗ - Logic operator "EXCEPT OR".

2.1 Macromodel of Human Behavior (First Level)

The first level model shows the interrelation and interdependence of the four main subprocesses (Fig. 2): perception, information processing, decision making and behavioral act) and their results (intermediate and final events). A starting event that initiates human behavior is a situation - a set of objects and phenomena at a certain point in time emitting signals perceived by the senses. These signals trigger the first subprocess - "perception", which ends with the first intermediate event - a concept (a meaningful image of the situation). The senses and corresponding structures of the brain only scan information relevant to the individual's needs and motivations. A person's perception is influenced by internal factors: perception thresholds, personal motives, and stress levels. The concept initiates the next subprocess - "information processing" which ends with an event called the "mental model of a situation".

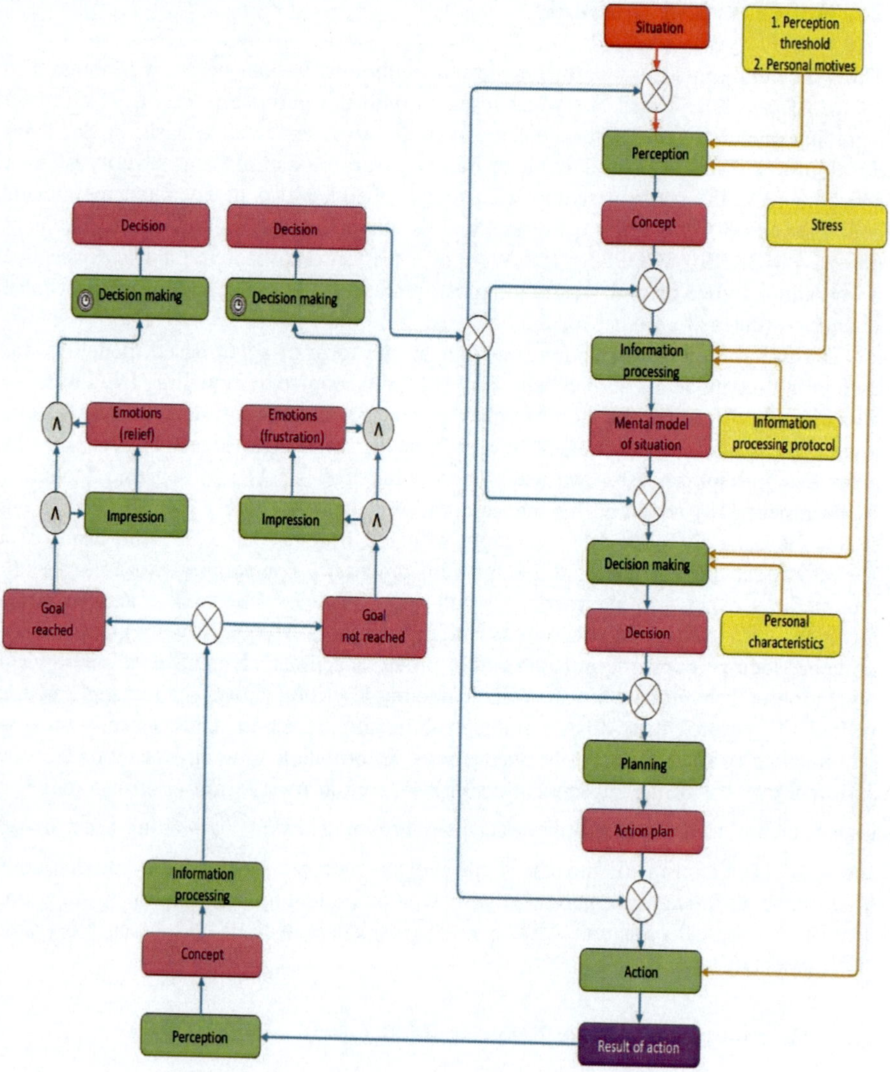

Fig. 2. Macro level of the Human Behavioral model

The mental model of a situation is a meaningful image in cause-effect relationships. Information processing is influenced by stress levels and internal processing protocols in the brain. The mental model initiates the next subprocess - "decision making" which ends with "decision made" event. The decision includes a general goal (what is needed) and a decision method (how to do it). The decision taken initiates a "planning" action which ends with an "action plan" event. The action plan (tasks, sequence, conditions of the action (executors, time, place, resources, etc.) can be very short and can be formed in seconds, such as a treatment plan. Decision making is influenced by the individual's personality characteristics (psychotype, values, emotional intelligence) and stress level. The action plan initiates the next subprocess - "behavioral act" which ends with the event "action result". The behavioral act and its result are the visible parts of behavior. The behavioral act is influenced by the stress level. The action result is perceived as a signal or feedback. Signals again initiate processes of perception and processing of information as a result two alternative events "target reached" and "target not reached" can occur. The first event initiates the action "impression" which ends with the event "emotion - relief". "Emotion-relieve" and the event "goal achieved" initiate a "decision making" action which ends with the event "decision - perception in background scan mode". The made decision initiates an active perception subprocess (active search for relevant information) in the form of a background scan. The second event (unreached goal) also initiates the "impression" action, but which ends with the "emotion-frustration" event. "Emotion-frustration" and the "unreached goal" event initiate the "decision making" action which ends with the "decision - repeated perception (either information processing or decision making or repeated behavioral act" (depending on the result) event. The made decision initiates the beginning of the corresponding sub-process of human behavior.

2.2 Subprocess Model "Perception" and "Information Processing" (Second Level Detalization)

Perception is the integration of the received sensory information with the previously assimilated information and other sensory inputs in order to form a cognitive surrounding world model - mental model. Perception allows a person to navigate in the environment. Perception is always subjective, it includes three consecutive actions: registration, identification and interpretation [5, 13, 14]. Start events for perception are external stimuli: the situation (atypical signals) either signals from the previous behavioral act result or a decision made. These external signals may come to the senses regardless of the person's will or the person actively seeks them out as a result of the made decision (information retrieval and data collection). The registration ends with an event called a "sensation". The sensation is the record result of not only one signal type. At the sensory organs level the basic filter works that reduces an abundance of sensory inputs. Absolute and relative thresholds affect the registration process. Absolute thresholds determine the range within which environmental stimuli (e.g. sound, light) can actually be detected. Absolute thresholds are determined by biology and may change with age, adaptation or fatigue. The relative threshold determines how different the two stimuli must be in order to be distinguished - the threshold of increment to the background (duration, wavelength, strength, etc.), which according to Weber's law is a constant value. For example, it means that in noisy environments the medical monitor's sound signals should be much louder than in absolute silence [5, 13]. Sensation initiates an "attention" action which ends with the "focalization (concentration) of associative thinking" event. The latter initiates a "recollection" action. Memory is a mechanism for extracting available information from memory. Recollection ends with the "mind sets and hypotheses extracted from memory" event. Mind sets are predispositions to a certain activity in a certain situation which are formed on the basis of expectations, experienced emotions and beliefs about the perceived object or event (we see what we are used to seeing and hearing what we are used to hearing). A hypothesis is an assumption about the essence of a given object and events which are formed on the basis of available knowledge in the memory (known objects are recognized faster). Mind sets and hypotheses define an individual pattern of sensations recognition. Sensations, mind sets and hypotheses initiate the "identification" (recognition) action. Identification ends with an event called "Gestalt". Formation of a complete image (gestalt) is a fundamental feature of the perception process which indicates that a person perceives not objects' individual parts but the entire object [5, 13, 18]. The identification process is strongly influenced by personal motives. So-called competing motives can form a Gestalt that best matches an unmet need [5, 13]. The formed Gestalt initiates an "attention" action as a result of which associative thinking is focused and hypotheses and mind sets are again extracted from memory. Gestalt, mind sets and hypotheses extracted from memory initiate an "interpretation" action, which is to give meaning to an object. At this stage, perception enters consciousness for the first time: the figure becomes a conscious object of the real world and the background becomes a semantic context. The interpretation ends with the "concept" event (a meaningful image). The concept answers the question "what is it?". Existing mind sets, especially expectations, have a very strong influence on the recognition vector ("what?" is what we think is more likely) [5, 13]. The concept initially initiates three actions: "impression", "attention" and "registration of a new need". Impression ends with the "emotion" event (excitement, relief or frustration). The registration of the new need ends with the "new motive" event. In turn, the focalization of associative thinking initiates the "memory" action, which ends with the "mind sets extracted from memory" event (as a result of previous experience with similar situations). Focalization of analytical thinking initiates a "memory" action, which ends with the "knowledge, rules, skills and other relevant information extracted from memory" event. The "attention" action is influenced by the stress level. Concept, new motive, emotions, extracted from memory mind sets, knowledge, rules, skills and relevant information initiate a sub-process called "information processing" (interpretation). In the information processing process a meaningful image (concept) is integrated with the information and knowledge extracted from memory. Information processing ends with the "mental model of a situation" event.

The mental model presents a cognitive image of the real world sector in causal relationships that allow us to explain why it happened and to predict what will happen next. Information processing is affected by stress levels and brain processing protocols. The latter follow three basic principles: economy principle (in the process of information processing, the existing mental model is first of all extracted from the memory), guarding the feeling of own competence(a person does not like to admit that he or she does not know something and always tries to find a suitable mental model in his or her memory, closing his or her eyes to the fact that it may not be quite relevant to the current situation), avoiding ambiguity and uncertainties (the most probable mental model is chosen from the list of alternatives, where the evaluation of the process of information processing is performed). Let's consider the main events that take place during the perception and information processing by the example of the resuscitator's assessment of the capnography curve on the monitor of the following medical equipment (one of the efficiency assessment of artificial lung ventilation parameters): feeling

- lines and angles on the monitor display; Gestalt - capnography curve form; concept - normal capnography curve; emotions - relief; mental model situation - "normal capnography curve along with other ventilation parameters and data of clinical examination and laboratory". Subprocess model "Decision-making" and "Behavioral Act" (second level detalization) as a diagram of subprocesses "Decision making", "Planning" and "Behavioral act". The decision ends with the "decision taken" event which includes the overall objective and decision method. The planning ends with the action plan event, the behavior act ends with the action result event. Decision making and planning is as much a part of associative and analytical thinking as information processing are studied in detail in J. Rasmussen's works [22] (Table 1).

Table 1. Three levels of behavioral abstraction.

Characteristics	Behavior abstraction level		
	Skills-based behavior	Rules-based behavior	Knowledge-based behavior
Tasks to be resolved	*Routine tasks*	*Familiar task*	*Unknown task*
Consciousness inclusion	*Minimal*	*Moderate*	*Full*
Attention	*Background Control + Focalization on intermediate indicators*	*Background Control + Focalization on actions and events*	*Full focalization in the form of concentration and vigilance (background check minimum)*

Subprocess model in "skills-based behavior" model. The situation mental model initiates the "attention" action which ends with the "associative thinking focalization" events and "analytical thinking focalization". The mental model, associative thinking focalization, and analytical thinking focalization initiate a "problem-solving" action. A problem is a mental model that shows deviations from the existing order of things [5, 20, 21]. The problems search can end with two alternative events: "having a problem" and "no problem". The absence of a problem initiates a "goal setting" action, which ends with the "chosen goal" event (as expected result). The goal initiates an "attention" action which ends with an "associative thinking focalization" exercise. The purpose and skills extracted from memory initiate a "response choice" action which ends with a "response choice" event. The response method initiates an "attention" action which ends with the "associative thinking focalization" event. Associative thinking focalization and response method initiates a "planning" action which ends with an "automatic sensorimotor pattern of behavior" event. The pattern initiates the "attention" action which ends with the "background control" event. The "goal reached" initiates a perception subprocess in the background. The "goal not achieved" event or the "problem presence" event takes the decision making and subsequent actions to the next level of abstraction associated with the rules.

Subprocess model in "rule-based behavior" model. The presence of a problem or failure to achieve a goal in a skill-based behavioral act initiates an 'attention' action which ends with an 'analytical thinking focalization' event. The latter initiates the "memory" action which ends with the "knowledge extracted from memory" event. A familiar problem initiates a "goal setting" action which ends with the "chosen goal" event (as the expected result of problem solving). The goal initiates an "attention" action which ends with the "focus on associative and analytical thinking" event. The latter initiates a "memory" action which ends with the "rules taken from memory" event. The purpose and rules extracted from memory initiate "rule selection" are extracted from memory. The rule selection ends with the "selected rule" event. The selected rule initiates an "attention" action, which ends with the event "associative and analytical thinking focalization". The selected rule, thinking focalization and selected scripts initiate the "planning" action which ends with the "scenario" event, which is a formalized action plan. The scenario initiates the "attention" action which ends with the event "associative and analytical focalization + background control". Scenario and focus thinking with background control initiate a "rule-controlled behavioral act" that can end with two alternative events: "goal achieved" and "goal not achieved". "Goal achieved" initiates a perception sub-process in the background. The "goal not achieved" event or the event "unfamiliar problem" transfers the decision making and subsequent actions to the highest level of abstraction with the highest level of uncertainty and requiring only analytical thinking - the level associated with knowledge.

Subprocess model in "knowledge-based behavior" model is an unfamiliar problem or failure to achieve a goal in a rule-based behavioral act initiates an "attention" action which ends with an "analytical thinking focalization" event. It initiates a "memory" process. An unfamiliar problem and extracted knowledge initiates a "problem identification" process is a complex mental model with identified cause-effect relationships taking into account alternative causes and uncertainty [5, 21]. The latter initiates the "goal setting" process, which ends with the "goal" event (as an expected result of the problem solving). The goal initiates the "attention" action which ends with the event "analytical thinking focalization". The latter initiates a "memory" process the final event of which is knowledge extracted from memory. The goal and the knowledge extracted from memory initiate a "problem-solving" action which is an analysis of alternatives with an assessment of risks and uncertainties. At its best, the search for a solution ends with two events: "main alternative selected" and "reserve alternative selected". The chosen alternatives, analytical thinking focalization and the knowledge extracted from memory initiate "problem-solving planning" actions, which culminate in the construction of based and back-up plans. The action plan in this case includes a detailed elaboration of the overall goal in the form of intermediate goals, as well as objectives, timelines, implementers, information and resources. The based plan initiates an "attention" action which ends with the "focalization on associative and analytical thinking + background control" event. Based plan and focus thinking with background control initiates the "knowledge-controlled behavioral act number 1" action in accordance with the master plan. The results of a knowledge-based behavior act can be two alternatives: 'goal achieved' and 'goal not achieved'. "Goal achieved" initiates a perception sub-process in the background. "Goal not achieved" initiates an "attention" action which ends with a focus on analytical and associative thinking with background control. Focalization on thinking with background control and unachieved goal initiate the action "behavioral act number 2" in accordance with the back-up plan. Failure to achieve the goal again indicates that the original mental model is not correct, resulting in the re-initiation of the "problem identification" action and all subsequent actions described above. Good plans should have early switchover points (from based to back-up plan, from back-up to "problem identification" action) before an unwanted outcome (when it becomes obvious that the goal has not been achieved). These switch points are called intermediate indicators which are intermediate events that have a predictive value for the final event. Strategies to prevent human error. The human behavior model presented by the authors demonstrates the most vulnerable places of psychological regulation of behavior, where failures can occur leading to human errors divided into two large groups [2, 4, 5, 23, 24]:

- Mistakes caused by a wrong decision which can be caused by both inadequate goals, methods and plans, and defects in the previous stages of behaior - distortions (illusion) and hallucinations (delusion) and information processing (inadequate mental model);
- Slips, lapses and somatic tripping due to the wrong execution of the right decision (wrong behavioral act).

Obviously, the share of the first error group is prevailing, since the possibilities of external control and self-monitoring in this case are extremely limited. The probability of human error increases when individuals are combined into groups (additive effect) and decreases when a complementary command is formed from a group (multiplying effect). Taking into account the progressive increase in the probability of human error under the constant impact of the root causes described in the introduction, the risk of an adverse event can be significantly reduced by reducing the impact of organizational and non-organizational latent failures. The benefits of such measures are undoubtedly higher, because, unlike randomly occurring and poorly managed human errors, latent failures exist constantly, can be timely detected and effectively managed [2, 4–6].

3 Conclusion

The human behavior model described in three algorithms diagrams visually demonstrates that fact that person's obvious action is predefined by rather complex mechanisms of perception, information processing, decision-making and planning which are not controlled by consciousness and regulated at the associative thinking level, mind sets, emotions and personal motives, and also are influenced by a number of human organism factors itself (stress condition, etc.). All aforesaid speaks about inevitability and, the most important, about considerable complexity of human errors management at the individual behavior level, as they, as well as normal behavioral acts are a consequence of not aberrant processes in a brain, as a result of normal psychological regulation of behavior which is poorly controlled by the individual and, moreover, by third parties and even modern information systems.

References

1. Zegers, M., Bruijne, M.C., Wagner, C., et al.: Adverse events and potentially preventable deaths in Dutch hospitals: results of a retrospective patient record review study. Qual. Saf. Health Care. **18**, 297–302 (2009)
2. Voskanyan, Y., Shikina, I., Kidalov, F., Davidov, D.: Medical care safety - problems and perspectives. In: Integrated Science in Digital Age. Lecture Notes in Networks and Systems, vol. 78, pp. 291–304 (2020). https://doi.org/10.1007/978-3-030-22493-6_26
3. Makary, M.A., Daniel, M.: Medical error—the third leading cause of death in the US. BMJ **353**(3), 1–5 (2016)
4. Reason, J.: Human error: models and management. Br. Med. J. **320**, 768–770 (2000)
5. St. Pierre, M., Hofinger, G., Buerschaper, C.: Crisis Management in Acute Care Settings Human Factors and Team Psychology in a High Stakes Environment. Springer, Heidelberg (2008)
6. van Vuuren, W.: Organisational failure: an exploratory study in the steel industry and the medical domain. Technische Universiteit Eindhoven, Eindhoven (1998). https://doi.org/10.6100/IR509397
7. Voskanyan, Y., Shikina, I., Kidalov, F., Andreeva, O., Makhovskaya, T.: Impact of macro factors on effectiveness of implementation of medical care safety management system. In: Integrated Science in Digital Age. LNNS, vol. 136, pp. 346–355. Springer (2021). https://doi.org/10.1007/978-3-030-49264-9_31
8. Voskanyan, Y., Shikina, I., Kidalov, F., Andreeva, O.: Medical safety management system from the perspective of a multifactorial model of adverse events in medicine. Am. J. Emerg. Crit. Care Med. **3**(1), 001–014 (2020). https://doi.org/10.37871/ajeccm.id14
9. Voskanyan, Y., Shikina, I., Andreeva, O., Kidalov, F., Davidov, D.: Multifactorial model of adverse events and medical safety management system. J. Digit. Sci. **2**(1), 29–39 (2020). https://doi.org/10.33847/2686-8296.2.1_3
10. Hoffmann, B., Rohe, J.: Patient safety and error management. Dtsch Arztebl Int **107**(6), 92–99 (2010). https://doi.org/10.3238/arztebl.2010.0092
11. Beuzekom, M., Boer, F., Akerboom, S., Hudson, P.: Patient safety: latent risk factors. Br. J. Anaesth. **105**(1), 52–59 (2010)
12. Mattsson, S., Fast-Berglund, A., Stahre, J.: Managing production complexity by supporting cognitive process in final assembly. Proc.Cirp **17**, 212–217 (2014)
13. Akintunde, E.A.: Theories and concept for human behavior in environmental preservation. J. Environ. Public Health. **1**(2), 120–133 (2017)
14. Dörner, D.: Bauplan für eine Seele. Rowohlt, Reinbek (2001). ISBN 3499611937
15. Van der Schaaf, T.W., Kanse, L.: Error recovery in socio-technical systems. Paper Presented at CSAPC 1999, 7th European Conference on Cognitive Science Approaches to Process Control, Villeneuve d'Ascq, France, 21–24 September 1999 (1999)
16. Mitchell, R.J., Williamson, A.M., Molesworth, B., Chung, A.Z.Q.: A review of the use of human factors classification frameworks that identify causal factors for adverse events in the hospital setting. Ergonomics **57**(10), 1443–1472 (2014). https://doi.org/10.1080/00140139.2014.933886
17. Shorrock, S.T., Kirwan, B.: Development and application of a human error identification tool for air traffic control. Appl. Ergon. **33**, 319–336 (2002)
18. Rudomin, H.B., Millan, E.: Fragment shaders for agent animation using finite state machinesю. Simul. Modell. Pract. Theory J. **8**(13), 741–751 (2005)
19. Davis, R., Brabaender, E.: ARIS Design Platform: Getting Started with BPM. Springer, London (2007). ISBN 1-84628-612-3

20. Kahneman, D., Slovic, P., Tversky, A.: Judgement Under Uncertainty: Heuristics and Biases. Cambridge University Press, Cambridge (1982)
21. Klein, G.: A recognition-primed decision (RPD) model of rapid decision-making. In: Klein, G., Orasanu, J., Calderwood, R., Zsamboka, E. (eds.) Decision-Making in Action: Models and Methods. Ablex, Norwood (1992)
22. Rasmussen, J.: Skills, rules, knowledge: signals, signs and symbols and other distinctions in human performance models. IEEE Trans. Syst. Man Cybern. **13**(3), 257–267 (1983)
23. Rozhkova, N., Blinova, U., Rozhkova, D.: The concept of management accounting based on the information technologies application. In: Antipova, T., Rocha, Á. (eds.) Information Technology Science. MOSITS 2017. Advances in Intelligent Systems and Computing, vol. 724, pp. 89–95. Springer, Cham (2018). https://doi.org/10.1007/978-3-319-74980-8_8
24. van der Snijders, C., Schaaf, T.W., Klip, H., van Lingen, R.A., Fetter, W.P.F., Molendijk, A.: On behalf of the NEOSAFE study group feasibility and reliability of PRISMA-Medical for specialty-based incident analysis Qual. Saf. Health Care **18**, 486–491 (2009). https://doi.org/10.1136/qshc.2008.028068

Physical and Material Sciences

Change of GeTe/Sb2Te3 Thin-Film Memory Elements Resistance R_{ON} Under External Pressure

Evgeniy Troyan[1(✉)] and Alexander Doronin[2]

[1] New Europe Innovative Technologies, Inc., Minsk, Belarus
troevfe@gmail.com
[2] Victoria University of Wellington, Wellington, New Zealand
aledoronin@gmail.com

Abstract. The $Ge_2Sb_2Te_5$ is a phase change material (PCM) widely used in non-volatile random-access memory (PC-RAM) devices since 2017. Despite the compound is under intense investigation its electronic structure and nature of a switching mechanism is not fully understood. The present work sheds new light on the nature of memory and switching effects observed in such compounds. Results of our research demonstrate that crystal structure of $Ge_2Sb_2Te_5$ possess different topological quantum phases: a 2D topological quantum phase is realized at the switching effect and a Weyl semimetal phase or 3D topological crystalline phase is a characteristic of the non-volatile memory effect. Since the structural phase transition can be caused by the external electric field, we discovered that the switching between different topologically non-trivial phases can be driven by variation of electrical potential on the top gate of topological field effect transistor. A similar pattern is observed when the thickness of the layers $GeTe/Sb_2Te_3$ changes. The obtained results have been possible to develop a unified physical model that explains the simultaneously observed effects of non-volatile memory and switching.

Keywords: Chalcogenide films and alloys · Effects memory and switching · Interfaces · Weyl and Dirac semimetals

1 Introduction

In the beginning of the 1960s it has been discovered that amorphous PCMS can undergo rapid and reversible phase transition upon heating (T°), external voltage (V), or current pulse (I). This transition in PCMS has been explained by a large property contrast between amorphous and crystalline phases. This idea is popular now. The basis of this idea has been believed to involve the melt-quench mechanism. However, for first time it has been discovered by us [1] and later confirmed by other researchers [2] that such idea does is not aligned well with several research results. Therefore, in 1990s we have been conducting more careful structural analysis of amorphous chalcogenide (Ch) alloy $Ge_2Sb_2Te_5$ and its connections with electrical properties. More than 50 years ago Petrov, etc. [3] discovered that amorphous chalcogenide alloy $Ge_2Sb_2Te_5$ (GTS) represents the periodic system consisting of nano-layers

T. Antipova (Ed.): ICCS 2020, LNNS 186, pp. 427–433, 2021.
https://doi.org/10.1007/978-3-030-66093-2_41

$[GeTe]_{n=2}[Sb2Te3]_{m=1}$ which in a consequence received the name "Petrov's structure" [4]. The quintuple layers (QLS) Sb_2Te_3 chaotically is located between nano-blocks $[GeTe]_2$ in this structure. We paid specific attention to characteristics of "Petrov's structure".

Firstly, amorphous chalcogenide "Petrov's structure" has a large number of the repeating nano-interfaces, namely: (GeTe/Sb2Te3)n. Therefore, we created a thin-film element with one GeTe/Sb_2Te_3 interface (with one superlattice fragment) using Molecular-beam epitaxy (MBE) and arranged it between two metallic electrodes. As a result, the steadily working thin-film element of memory was created, depending on polarity of the applied external voltage. The element of memory switched from a state with a high resistance (OFF-state) in a state with a low resistance (ON-state) only at a negative potential of external voltage on an electrode contacting to Sb_2Te_3 film. The element reverted to the original state with a high resistance (OFF-state) only when changing polarity of external voltage applied to electrodes. Thus I-V characteristic in thin film GeTe/Sb_2Te_3 sample in the present work exhibit asymmetry, i.e. it is voltage-controlled and bias polarity-dependant. Direct correlation between electrical parameters of the produced GeTe/Sb_2Te_3 samples and the polarity of the applied voltage places in question widely used models of the switching effects in chalcogenide semiconductors.

Secondly, such separate nano-block "Petrov's structure" as Sb_2Te_3 is the three-dimensional (3D) topological insulator [7]. On other hand, GeTe is a high-temperature ferroelectric material. It is known [8] that upon TC = 630−700 K in the ferroelectric material GeTe induced internal electric field. Therefore, in the beginning of 2000, we created the thin-film structure with three electrodes (FET or transistor) in which one electrode ("gate") was located above GeTe film, and two others contacted to Sb2Te3 film. We would like to bring your attention to that, that the main objective of this experiment consisted in a research of influence of the alternating voltage to the "gate", on such key parameters of a GeTe/Sb_2Te_3 memory cell as value of a threshold current (Ith) and voltage (Uth), switching time of (t_{SW}), etc. We observed that the minor change of voltage on "gate" reduce of Uth and t_{SW}. Besides, only the effect of memory was observed at a positive voltage on "gate" in thin film structure (Fig. 1,a). In the structure was observed a transition from effect of memory to the effect of asymmetric threshold switching (Fig. 1b, c) when a polarity of voltage on "gate" was changed from positive to negative (at the size of negative voltage of U = −1.1−1.5 V). This transition was reversible and repeated repeatedly with the change of voltage polarity at the "gate." Hence, we concluded that in GeTe/Sb_2Te_3 films, memory and switching effects have a common electric nature and occur within of the GeTe/Sb_2Te_3 interface [1] Temperature (T°) also has significant effect on these switching effects that is not surprising since GeTe is a high-temperature ferroelectric material and the increase of T° can cause ferroelectric phase transition into GeTe. It is possible for this reason, researchers often connect effects of memory in the GTS alloys with thermal processes, i.e. with crystallization of an amorphous state. In addition, with little effect of the external pressure (p), we observed a huge change in the resistance RON (ON-state or SET phase) (2−3 orders) of the memory elements. At the same time, analogical value of the external pressure did not influence, or have very weakly affect for the RON of the switching elements, what excluded the effect of the mistake or so-called "laboratory rack". Due to the received results, we have developed a more complete unified physical model of memory and threefold switching effects in iPCM [8]. Contrary to the previously proposed thermal model, it was concluded that the phase-change occur within of the interface GeTe/Sb_2Te_3, not requiring the melt-quench thermal cycle. Our research is explained in terms of the binding anisotropy of GeTe and Sb_2Te_3 and the strong tendency of these materials to intermix and atomic migration [9].

2 Research Findings

We determined that as a result of the process of polar electro-migration of components under the influence of external voltage (at U = Uth), with the participation of surface states of van der Waals (dvdW) gap has formed on the GeTe/Sb_2Te_3 interface [14] where indirect redistribution of electron density in cation ions of different valence (for example, Ge and Sb) has been occurring. Some part of the interface changes to highly ordered, stable layered superstructure $GeSbTe_4$ with semi-metallic type of conductivity [5, 10]. In contrast to crystalline $Ge_1Sb_2Te_4$ trigonal structure a process of the forming the highly ordered superstructure $GeSbTe_4$ is connected with an effect of bands inversion [23]. Thus, $GeSbTe_4$ superstructure is created within of vdW gap under the external electric field. It is so-called "filament" of conductivity. The "filament" has a high conductive state what responsible for the ON-state in the GeTe/Sb_2Te_3 superlattice. The time stability of this superstructure formed within of vdW gap can be maintained by of an internal ferroelectric field at ambient T° [23].

However, such a physical scenario is not sufficiently complete since it is important to note that the stable layered $GeSbTe_4$ superstructure ("filament") comprise from two types of two dimensional dichalcogenide metastable atomic blocks: $GeTe_2$ and $SbTe_2$. These two-dimensional dichalcogenide metastable atomic blocks sandwiched between two chalcogenide layers: GeTe and Sb_2Te_3. If $GeTe_2$ dichalcogenide atomic block is situated near to the GeTe layer and interacts with him, so $SbTe_2$ dichalcogenide atomic block is in contact with the Sb_2Te_3 one. The internal ferroelectric field can created in the first metastable $GeTe_2$ dichalcogenide atomic block ($GeTe_2$ characterized by distorted tetrahedral coordination of Ge ions situated in cation sites [10]). The second metastable $SbTe_2$ degenerate metal dichalcogenide atomic layer takes direct part in a forming of 3D Weyl semimetal phase [12, 14].

Recently, it was been discovered that in transition metal dichalcogenides there is a new topological state – a phase Weyl semimetal [13]. As the ferroelectric polarization of $GeTe_2$ block lifts not only the degeneration but also breaks the inversion symmetry of $SbTe_2$ block, so it can possibly be claimed that the SET phase (or ON-state) of the GeTe/Sb_2Te_3 memory cell is a phase of 3D Weyl semimetal (WSM) [14, 15]. WSM is a three-dimensional crystal, which break either time-reversal or inversion symmetry, where pairs of energy bands touch at k points, referred to as Weyl nodes [16]. The Weyl nodes have chiralities (topological protected chiral charges) and act as sources or sinks of Berry curvature flux in the Brillouin zone (BZ) [12]. Weyl hosts unusual surface states which form disjoint open Fermi arcs connecting the projection of the Weyl nodes onto the surface BZ where the Fermi arcs on opposite surfaces of sample are linked through the bulk states [17].

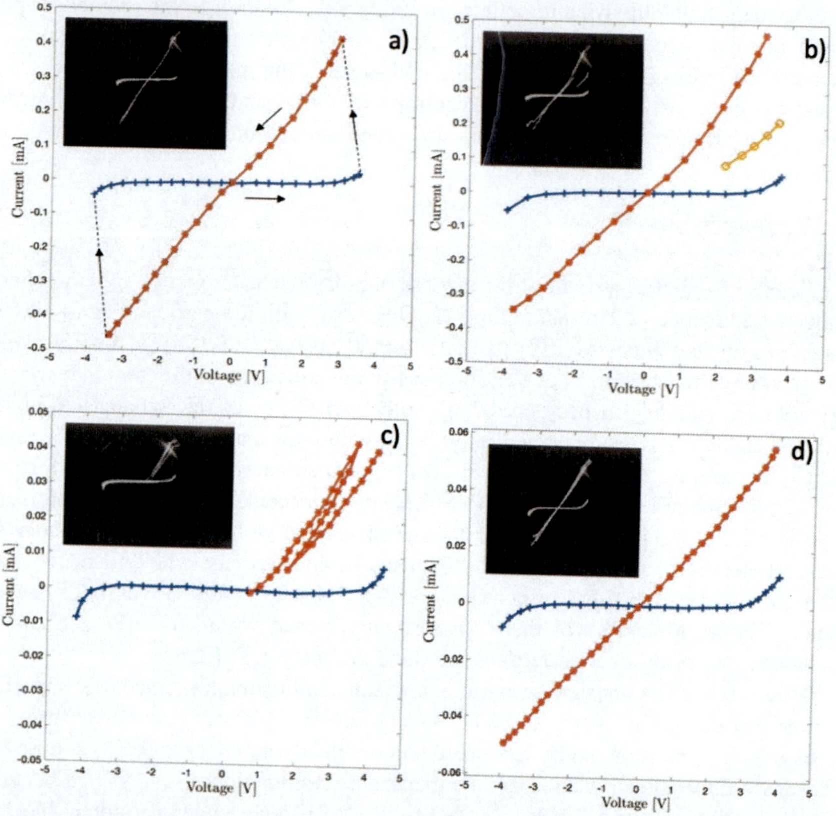

Fig. 1. Polarity-depend I-V characteristic in $GeTe/Sb_2Te_3$ thin film FET with threshold switching and memory effects, f = 100 Hz: a). memory effect is observed only, Ugate = 0 V, arrows show dynamics of I-V characteristic for horizontal sweep of variable voltage; b). first signs of threshold switching effect are there. Memory effect is insignificant on the I-V characteristic, Ugate = −0,9 V; c). effect of threshold switching is observed only, Ugate = −1,1 V; d). a return to net effect of memory, Ugate towards 0 V; on inserts – oscillograms of threshold switching and memory effects.

Our findings demonstrate that in the case $GeTe/Sb_2Te_3$ super-lattice undergoes the phase transition from OFF-state to ON-state (memory effect) a mechanism of giant enhancement (by up to five orders of magnitude and more it) of electrical conductivity are connected with an emergence of the gapless chiral monopole-antimonopole Weyl pairs on the surface states of $GeSbTe_4$ superstructure. The shape of the Fermi surface depends sensitively on the surface termination exhibiting different topology. These is two types of chalcogenide atomic layers in the memory cell are $GeTe_2$ and $SbTe_2$ and they have two different types of surface states which take direct part in the processes of the bands inversion [14]. Therefore, the $GeTe_2$ - terminated and $SbTe_2$-terminated surfaces have different pairs of the topological Fermi arcs [14]. These Fermi arcs linked together through the bulk states of the $GeSbTe_4$ superstructure. The pairs of the Fermi arcs are situated on the surfaces asymmetrically one above another, i.e. they are specularly reflect each other. Weyl's node with a negative charge (chirality) on the $GeTe_2$ surface is located over Weyl's node with a positive charge (chirality) on the $SbTe_2$ surface and vice versa, i.e. these Weyl's nodes have the opposite chirality. The electrical conductivity of SET phase (low-resistive ON-state) of the $GeTe/Sb_2Te_3$ superlattice is conditioned by a constructive interference on the Weyl nodes of at least two waves of the polarized electrons moving on the pairs of Fermi arcs with opposite chirality. Though the chirality of Weyl's nodes prohibits the back-scattering of conduction electrons from each chiral channel, we speculate (assume) here that the size of resistance ON-state of the $GeTe/Sb_2Te_3$ super-lattice is defined by the sizes of disjoint open Fermi arcs connecting the projections of the Weyl's nodes onto the BZ, i.e. by of the lengths of the arcs. The sizes of the lengths of disjoint open Fermi arcs depend from a distance q connecting opposite arcs, i.e. q is defined by $GeSbTe_4$ layer thickness [14]. The scattering vector q connects wave vectors of initial and final states with opposite group velocities, supporting constructive interference [18]. Therefore, a magnitude of the resistance (RON) of the SET phase (memory state) must be very dependant sensitively on compressive strain [14].

Our findings clearly demonstrate that appearance of the threshold switching effect in the $GeTe/Sb_2Te_3$ super-lattice can be explained by the phase transition from Weyl semimetal phase to a metastable phase of 2D topological insulator which may be identified as Dirac semimetal at a negative voltage on the "gate" of the thin film transistor. The negative potential on the gate of the FET reduces not only the magnitude of the ferroelectric polarization of $GeTe_2$ and GeTe blocks but it changes also the band inversion strength of the $SbTe_2$ and Sb_2Te_3 building blocks. The pairs of the Weyl nodes merge in a degenerate Dirac points on the Fermi surfaces both the GeTe2 and $SbTe_2$ building blocks [19]. If a spacer $GeSbTe_4$ layer is sufficiently thin, these the Dirac points can couple in a gap closing points through hybridization [11]. The constructive interference of polarized electrons transformed to a destructive one on these Dirac points. In result, a magnitude of the resistance (RON) of the SET phase (in the threshold switching state) sharp increased by up to one-two orders of magnitude (Fig. 1, b,c). The stability in the time of this new superstructure formed within of the vdW gap can be retained and fixed at ambient temperature by external electric field only [23]. Moreover, a magnitude of the resistance (RON) of the SET phase (for switching effect) not must be to depend sensitively on compressive strain.

3 Conclusions

Thus, our findings show that using this model, one can explain not only a nature of threshold switching and memory effects in the $GeTe/Sb_2Te_3$ super-lattices but appearance of high temperature topological superconductivity in these and another chalcogenide semiconductors under pressure [19–23]. Ultimately, the unified model can be used not only for $GeTe/Sb_2Te_3$ superlattices, but also and for $Ge_2Sb_2Te_5$ compounds. For example: a) threshold switching effect in these $Ge_2Sb_2Te_5$ compounds is related to giant spin-orbit interaction (SOI) and to inversion of zones of the conductive nano-filaments formed within of the nano-interfaces ($GeTe/Sb_2Te_3$)n: i.e. with phase change from a normal insulator - to 2D topological insulator; b) memory effect in the $Ge_2Sb_2Te_5$ compounds is due to induction in some constituent regions of the complex GTS alloy (for example, in amorphous $Ge_2Sb_2Te_5$ compound it can be the regions of GeTex) of the internal ferroelectric field (Uind.). At the absence of external voltage, the internal field Uind maintains the field balance. That stabilizes in time the highly ordered nano-filaments ($GeSbTe_4$) with degenerate 3D Weyl semi-metallic phases that formed within of the nano-interfaces ($GeTe/Sb_2Te_3$)n at ambient T°; c) decrease of the value of the resistance RON in the GTS with increasing of the value of external pressure, is caused by the change the thicknesses of the conductive nano-parts ($GeSbTe_4$)n that are formed within of the nano-interfaces ($GeTe/Sb_2Te_3$)n after the switching of memory elements in ON-state.

Thus, obtained results reveal the rich and exotic physics of the phase changing materials opening previously unexplored applications of these materials for topological superconductivity and spintronics. On the basis of the unified model, we have created novel thin-film elements and were granted patents for their designs in the Russian Federation, EU and USA [23]. The test methodology and obtained data are opened there are in these patents.

References

1. Troyan, E.F.: Candidate's thesis by E.F. Troyan on the issue in dec.4 (1997)
2. Sun, H.-J., Hou, L.-S., Wu, Y.-Q., Tang, X.-D.: Reversible resistance switching effect in amorphous Ge1Sb4Te7 thin films without phase transition. Chin. Phys. Lett. **2**(2), p. 024203-1-3 (2009)
3. Petrov, I.I., Imamov, R.M., Pinsker, Z.G.: Electron-diffraction determination of the structures of $Ge_2Sb_2Te_5$ and $GeSb_4Te_7$. Sov. Phys. Crystallogr. **13**, 339–342 (1968)
4. Tominaga, J., Simson, R., Fons, P., Kolobow, A.: Electrical-field induced giant magnetoresistivitiy in (non-magnetic) phase change material. Appl. Phys. Lett. **99**, 152105 (2011)
5. Troyan, E.F.: patent US 5,363,329 (1994)
6. Kvyatkovskii, O.E., Maksimov, E.G.: Microscopic theory of the lattice dynamics and the nature of the ferroelectric instability in crystals. Sov.,Usp. Fiz. Nauk, **154**(1), pp. 3–48 (1988)
7. Zhang, H., Liu, C.-X., Qi, X.-L., Dai, X., Fang, Z., Zhang, S.-C.: Topological insulators Bi_2Se_3, Bi_2Te_3 and Sb_2Te_3 with a single Dirac cone on the surface. Nat. Phys. **5**, 438 (2009)
8. Kolostitcin, B.S., Troyan, E.F.: Switching and memory effects in thin film disorder chalcogenide semiconductors. Doklady BGYIR **104**(2), 25–30 (2017)
9. Momand, J.: Structure and reconfiguration of epitaxy $GeTe/Sb_2Te_3$ super-lattices. University of Groningen, pp. 5–86 (2017)
10. Lontyk, A., Behrens, M., Rauschenbach, B.: Phase change thin films for memory applications. Nanoscale Adv. **1**, 3836 (2019)
11. Kim, M., Kim, C.H., Kim, H.-J., Ihm, J.: Topological quantum phase transitions driven by external fields in Sb_2Te_3 thin films. PNAS **109**(3), 671–674 (2012)
12. Burkov, A.A., Hook, M.D., Balents, L.: Topological nodal semimetals. Phys. Rev., B **84**, pp. 235126-1-5 (2011)

13. Xu, S.-Y., Belopolski, I., Alidoust, N., Neupane, M., Bian, G., Zhang, C., Sankar, R., Chang, G., Yuan, Z., Lee, C.-C., Hyang, S.-M., Zheng, H., Ma, J., Sanchez, D.S., Wang, B., Bansil, A., Chou, F., Shibayev, P.P., Lin, H., Jia, S., Hasan, M.Z.: Discovery of a Weyl fermion semimetal and topological Fermi arks. Science **349**, 613 (2015)
14. Kim, J., Kioussis, M.: Weyl node assisted conductivity switch in interfacial phase change memory with van der Waals interfaces. arXiv: 1702.05579v1, Phys. Rev. B **96**, p. 235304 (2017)
15. Eremeev, S.V., Rusinov, I.P., Echenique, P.M., Chulcov, E.V.: Temperature driven topological quantum phase transitions in a phase change material $Ge_2Sb_2Te_5$. Nature, Scientific Reports, **6**, p. 38799 (2016)
16. Nielsen, H., Ninomiya, M.: The Adler-Bell-Jackiw anomaly and Weyl fermions in crystal. Phys. Lett. B **130B**, 389 (1983)
17. Wan, X., Turner, A.N., Vishwanach, A., Savrasov, S.Y.: Topological semimetal and Fermi-arc surface states in the electron structure of pyrochlore iridates. Phys.Rev. B **83**, pp. 25101-1-6 (2011)
18. Rubmann, P., Weber, A.P., Glott, F., Xu, N., Fanciulli, M., Muff, S., Magrez, A., Bugnon, P., Berger, H., Bode, M.: Universal scattering response across the type-II Weyl semimetal phase diagram. Phys. Rev. **B97**, pp. 075106-1-8 (2018)
19. Hasan, M.Z., Xu, S.-Y., Bian, G.: Topological insulators, topological superconductors and Weyl fermion semimetals: discovery, perspectives and outlooks. Phys. Scr. **164**, 1–22 (2015)
20. Wang, S., Lin, B.-C., Wang, A.-Q., Yu, D.-P., Liao, Z.-M.: Quantum transport in Dirac and Weyl semimetal: a review. Adv. Phys. **2**(3), 518–544 (2017)
21. Yang, H., Kim, S.W., Chhhowalla, M., Lee, Y.H.: Structural and quantum-state phase transitions in van der Waals layered materials. Nat. Phys. **13**, 931–937 (2017)
22. Grenberg, E., Hen, B., Layek, S., Pozin, I., Friedman, R., Shelukhin, V., Rosenberg, Y., Karpovski, M., Pasternak, M., Steren, E., Dagan, Y., Rosenberg, G., Palevski, A.: Superconductivity in multiple phase of compressed GeSb2Te4. arXiv: 1607.04795v1 (2016)
23. Troyan, E.F.: patent US 9,865,811, jan.9, 2018, patent RF2618959, 2017; patent EP3151294 (2019)

Author Index

T. Antipova (Ed.): ICCS 2020, LNNS 186, pp. 435–436, 2021.
https://doi.org/10.1007/978-3-030-66093-2